TAKING SIDES

Clashing Views on

Global Issues

FOURTH EDITION

McGraw Hill **Contemporary Learning Series**

A Division of The McGraw-Hill Companies

TAKING SIDES

Clashing Views on

Global Issues

FOURTH EDITION

Selected, Edited, and with Introductions by

James E. Harf
University of Tampa

and

Mark Owen Lombardi
College of Santa Fe

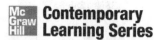

Mc Graw Hill **Contemporary Learning Series**

A Division of The McGraw-Hill Companies

*To my daughter, Marie: May your world
conquer those global issues left unresolved
by my generation. (J.E.H.)*

*For Betty and Marty, who instilled a love of
education and need to explore the world. (M.O.L.)*

Photo Acknowledgment
Cover image: Brand X Pictures/PunchStock

Cover Acknowledgment
Maggie Lytle

Manufactured in the United States of America

Fourth Edition

123456789DOCDOC9876

Library of Congress Cataloging-in-Publication Data
Main entry under title:
Taking sides: clashing views on controversial global issues/selected, edited, and with
introductions by James E. Harf and Mark Owen Lombardi.—4th ed.
Includes bibliographical references and index.
1. Globalization. 2. International relations. 3. Global environmental change. 4. Population.
5. Emigration and immigration. I. Harf, James E., *ed.* II. Lombardi, Mark Owen, *ed.* III. Series.
303

0-07-352724-6
978-0-07-352724-6
ISSN: 1536-3317

Printed on Recycled Paper

Preface

This volume reflects the changing nature of the international system. The personal reflections of its two editors bring this into sharp focus. We are both products of the separate environments in which we have been raised. Our 20-year age difference suggests that each of us brings to the study of international affairs a particular lens through which two distinct yet interconnected generations view contemporary world issues.

The older editor began his formal schooling a few weeks after the United States dropped two atomic bombs on Japan. His precollegiate experience took place amid the fear that an evil enemy was lurking throughout the world, poised to conquer America and enslave its occupants. During his initial month in college, the Soviet Union launched the first space vehicle, *Sputnik*, forever changing the global landscape. Yet his 1961 international relations textbook assumed that, despite the bomb and the ability to exploit outer space, the driving forces behind national behavior had not changed very much. By the time he accepted his first university teaching position, the world had been locked in a protracted Cold War for two decades. Yet evidence of a new global agenda of problems totally unrelated to the American-Soviet rivalry was beginning to appear. Soon this new set of issues would challenge the traditional agenda for the world's attention and also occupy the older editor's research agenda during much of his career.

When the older editor moved on to a new stage of his long career in summer 2001, the Cold War existed no more. And the set of problems that had occupied his and others' attention for some time—population growth, resource scarcities, environmental degradation, and the like—were being transformed by the overarching presence of globalization. The Internet and other advances in communication and transportation signal an entirely new environment in which global actors design and carry out their policies. This was brought home in September 2001 with the terrorist attacks on the United States that have changed long-held assumptions regarding violence, war, and the application of power in the post–Cold War world.

This older editor's world has now undergone three major changes during his professional career, from a traditionalist, realist paradigm where military power dominated, to a new global agenda where problems demanded the attention and cooperation of a kaleidoscope of actors, to yet another set of problems and a transformation of existing issues brought on by globalization, where the ease with which people, information, ideas, money, and goods cross national boundaries yield both positive and negative consequences for the world at large.

The younger editor began his formal schooling just after the United States and the Soviet Union had settled "comfortably" into the era of mutual assured destruction (MAD). He entered his university studies as increased tensions and U.S.-Soviet rivalry threatened to tear apart the fragile fabric of détente. As he embarked on his teaching career, the communist empire was taking its first steps toward total disintegration. Amid this revolutionary change, deeper and more complex issues of security, technology, development, and ethnic conflict became more pronounced and increasingly present in the consciousness of actors and scholars alike. As he, too, embarked on a new academic career, September 11 woke all Americans up to the harsh reality that the paradigms of violence and war were changing. Now at the midpoint of his academic life, this younger editor contemplates the speed and force of globalization, and how the issues that this phenomenon has brought forth will affect the lens through which he observes world affairs, influencing his and other research agendas over the next few decades.

The decision of this book's publisher to embark on a text on global issues reflects how the nature of international affairs has changed. Many textbooks on world politics and American foreign policy address longstanding, important concerns of the student of world events, employing for the most part the traditional paradigm of the past century. Simply put, this traditional scholarly view assumed a world where power, particular military might, dominated the world picture. Nation-states, the only really important world actors, focused on pursuing power and then using it in rather conventional ways to become a larger presence on the international scene. The specific problems of the day might have changed, but the fundamental principles that guided the international system remained the same. Most books in these fields reflect this outlook.

But the field of international affairs is ever changing, and textbooks are beginning to reflect this evolution. This new volume takes into account the fact that the age of globalization has accelerated, transforming trends that began over three decades ago. No longer are nation-states the only actors on the global stage. Moreover, their position of dominance is increasingly challenged by an array of other actors—international governmental organizations, private international groups, multinational corporations, and important individuals—who might be better equipped to address newly emerging issues (or who might also serve as the source of yet other problems). This new agenda took root in the late 1960s where astute observers began to identify disquieting trends: quickening population growth in the poorer sectors of the globe, growing disruptions in the world's ability to feed its population, increasing shortfalls in required resources, and expanding evidence of negative environmental impacts, such as a variety of pollution evils.

An even more recent phenomenon is the unleashing of ethnic pride, which manifests itself in both positive and negative ways. The history of post–Cold War conflict is a history of intrastate, ethnically driven violence that has torn apart those countries that have been unable to deal effectively with the changes brought on by the end of the Cold War. The most insidious manifestation of this emphasis on ethnicity is the emergence of terrorist groups who use religion and other aspects of ethnicity to justify bringing death and destruction on their perceived enemies. As national governments attempt to cope with this latest phenomenon, they too are changing the nature of war and violence. The global agenda's current transformation, brought about by globalization, demands that our attention turn toward the latter's consequences.

The format of *Taking Sides: Clashing Views on Global Issues*, 4th edition, follows the successful formula of other books in the Taking Sides series. The book begins with an introduction to the emergence of global issues, the new age of globalization, and the effect of 9/11 and the international community's response to the world that ushered in the twentieth century. It then addresses twenty current global issues grouped into four parts. Population takes center stage in Part 1 because it not only represents a global issue by itself, but it also affects the parameters of most other global issues. Part 2 addresses a range of problems associated with global resources and their environmental impact. Parts 3 and 4 feature issues borne out of the emerging agenda of the twenty-first century. The former part examines widely disparate expanding forces and movements across national boundaries, such as global pandemics as well as international financial institutions. Part 4 focuses on new security issues in the post–Cold War and post–September 11 eras, such as whether the world is headed for a nuclear 9/11 or has U.S. hegemony rendered the United Nations irrelevant.

Each issue has two readings, one pro and one con. The readings are preceded by an issue *introduction* that sets the stage for the debate and briefly describes the two readings. Each issue concludes with a *postscript* that summarizes the issues, suggests further avenues for thought, and provides *suggestions for further reading*. At the back of the book is a listing of all the *contributors to this volume* with a brief biographical sketch of each of the prominent figures whose views are debated here.

Changes to this edition This fourth edition represents a significant revision. It contains twenty issues, an increase of one issue from the previous edition. Six of the issues are completely new: "Are Declining Growth Rates Rather than Rapid Population Growth Today's Major Global Population Problem?" (Issue 1); "Is the International Community Adequately Prepared to Address Global Health Pandemics?" (Issue 11); "Has the International Community Designed an Adequate Strategy to Address Human Trafficking?" (Issue 12); "Does Immigration Policy Affect Terrorism?" (Issue 16); "Can Nuclear Proliferation Be Stopped?" (Issue 19); and "Has U.S. Hegemony Rendered the United Nations Irrelevant?" (Issue 20). In addition, the scope of one issue was enlarged to include consequences for the entire planet rather than for just part of it: "Does Global Urbanization Lead Primarily to Undesirable Consequences?" (Issue 4). In nine of the remaining thirteen issues, one or both selections were replaced to bring the issues up to date. Only three of the twenty issues in the book have both original selections.

A word to the instructor An *Instructor's Manual With Test Questions* (multiple-choice and essay) is available through the publisher for the instructor using *Taking Sides* in the classroom. A general guidebook, *Using Taking Sides in the Classroom,* which discusses methods and techniques for integrating the pro-con approach into any classroom setting, is also available. An online version of *Using Taking Sides in the Classroom* and a correspondence service for *Taking Sides* adopters can be found at http://www.mhcls.com/usingts/.

Taking Sides: Clashing Views on Global Issues is only one title in the Taking Sides series. If you are interested in seeing the table of contents for any of the other titles, please visit the Taking Sides Web site at http://www.mhcls.com/takingsides/.

James E. Harf
University of Tampa

Mark Owen Lombardi
College of Santa Fe

Contents In Brief

Contents

Michael Meyer, a writer for *Newsweek International*, argues that the
new global population threat is not too many people as a consequence
of continuing high growth rates. On the contrary, declining birth rates
will ultimately lead to depopulation in many places on Earth, a virtual
population implosion, in both the developed and developing worlds.
Danielle Nierenberg, a research associate at the Worldwatch Institute,
and Mia MacDonald, a policy analyst and Worldwatch Institute senior
fellow, argue that the consequences of a still-rising population have
worsened in some ways because of the simultaneous existence of fast-
rising consumption patterns, creating a new set of concerns.

Robert McNamara, former president of the World Bank, argues in this
piece written during his presidency that the developed countries of the
world and international organizations should help the countries of the
developing world reduce their population growth rates. Steven W.
Mosher, president of the Population Research Institute, an organization
dedicated to debunking the idea that the world is overpopulated, argues
that McNamara's World Bank and other international financial lending
agencies have served for over a decade as "loan sharks" for those
groups and individuals who were pressuring developing countries to
adopt fertility reduction programs for self-interest reasons.

doomsdayers to task for their shoddy use of science. Bioscientist David Pimentel takes to task Lomborg's findings, accusing him of selective use of data to support his conclusions.

Red Cavaney, president and chief executive officer of the American Petroleum Institute, argues that recent revolutionary advances in technology will yield sufficient quantities of available oil for the foreseeable future. James Howard Kunster, author of *The Long Emergency 2005,* suggests that simply passing the all-time production peak of oil and heading toward its steady depletion will result in a global energy predicament that will substantially change our lives.

Sylvie Brunel, former president of Action Against Hunger, argues that "there is no doubt that world food production . . . is enough to meet the needs of" all the world's peoples. Janet Raloff, a writer for *Science News,* looks at a number of factors—declining per capita grain harvests, world's growing appetite for meat, the declining availability of fish for the developing world, and continuing individual poverty.

The summary of the most recent assessment of climatic change by a UN-sponsored group of scientists concludes that an increasing set of observations reveals that the world is warming and much of it is due to human intervention. Christopher Essex and Ross McKitrick, Canadian university professors of applied mathematics and economics, respectively, attempt to prove wrong the popularly held assumption that scientists know what is happening with respect to climate and weather, and thus understand the phenomenon of global warming.

Rosegrant and colleagues conclude that if current water policies
continue, farmers will find it difficult to grow sufficient food to meet the
world's needs. Water is not only plentiful but also a renewable
resource that, if properly treated as valuable, should not pose a future
problem.

PART 3   EXPANDING GLOBAL FORCES AND MOVEMENTS   161

Mr. Mayor, former director-general of UNESCO, suggests that drug
trafficking and consumption "constitute one of the most serious threats
to our planet," and the world must dry up the demand and attack the
financial power of organized crime. Harry G. Levine, professor of
sociology at Queens College, City University of New York, argues that
the emphasis on drug prohibition should be replaced by a focus on
"harm reduction," creating mechanisms to address tolerance, regulation,
and public health.

The document from the World Health Organization lays out a
comprehensive program of action for individual countries, the international
community, and WHO to address the next influenza pandemic. H. T.
Goranson, a former top national scientist with the U.S. government,
describes the grave dangers posed by global pandemics and highlights
flaws in the international community's ability to respond.

Janie Chuang, practitioner-in-residence at the American University Washington College of Law, suggests that governments have been finally motivated to take action against human traffickers as a consequence of the concern over national security implications of forced human labor movement and the involvement of transnational criminal syndicates. Dina Francesca Haynes, associate professor of law at New England School of Law, argues that none of the models underlying domestic legislation to deal with human traffickers is "terribly effective" in addressing the issue effectively.

Norberg argues that throughout history the expansion of trade through the capitalist system has created wealth in nations. He argues that developing countries need only overcome their own shortcomings and adopt this model within a globalizing world to take advantage of this reality. Tynes contends that globalization not only creates economic poverty among developing states but first world countries through mass media control, corporate action, and policies that exacerbate this problem and create deeper gaps between rich and poor.

George Monbiot argues that U.S. media control and the desire for profits have not created a political or social awakening around the globe but rather a quest for profits that spreads American culture in its basest forms without any inherent benefits. Thus, by implication, the world is suffering under the yoke of American cultural imperialism. Philippe Legrain is a British economist who presents two views of cultural

Political scientist Samuel P. Huntington argues that the emerging
conflicts of the twenty-first century will be cultural and not ideological.
He identifies the key fault lines of conflict and discusses how these
conflicts will reshape global policy. Wendell Bell, professor emeritus of
sociology at Yale University, argues that by emphasizing our common
humanity and shared values, cultural divisions will not be the defining
wave but rather shared mission and vision will characterize human
experience in this century.

Ira Straus, U.S. coordinator of the Independent International Committee on
Eastern Europe and Russia argues that non-proliferation has historically been a key component of U.S. policy but that it has been
subsumed with terrorism of late, and as such, we have lost precious
ground in the fight to maintain non-proliferation. As a result, the threat of
states such as North Korea and Iran is dangerous and runs the risk of
breaking the proliferation regime's back. Mirza Aslam Beg, former
chief of staff of the Pakistani army argues that proliferation combined
with deterrence leads to stability and not instability. He contends that
proliferation will happen naturally and therefore, the world community
should deal with its realities and not try and prevent it, thus creating a
set of regional balances that will lead to peace.

Tom DeWeese argues that the United Nations is principally responsible
for much of the chaos in the world and that its ability to positively impact
change is negligible. Therefore, it is irrelevant to U.S. policy, and thus
global interests, and should be dismantled. Shashi Tharoor argues that
the United Nations does a great deal of good and in this age of terrorism
and U.S. global hegemony, it is necessary to peacefully negotiate and
be a force for reason around the world.

Introduction

Global Issues in the Twenty-First Century

James E. Harf

Mark Owen Lombardi

Threats of the New Millennium

As the new millennium dawned, the world witnessed two very different events whose impacts were deeply felt immediately and that are still being experienced far and wide. One such episode was the tragedy of 9/11, a series of incidents that ushered in a new era of terrorism. It burst upon the international scene with the force of a mega-catastrophe, occupying virtually every waking moment of national and global leaders throughout the world and seizing the attention of the rest of the planet's citizens who had access to a television set for months to come. The focused interest of national policymakers was soon transformed into a war on terrorism, while average citizens sought to cope with changes brought on by both the tragic events of September 2001 and the global community's response to them. Both governmental leaders and citizens continue to address the consequences of this first intrusion of the new millennium on a world now far different in many ways since the pre-9/11 era.

The second event was the creation of a set of ambitious millennium development goals by the United Nations. In September 2000, with much fanfare, 189 national governments committed to eight major goals in an initiative known as the UN Millennium Development Goals (MDG): eradicate extreme poverty and hunger; achieve universal primary education; promote gender equality and empower women; reduce child mortality; improve maternal health; combat HIV/AIDS, malaria, and other diseases; ensure environmental sustainability, and develop a global partnership for development. This initiative was important not only because the UN was setting an actionable 15-year agenda against a relatively new set of global issues, but also because it signified a major change in how the international community would henceforth address such problems confronting humankind. The new initiative represented a recognition of: (1) shared responsibility between rich and poor nations for solving such problems; (2) a link between goals; (3) the paramount role to be played by national governments in the process: and (4) the need for measurable outcome indicators of success. The UN Millennium Development Goals initiative went virtually unnoticed by much of the public, although governmental decision-makers involved with the UN understood its significance. Jay Leno of NBC's *Tonight* show would have a field day questioning passersby on the street about their knowledge of this UN initiative.

These two major events, although vastly different, symbolize the world in which we now find ourselves, a world far more complex and more violent than either the earlier one characterized by the Cold War struggle between the United States and the Soviet Union, or the post–Cold War era of the 1990s, where global and national leaders struggled to identify and then find their proper place in the post–Cold War world order. Consider the first event, the 9/11 tragedy. It reminds us all that the use and abuse of power in pursuit of political goals in earlier centuries is still a viable option for those throughout the world who believe themselves disadvantaged because of various political, economic, or social conditions and structures. The only difference is the perpetrators' choice of military hardware and strategy. Formally declared wars fought by regular national military forces committed (at least on paper) to the tenets of just war theory have now been replaced by a plethora of military actions and other violent events whose defining characteristics conjure up terrorism, perpetrated by individuals without attachments to a regular military and/or without allegiance to a national government and country, and who do not hesitate to put ordinary citizens in harm's way.

On the other hand, the second event of the new century, the UN Millennium Goals initiative, symbolizes the other side of the global coin, the recognition that the international community is also beset with a number of problems unrelated to military actions or national security, at least in a direct sense. Rather, the past three or four decades have witnessed the emergence and thrust to prominence of a number of new problems relating to social, economic, and environment characteristics of the citizens that inhabit this planet. These problems impact the basic quality of life of global inhabitants in ways very different from the scourges of military violence. Yet they are just as dangerous and just as threatening. At the heart of this global change affecting not only its character but also its prevalence is a phenomenon called globalization.

The Emergence of the Age of Globalization

The Cold War era, marked by the domination of two superpowers in the decades following the end of World War II, has given way to a new era called "globalization." This new epoch is characterized by a dramatic shrinking of the globe in terms of travel and communication, increased participation in global policymaking by an expanding array of national and nonstate actors, and an exploding volume of problems with ever-growing consequences. While the tearing down of the Berlin Wall almost two decades ago dramatically symbolized the end of the Cold War era, the creation of the Internet graphically illustrates the emergence of the globalization era, while the fallen World Trade Center symbolizes a new paradigm for conflict and violence.

Early signs of this transition were manifested during the latter part of the twentieth century by a series of problems that transcended national boundaries. These problems caught the attention of policymakers who demanded comprehensive analyses and solutions. The effects of significant population growth in the developing world, for example, called for multilateral action by leaders in developed and developing nations alike. Acid rain, created by emissions from smokestacks, was the precursor to a host of environmental crises that appeared with increasing frequency, challenging the traditional political order and those that commanded it. Finally, much of the world began to sense that the planet is in reality "Spaceship Earth," with finite resources in danger of being exhausted and which require careful stewardship.

These global concerns remain today. Some are being addressed successfully, while others are languishing amid a lack of consensus about the nature of the problems and how to solve them. Also, with the shattering of the Cold War system and the advent of the globalization age, new issues have emerged that broaden our conception of the global agenda. In the past ten years, national borders have shrunk or disappeared, with a resultant increase in the movement of ideas, products, resources, finances and people throughout the globe. The ease with which people and objects move throughout the globe has greatly magnified another fear, the spread of disease. The term "epidemic" has been replaced by the phrase "global pandemic" as virulent scourges unleashed in one part of the globe now have greater potential to find their way to the far corners of the planet. The world has also come to fear an expanded potential for terrorism, as new technologies combined with old but strong hatreds have conspired to make the world far less safe than it had been. The pistol that killed the Austrian Archduke in Sarajevo in 1914 ushering in World War I has been replaced by the jumbo jet used as a missile to bring down the World Trade Center and with it, to snuff out the lives of thousands of innocent victims.

This increase in the movement of information and ideas has ushered in global concerns over cultural imperialism and religious/ethnic wars. The ability both to retrieve and disseminate information in the contemporary era will have an impact in this century as great as, if not greater than, the telephone, radio, and television in the last century. The potential for global good or ill is mindboggling. Finally, traditional notions of great-power security embodied in the Cold War rivalry have given way to concerns about terrorism, genocide, nuclear proliferation, cultural conflict, and the diminishing of international law.

Globalization heightens our awareness of a vast array of global issues that will challenge individuals as well as governmental and nongovernmental actors. Since the demise of the Cold War world, analysts and other observers have become free to define, examine, and explore solutions to such issues on a truly global scale. This text seeks to identify those issues that are central to the discourse on the impact of globalization. The issues in this volume provide a broad overview of the mosaic of global issues that will affect students' daily lives.

What Is a Global Issue?

We begin by addressing the basic characteristics of a *global issue*.[1] By definition, the word *issue* suggests disagreement along several related dimensions:

1. whether a problem exists and how it comes about;
2. the characteristics of the problem;

---

[1] The characteristics are extracted from James E. Harf and B. Thomas Trout, *The Politics of Global Resources,* Duke University Press, 1986, pp. 12–28.

3. the preferred future alternatives or solutions; and/or
4. how these preferred futures are to be obtained.

These problems are real, vexing, and controversial because policymakers bring to their analysis different historical experiences, values, goals, and objectives. These differences impede and may even prevent successful problem solving. In short, the key ingredient of an issue is disagreement.

The word *global* in the phrase *global issue* is what makes the set of problems confronting the human race today far different from those that challenged earlier generations. Historically, problems were confined to a village, city, or region. The capacity of the human race to fulfill its daily needs was limited to a much smaller space: the immediate environment. When walking and the horse were the principal modes of transportation, people pursued their food, fuel, and other necessities near their homes. When local resources were exhausted, people moved elsewhere in search of a more inviting environment. With the invention of the locomotive and the automobile, however, humans were able to travel greater distances on a daily basis. National energy and agricultural systems emerged, freeing humans from a reliance on their local community. Problems then became national, as disruptions in the production or distribution system in one part of a country had repercussions in some other part.

With the advent of transoceanic transportation capabilities, national distribution systems gave way to global systems. The "breadbasket" countries of the world—those best able to grow food—expanded production in order to serve an ever-larger global market. Oil-producing countries with vast reservoirs eagerly increased production in order to satisfy the needs of an ever-growing worldwide clientele. Countries that were best able to provide other resources to a larger part of the world entered the global trade system as well. These spatial changes were a consequence of significant increases in population and per capita consumption, both of which dramatically heightened the demand for resources.

Additionally, the larger machines of the Industrial Revolution were capable of moving greater quantities of dirt and of emitting greater amounts of pollutants into the atmosphere. With these new production and distribution capabilities came more prevalent manifestations of existing problems as well as the emergence of new ones. Providers either engaged in greater exploitation of their own environment or searched further afield in order to satisfy additional resource requirements.

The character of these new problems is thus different from those of earlier eras. First, they transcend national boundaries and impact virtually every corner of the globe. In effect, these issues help make national borders increasingly meaningless. Environmental pollution or poisonous gases do not recognize or respect national borders. Birds carrying the avian flu have no knowledge of political boundaries.

Second, these new issues cannot be resolved by the autonomous action of a single actor, be it a national government, international organization, or multinational corporation. A country cannot guarantee its own energy or food security without participating in a global energy or food system.

Third, these issues are characterized by a wide array of value systems. To a family in the developing world, giving birth to a fifth or sixth child may contribute to the family's immediate economic well-being. But to a research scholar at the United Nations Population Fund, the consequence of such an action multiplied across the entire developing world leads to expanding poverty and resource depletion.

Fourth, these issues will not go away. They require specific policy action by a consortium of local, national, and international leaders. Simply ignoring the issue cannot eliminate the threat of chemical or biological terrorism, for example. If global warming does exist, it will not disappear unless specific policies are developed and implemented.

These issues are also characterized by their persistence over time. The human race has developed the capacity to manipulate its external environment and, in so doing, has created a host of opportunities and challenges. The accelerating pace of technological change suggests that global issues will proliferate and will continue to challenge human beings throughout the next millennium.

In the final analysis, however, a global issue is defined as such only through mutual agreement by a host of actors within the international community. Some may disagree about the nature, severity, or presence of a given issue. These concerns then become areas of focus after a significant number of actors (states, international organizations, the United Nations, and others) begin to focus systematic and organized attention on the issue itself.

The Creation of the Global Issues Agenda

Throughout the first part of the Cold War, the international community found itself more often as an observer of, rather than a player in, the superpower conflict for global domination being waged by the United States and the Soviet Union. Other forces and trends competed for the attention of world leaders during this period. Economic development exploded in many parts of the globe, while other areas floundered in poverty. Resource consumption expanded alongside economic growth throughout the developed world, creating both pollution and scarcity. Population growth rates increased dramatically in societies that were unable to cope with the consequences of a burgeoning population. New forms of violence replaced traditional ways of employing force. And nonstate actors became a larger part of the conflict problem.

As the major participants in the Cold War settled into a system of routinized behavior designed to avoid catastrophic nuclear war, other actors in the international system began to worry about the disquieting effects associated with exploding population and consumption. Nation-states, international governmental organizations, and private groups sought to identify these growth-related problems and to seek solutions. The United Nations (UN) soon became the principal force to which those affected by such global behavior turned.

International community members began to join forces under the auspices of the UN to design comprehensive strategies. Such plans became known as international regimes—a set of agreements, structures, and plans of action involving all relevant global actors in addressing specific issues before them.

For example, in response to concerns of the Swedish government in the early 1970s about the discovery of a significant amount of dead fish in its lakes, the UN brought together relevant parties to discuss the problem, later acknowledged to be caused by acid rain. This 1972 conference, called the United Nations Conference on the Human Environment, represented the first major attempt of the international community to address the global ecological agenda. Representatives from 113 countries appeared in Stockholm, Sweden (the Soviet-bloc countries boycotted the conference for reasons unrelated to the environmental issue). Over 500 international nongovernmental organizations from such disparate interests such as the environment, science, women, religion, and business also sent delegates. Members of the United Nations Secretariat and specialized international governmental organizations such as the World Bank and the Food and Agricultural Organization (FAO) were present, as well. Experts from all walks of life—scientists, environmentalists, and others—attended.

The conference in Stockholm ushered in a new age of thinking about issues in their global context and became the model for future conferences on global issues. Position papers were prepared for delegates, and preparatory committees hammered out draft declarations and plans of action. These served as the starting points for conference deliberations. The final plan of action included the establishment of a formal mechanism, termed the United Nations Environmental Programme or UNEP, for addressing environmental issues in the future.

The most significant outcome of the conference was the creation of awareness among national government officials regarding the severity of ecological issues, the extent to which they cross national boundaries, and the need for international cooperation in order to solve these issues. The era of global issues identification, analysis, and problem solving was born.

Twenty years later, the global community reconvened in Rio de Janeiro, Brazil, at the United Nations Conference on Environment and Development, more popularly known as the Earth Summit. This meeting produced "Agenda 21," a report that outlined a plan of action for achieving development throughout the world while at the same time avoiding environmental degradation.

Another example of the international community's venture into this emerging agenda focused on the global population problem. In the early 1970s, the UN responded to the pleas of a variety of groups to consider the negative consequences of exploding population birth rates in many developing countries. The reason for such concern grew out of the belief that such growth was deterring economic growth in those areas of the world that were the most poverty stricken.

The result was the United Nations World Population Conference held in Bucharest, Romania, in 1974, and attended by delegates from 137 countries as well as by representatives of numerous governmental and nongovernmental organizations. The principal result of the conference was the placement of population on the global agenda and the development of a formal plan of action. The international community has continued its formal attention to the population issue, and has convened every ten years since Bucharest—in Mexico City in 1984 and in

Cairo in 1994—to evaluate strategy and assume new tasks to lower growth rates, as well as to address other population-related issues. A decade later, it is currently assessing the strategies agreed upon at Cairo. One thing is clear, however. High growth rates in developing countries have dropped dramatically, suggesting that international conferencing worked in this instance.

These kinds of conferences, where a disparate group of experts join a wide range of governments, private groups, and individuals, is now characteristic of how the global community is addressing the emerging global agenda. But the international community was far less certain of the wisdom of this strategy back in 1972. However, the success of the United Nations Conference on the Human Environment (now popularly known as the Stockholm Conference) led to similar strategies in a host of other environmental areas and provided the blueprint for the analysis of unforeseen global issues. For most of the issues in this text, the international community is either now formally addressing the problem or is considering its first steps.

The Nexus of Global Issues and Globalization

In 1989, the Berlin Wall fell, and with it a variety of assumptions, attitudes and expectations of the international system. This event did not usher in a utopia, nor did it irrevocably change all existing ideas. It did, however, create a void. As the dominant Cold War system entrenched since the end of World War II disintegrated, alternative views, issues, actors, and perspectives began to emerge to fill the void left by the collapse of that old system. What helped to bring down the previous era and what has emerged as the new dominant international construct is what we now call globalization.

Throughout the 1990s and into the twenty-first century, scholars and policymakers have struggled to define this new era. As the early years of the new century ushered in a different and heightened level of violence, a sense of urgency emerged. Some have analyzed the new era in terms of the victory of Western or American ideals, the dominance of global capitalism, and the spread of democracy vs. the have-nots of the world who use religious fanaticism as a ploy to rearrange power within the international system. Others have defined it simply in terms of the multiplicity of actors now performing on the world stage, and how states and their sovereignty have declined in importance and impact vis-à-vis others, such as multinational corporations, and nongovernmental groups like Greenpeace and Amnesty International. Still others have focused on the vital element of technology and its impact on communications, information storage and retrieval, and global exchange.

Whether globalization reflects one, two, or all of these characteristics is not as important as the fundamental realization that globalization is the dominant element of a new era in international politics. This new period is characterized by several basic traits that greatly impact the definition, analysis, and solution of global issues. They include the following:

- an emphasis on information technology;
- the increasing speed of information and idea flows;

- a need for greater sophistication and expertise to manage such flows;
- the control and dissemination of technology; and
- the cultural diffusion and interaction that come with information expansion and dissemination.

Each of these areas has helped shape a new emerging global issues agenda. Current issues remain important and, indeed, these factors help us to understand them on a much deeper level. Yet globalization has also created new problems and has brought them to the fore such that significant numbers of actors now recognize their salience in the international system.

For example, the spread of information technology has made ideas, attitudes, and information more available to people throughout the world. Americans in Columbus, Ohio, had the ability to log onto the Internet and speak with their counterparts in Kosovo to discover when NATO bombing had begun and to gauge the accuracy of later news reports on the bombing. Norwegian students can share values and customs directly with their counterparts in South Africa, thereby experiencing cultural attitudes firsthand without the filtering mechanisms of governments or even parents and teachers. Scientific information that is available through computer technology can now be used to build sophisticated biological and chemical weapons of immense destructive capability. Ethnic conflicts and genocide between groups of people are now global news, forcing millions to come to grips with issues of intervention, prevention, and punishment. And terrorists in different parts of the globe can communicate readily with one another, transferring plans and even money across national and continental boundaries with ease.

Globalization is an international system and it is also rapidly changing. Because of the fluid nature of this system and the fact that it is both relatively new and largely fueled by the amazing speed of technology, continuing issues are constantly being transformed and new issues are emerging regularly. The nexus of globalization and global issues has now become, in many ways, the defining dynamic of understanding global issues. Whether it is new forms of terrorism, new conceptions of security, expanding international law, solving ethnic conflicts, dealing with mass migration, coping with individual freedom and access to information, or addressing cultural clash and cultural imperialism, the transition from a Cold War world to a globalized world helps us understand in part what these issues are and why they are important.

Identifying the New Global Issues Agenda

The analysis of global issues by scholars and policymakers has changed. Our assumptions, ideas, and conceptions of what makes a problem a global issue have expanded. New technologies, new actors, and new strategies within a changing global order have made the study of global issues a broader and more complex undertaking.

The organization of this text reflects these phenomena. Parts 1 and 2 focus on the continuing global agenda of the post–Cold War era. The emphasis is on global population and environmental issues and the nexus between these two phenomena. Has the threat of uncontrolled world population growth subsided or will the built-in momentum of the past thirty years override any recent strides in slowing birth rates in the developing world? Should the international community continue to address this problem? Is global aging about to unleash a host of problems for governments of the developed world? Is rapid urbanization creating a whole new set of problems unique to such urban settings? Do environmentalists overstate their case or is the charge of "crying wolf" by environmental conservatives a misplaced attack? Is the world running out of natural resources, or is the concern of many about resource availability, be it food, oil, water, air, and/or pristine land, simply misguided? Should the world continue to rely on oil or should the search for viable alternatives take on a new urgency? Will the world be able to feed itself or provide enough water in the foreseeable future? Is global warming for real?

Part 3 addresses the consequences of the decline of national boundaries and the resultant increased international flow of information, ideas, money, and material things in this globalization age. Can the global community win the war on drugs? Is the international community prepared for the next global health pandemic? Has this community also designed an adequate strategy to address human trafficking? Is globalization a positive or negative development? Is America guilty of cultural imperialism? Do the actions of international financial institutions and multinational corporations enhance the capacity of poor countries to develop or are they simply another form of exploitation perpetrated on the developing world? Are the international media hurting global society?

Part 4 addresses the new global security dilemma. Does immigration policy affect terrorism? Are we headed for a nuclear 9/11? Are cultural and ethnic wars the defining dimensions of conflict in this century? Can nuclear proliferation be stopped? Has U.S. hegemony rendered the United Nations irrelevant?

The revolutionary changes of the last few decades present us with serious challenges unlike any others in human history. However, as in all periods of historic change, we possess significant opportunities to overcome problems. The task ahead is to define these issues, explore their context, and develop solutions that are comprehensive in scope and effect. The role of all researchers in this field, or any other, is to analyze objectively such problems and search for workable solutions. As students of global issues, your task is to educate yourselves about these issues and become part of the solution.

On the Internet . . .

Population Reference Bureau

The Population Reference Bureau provides current information on international population trends and their implications from an objective viewpoint. The PopNet section of this Web site offers maps with regional and country-specific population information as well as information divided by selected topics.

http://www.prb.org

United Nations Population Fund (UNFPA)

The United Nations Population Fund (UNFPA) was established in 1969 and was originally called the United Nations Fund for Population Activities. This organization works with developing countries to educate people about reproductive and sexual health as well as about family planning. The UNFPA also supports population and development strategies that will benefit developing countries and advocates for the resources needed to accomplish these tasks.

http://www.unfpa.org

Population Connection

Population Connection (formerly Zero Population Growth) is a national, nonprofit organization working to slow population growth and to achieve a sustainable balance between Earth's people and its resources. The organization seeks to protect the environment and to ensure a high quality of life for present and future generations.

http://www.populationconnection.org

The CSIS Global Aging Initiative

The Center for Strategic and International Studies (CSIS) is a public policy research institution that approaches the issue of the aging population in developed countries in a bipartisan manner. The CSIS is involved in a two-year project to explore the global implications of aging in developed nations and to seek strategies on dealing with this issue. This site includes a list of publications that were presented at previous events.

http://www.csis.org/gai/

The Population Council

The Population Council is an international, nonprofit organization that conducts research on population matters from biological, social science, and public health perspectives. It was established in 1952 by John D. Rockefeller, III.

http://www.popcouncil.org

The Population Institute

The Population Institute is an international, educational, non-profit organization that seeks to voluntarily reduce excessive population growth. Established in 1989 and headquartered in Washington, D.C., it has members in 172 countries.

http://www.populationinstitute.org

PART 1

Global Population

*I*t is not a coincidence that many of the global issues in this book emerged at about the same time as world population growth exploded. No matter what the issue, the presence of a large and fast-growing population along-side it exacerbates the issue and transforms its basic characteristics. In the new millennium, declining growth rates, which first appeared in the devel-oped world but are now also evident in many parts of the developing world, pose a different set of problems. The emergence of a graying popula-tion throughout the globe, but particularly in the developed, world has the potential for significant impact. And the rapid growth within urban areas of the developing world continues to pose a different set of problems. The ability of the global community to respond to any given issue is diminished by certain population conditions, be it an extremely young consuming pop-ulation in a poor country in need of producers, an expanding urban popu-lation whose local public officials are unable to provide an appropriate infrastructure, a large working-age group in a nation without sufficient jobs, or an ever-growing senior population for whom additional services are needed.*

Thus we begin this text with a series of issues directly related to various aspects of world population. It serves as both a separate global agenda and as a context within which other issues are examined.

- Are Declining Growth Rates Rather Than Rapid Population Growth Today's Major Global Population Problem?

- Should the International Community Attempt to Curb Population Growth in the Developing World?

- Is Global Aging in the Developed World a Major Problem?

- Does Global Urbanization Lead Primarily to Undesirable Consequences?

ISSUE 1

Are Declining Growth Rates Rather than Rapid Population Growth Today's Major Global Population Problem?

YES: Michael Meyer, from "Birth Dearth," *Newsweek* (September 27, 2004)

NO: Danielle Nierenberg and Mia MacDonald, from "The Population Story . . . So Far," *World Watch Magazine* (September/October 2004)

ISSUE SUMMARY

YES: Michael Meyer, a writer for *Newsweek International*, argues that the new global population threat is not too many people as a consequence of continuing high growth rates. On the contrary, declining birth rates will ultimately lead to depopulation in many places on Earth, a virtual population implosion, in both the developed and developing worlds.

NO: Danielle Nierenberg, a research associate at the Worldwatch Institute, and Mia MacDonald, a policy analyst and Worldwatch Institute senior fellow, argue that the consequences of a still-rising population have worsened in some ways because of the simultaneous existence of fast-rising consumption patterns, creating a new set of concerns.

Beginning in the late 1960s, demographers began to observe dramatic increases in population growth, particularly in the developing world. As a consequence, both policymakers and scholars focused on projections of rising birth rates and declining death rates, the consequences of the resultant increased population levels, and strategies for combating such growth. By the mid-1970s, growth rates, particularly in the developing world, were such that the doubling of the world's population was predicted to occur in only a few decades. Consensus on both the resultant level of population growth and its implications was not immediately evident, however. And indeed, as the last millennium was coming to an end, a billion people were added to the world's population in just 12 short years. Contrast this time frame with the fact that it had taken all of recorded history until 1830 for the planet to reach a population of one billion and 100 years for the second billion.

In the last decade of the millennium, however, something unforeseen happened. Population growth slowed, not only in the developed sector of the globe but also in the developing world. Headlines proclaiming this sweeping change appeared with increasing regularity: "The birth dearth," "Overpopulation? Fiddlesticks!" "Population Growth Slows," and "End of the Boom." The UN Population Fund lowered its short-term and long-term projections. With the turn of the century, individual demographers began to change their calculations as the release of yearly figures suggested the need to do so.

More specifically, no longer could observers simply place all developing countries into a high-growth category and reserve the low-growth category for countries of the developed world. It had become clear that an increasing number of poorer countries had begun to experience a dramatic drop in fertility as well as growth rates.

These recent trends have been surprising, as most observers had long believed that the built-in momentum of population growth in the last third of the twentieth century would have major impacts well into the new century. The logic was understandable, as numbers do matter. They contain a potential built-in momentum that, if left unchecked, creates a geometric increase in births. For example, let us assume that a young third-world mother gives birth to three daughters before the age of 20. In turn, each of these three daughters has three daughters before she reaches the age of 20. The result of such fertility behavior is a nine-fold (900 percent) increase in the population (minus a much lower mortality-rate influence) within 40 years, or two generations. Contrast this pattern with that of a young mother in the developed world, who is currently reproducing at or slightly above/below replacement level. In the latter part of the globe, daughters simply replace mothers, and granddaughters replace daughters. Add a third generation of fertility, and the developing-world family will have increased 27-fold within 60 years, while the developed-world family's size would remain virtually unchanged.

Thus, problems relating to population growth may, in fact, be at the heart of most global issues. If growth is a problem, control it and half the ecological battle is won. Fail to control it, and global problem-solvers will be swimming upstream against an ever-increasing current. It is for this reason that the question of "out-of-control" population growth was selected to be the first issue in this volume.

Michael Meyer describes how families in both the developed and developing worlds are choosing to have fewer children and chronicles what he believes to be an array of negative consequences associated with this population transition. Danielle Nierenberg and Mia MacDonald argue that even though the rate of growth has declined, the latter is applied to a much larger base than at any time in history, including the last century. They suggest that since the "largest generation in history has arrived" (1.2 billion people between the ages of 10 and 19), population levels at mid-century will be far higher than today unless certain fertility choices are made by this group.

YES

Michael Meyer

Birth Dearth

Everyone knows there are too many people in the world. Whether we live in Lahore or Los Angeles, Shanghai or Sao Paulo, our lives are daily proof. We endure traffic gridlock, urban sprawl and environmental depredation. The evening news brings variations on Ramallah or Darfur—images of Third World famine, poverty, pestilence, war, global competition for jobs and increasingly scarce natural resources.

Just last week the United Nations warned that many of the world's cities are becoming hopelessly overcrowded. Lagos alone will grow from 6.5 million people in 1995 to 16 million by 2015, a miasma of slums and decay where a fifth of all children will die before they are 5. At a conference in London, the U.N. Population Fund weighed in with a similarly bleak report: unless something dramatically changes, the world's 50 poorest countries will triple in size by 2050, to 1.7 billion people.

Yet this is not the full story. To the contrary, in fact. Across the globe, people are having fewer and fewer children. Fertility rates have dropped by half since 1972, from six children per woman to 2.9. And demographers say they're still falling, faster than ever. The world's population will continue to grow—from today's 6.4 billion to around 9 billion in 2050. But after that, it will go sharply into decline. Indeed, a phenomenon that we're destined to learn much more about—depopulation—has already begun in a number of countries. Welcome to the New Demography. It will change everything about our world, from the absolute size and power of nations to global economic growth to the quality of our lives.

This revolutionary transformation will be led not so much by developed nations as by the developing ones. Most of us are familiar with demographic trends in Europe, where birthrates have been declining for years. To reproduce itself, a society's women must each bear 2.1 children. Europe's fertility rates fall far short of that, according to the 2002 U.N. population report. France and Ireland, at 1.8, top Europe's childbearing charts. Italy and Spain, at 1.2, bring up the rear. In between are countries such as Germany, whose fertility rate of 1.4 is exactly Europe's average. What does that mean? If the U.N. figures are right, Germany could shed nearly a fifth of its 82.5 million people over the next 40 years—roughly the equivalent of all of east Germany, a loss of population not seen in Europe since the Thirty Years' War.

From *Newsweek*, vol. 144, issue 13, September 27, 2004, pp. 54-61. Copyright © 2004 by Newsweek. Reprinted by permission.

And so it is across the Continent. Bulgaria will shrink by 38 percent, Romania by 27 percent, Estonia by 25 percent. "Parts of Eastern Europe, already sparsely populated, will just empty out," predicts Reiner Klingholz, director of the Berlin Institute for Population and Development. Russia is already losing close to 750,000 people yearly. (President Vladimir Putin calls it a "national crisis.") So is Western Europe, and that figure could grow to as much as 3 million a year by midcentury, if not more.

The surprise is how closely the less-developed world is following the same trajectory. In Asia it's well known that Japan will soon tip into population loss, if it hasn't already. With a fertility rate of 1.3 children per woman, the country stands to shed a quarter of its 127 million people over the next four decades, according to U.N. projections. But while the graying of Japan (average age: 42.3 years) has long been a staple of news headlines, what to make of China, whose fertility rate has declined from 5.8 in 1970 to 1.8 today, according to the U.N.? Chinese census data put the figure even lower, at 1.3. Coupled with increasing life spans, that means China's population will age as quickly in one generation as Europe's has over the past 100 years, reports the Center for Strategic and International Studies in Washington. With an expected median age of 44 in 2015, China will be older on average than the United States. By 2019 or soon after, its population will peak at 1.5 billion, then enter a steep decline. By midcentury, China could well lose 20 to 30 percent of its population every generation.

The picture is similar elsewhere in Asia, where birthrates are declining even in the absence of such stringent birth-control programs as China's. Indeed, it's happening despite often generous official incentives to procreate. The industrialized nations of Singapore, Hong Kong, Taiwan and South Korea all report subreplacement fertility, says Nicholas Eberstadt, a demographer at the American Enterprise Institute in Washington. To this list can be added Thailand, Burma, Australia and Sri Lanka, along with Cuba and many Caribbean nations, as well as Uruguay and Brazil. Mexico is aging so rapidly that within several decades it will not only stop growing but will have an older population than that of the United States. So much for the cliche of those Mexican youths swarming across the Rio Grande? "If these figures are accurate," says Eberstadt, "just about half of the world's population lives in subreplacement countries."

There are notable exceptions. In Europe, Albania and the outlier province of Kosovo are reproducing energetically. So are pockets of Asia: Mongolia, Pakistan and the Philippines. The United Nations projects that the Middle East will double in population over the next 20 years, growing from 326 million today to 649 million by 2050. Saudi Arabia has one of the highest fertility rates in the world, 5.7, after Palestinian territories at 5.9 and Yemen at 7.2. Yet there are surprises here, too. Tunisia has tipped below replacement. Lebanon and Iran are at the threshold. And though overall the region's population continues to grow, the increase is due mainly to lower infant mortality; fertility rates themselves are falling faster than in developed countries, indicating that over the coming decades the Middle East will age far more rapidly than other regions of the world. Birthrates in Africa remain high, and despite the AIDS epidemic its population is projected to keep growing. So is that of the United States.

We'll return to American exceptionalism, and what that might portend. But first, let's explore the causes of the birth dearth, as outlined in a pair of new books on the subject. "Never in the last 650 years, since the time of the Black Plague, have birth and fertility rates fallen so far, so fast, so low, for so long, in so many places," writes the sociologist Ben Wattenberg in "Fewer: How the New Demography of Depopulation Will Shape Our Future." Why? Wattenberg suggests that a variety of once independent trends have conjoined to produce a demographic tsunami. As the United Nations reported last week, people everywhere are leaving the countryside and moving to cities, which will be home to more than half the world's people by 2007. Once there, having a child becomes a cost rather than an asset. From 1970 to 2000, Nigeria's urban population climbed from 14 to 44 percent. South Korea went from 28 to 84 percent. So-called megacities, from Lagos to Mexico City, have exploded seemingly overnight. Birth rates have fallen in inverse correlation.

Other factors are at work. Increasing female literacy and enrollment in schools have tended to decrease fertility, as have divorce, abortion and the worldwide trend toward later marriage. Contraceptive use has risen dramatically over the past decade; according to U.N. data, 62 percent of married or "in union" women of reproductive age are now using some form of nonnatural birth control. In countries such as India, now the capital of global HIV, disease has become a factor. In Russia, the culprits include alcoholism, poor public health and industrial pollution that has whacked male sperm counts. Wealth discourages childbearing, as seen long ago in Europe and now in Asia. As Wattenberg puts it, "Capitalism is the best contraception."

The potential consequences of the population implosion are enormous. Consider the global economy, as Phillip Longman describes it in another recent book, "The Empty Cradle: How Falling Birthrates Threaten World Prosperity and What to Do About It." A population expert at the New America Foundation in Washington, he sees danger for global prosperity. Whether it's real estate or consumer spending, economic growth and population have always been closely linked. "There are people who cling to the hope that you can have a vibrant economy without a growing population, but mainstream economists are pessimistic," says Longman. You have only to look at Japan or Europe for a whiff of what the future might bring, he adds. In Italy, demographers forecast a 40 percent decline in the working-age population over the next four decades—accompanied by a commensurate drop in growth across the Continent, according to the European Commission. What happens when Europe's cohort of baby boomers begins to retire around 2020? Recent strikes and demonstrations in Germany, Italy, France and Austria over the most modest pension reforms are only the beginning of what promises to become a major sociological battle between Europe's older and younger generations.

That will be only a skirmish compared with the conflict brewing in China. There market reforms have removed the cradle-to-grave benefits of the planned economy, while the Communist Party hasn't constructed an adequate social safety net to take their place. Less than one quarter of the population is covered by retirement pensions, according to CSIS. That puts the burden of elder care almost entirely on what is now a generation of only children. The one-child policy has led to the so-called 4-2-1 problem, in which each child will be potentially responsible for caring for two parents and four grandparents.

Incomes in China aren't rising fast enough to offset this burden. In some rural villages, so many young people have fled to the cities that there may be nobody left to look after the elders. And the aging population could soon start to dull China's competitive edge, which depends on a seemingly endless supply of cheap labor. After 2015, this labor pool will begin to dry up, says economist Hu Angang. China will have little choice but to adopt a very Western-sounding solution, he says: it will have to raise the education level of its work force and make it more productive. Whether it can is an open question. Either way, this much is certain: among Asia's emerging economic powers, China will be the first to grow old before it gets rich.

Equally deep dislocations are becoming apparent in Japan. Akihiko Matsutani, an economist and author of a recent best seller, "The Economy of a Shrinking Population," predicts that by 2009 Japan's economy will enter an era of "negative growth." By 2030, national income will have shrunk by 15 percent. Speculating about the future is always dicey, but economists pose troubling questions. Take the legendarily high savings that have long buoyed the Japanese economy and financed borrowing worldwide, especially by the United States. As an aging Japan draws down those assets in retirement, will U.S. and global interest rates rise? At home, will Japanese businesses find themselves competing for increasingly scarce investment capital? And just what will they be investing in, as the country's consumers grow older, and demand for the latest in hot new products cools off? What of the effect on national infrastructure? With less tax revenue in state coffers, Matsutani predicts, governments will increasingly be forced to skimp on or delay repairs to the nation's roads, bridges, rail lines and the like. "Life will become less convenient," he says. Spanking-clean Tokyo might come to look more like New York City in the 1970s, when many urban dwellers decamped for the suburbs (taking their taxes with them) and city fathers could no longer afford the municipal upkeep. Can Japanese cope? "They will have to," says Matsutani. "There's no alternative."

Demographic change magnifies all of a country's problems, social as well as economic. An overburdened welfare state? Aging makes it collapse. Tensions over immigration? Differing birthrates intensify anxieties, just as the need for imported labor rises—perhaps the critical issue for the Europe of tomorrow. A poor education system, with too many kids left behind? Better fix it, because a shrinking work force requires higher productivity and greater flexibility, reflected in a new need for continuing job training, career switches and the health care needed to keep workers working into old age.

In an ideal world, perhaps, the growing gulf between the world's wealthy but shrinking countries and its poor, growing ones would create an opportunity. Labor would flow from the overpopulated, resource-poor south to the depopulating north, where jobs would continue to be plentiful. Capital and remittance income from the rich nations would flow along the reverse path, benefiting all. Will it happen? Perhaps, but that presupposes considerable labor mobility. Considering the resistance Europeans display toward large-scale immigration from North Africa, or Japan's almost zero-immigration policy, it's hard to be optimistic. Yes, attitudes are changing. Only a decade ago, for instance, Europeans also spoke of zero immigration. Today they recognize the need and, in bits and pieces, are beginning to plan for it. But will it happen on the scale required?

A more probable scenario may be an intensification of existing tensions between peoples determined to preserve their beleaguered national identities on the one hand, and immigrant groups on the other seeking to escape overcrowding and lack of opportunity at home. For countries such as the Philippines—still growing, and whose educated work force looks likely to break out of low-status jobs as nannies and gardeners and move up the global professional ladder—this may be less of a problem. It will be vastly more serious for the tens of millions of Arab youths who make up a majority of the population in the Middle East and North Africa, at least half of whom are unemployed.

America is the wild card in this global equation. While Europe and much of Asia shrinks, the United States' indigenous population looks likely to stay relatively constant, with fertility rates hovering almost precisely at replacement levels. Add in heavy immigration, and you quickly see that America is the only modern nation that will continue to grow. Over the next 45 years the United States will gain 100 million people, Wattenberg estimates, while Europe loses roughly as many.

This does not mean that Americans will escape the coming demographic whammy. They, too, face the problems of an aging work force and its burdens. (The cost of Medicare and Social Security will rise from 4.3 percent of GDP in 2000 to 11.5 percent in 2030 and 21 percent in 2050, according to the Congressional Budget Office.) They, too, face the prospect of increasing ethnic tensions, as a flat white population and a dwindling black one become gradually smaller minorities in a growing multicultural sea. And in our interdependent era, the troubles of America's major trading partners—Europe and Japan—will quickly become its own. To cite one example, what becomes of the vaunted "China market," invested in so heavily by U.S. companies, if by 2050 China loses an estimated 35 percent of its workers and the aged consume an ever-greater share of income?

America's demographic "unipolarity" has profound security implications as well. Washington worries about terrorism and failing states. Yet the chaos of today's fragmented world is likely to prove small in comparison to what could come. For U.S. leaders, Longman in "The Empty Cradle" sketches an unsettling prospect. Though the United States may have few military competitors, the technologies by which it projects geopolitical power—from laser-guided missiles and stealth bombers to a huge military infrastructure—may gradually become too expensive for a country facing massively rising social entitlements in an era of slowing global economic growth. If the war on terrorism turns out to be the "generational struggle" that national-security adviser Condoleezza Rice says it is, Longman concludes, then the United States might have difficulty paying for it.

None of this is writ, of course. Enlightened governments could help hold the line. France and the Netherlands have instituted family-friendly policies that help women combine work and motherhood, ranging from tax credits for kids to subsidized day care. Scandinavian countries have kept birthrates up with generous provisions for parental leave, health care and part-time employment. Still, similar programs offered by the shrinking city-state of Singapore—including a state-run dating service—have done little to reverse the birth dearth. Remember, too, that such prognoses have been wrong in the past. At the cusp of the postwar baby boom, demographers predicted a sharp fall in fertility and a global birth dearth. Yet even if this generation of seers turns out to be right, as seems likely, not all is bad. Environmentally, a smaller world is almost certainly a better world, whether in terms of cleaner air or, say, the return of wolves and rare flora to abandoned stretches of the East German countryside. And while people are living longer, they are also living healthier—at least in the developed world. That means they can (and probably should) work more years before retirement.

Yes, a younger generation will have to shoulder the burden of paying for their elders. But there will be compensations. As populations shrink, says economist Matsutani, national incomes may drop—but not necessarily per capita incomes. And in this realm of uncertainty, one mundane thing is probably sure: real-estate prices will fall. That will hurt seniors whose nest eggs are tied up in their homes, but it will be a boon to youngsters of the future. Who knows? Maybe the added space and cheap living will inspire them to, well, do whatever it takes to make more babies. Thus the cycle of life will restore its balance. . . .

**Danielle Nierenberg and
Mia MacDonald**

 **NO**

The Population Story . . . So Far

Forty years ago, the world's women bore an average of six children each. Today, that number is just below three. In 1960, 10–15 percent of married couples in developing countries used a modern method of contraception; now, 60 percent do.

To a considerable extent, these simple facts sum up the change in the Earth's human population prospects, then and now. In the mid-1960s, it was not uncommon to think about the human population as a time bomb. In 1971, population biologist Paul Ehrlich estimated that if human numbers kept increasing at the high rates of the time, by around 2900 the planet would be teeming with sixty million billion people (that's 60,000,000,000,000,000). But the rate of population rise actually peaked in the 1960s and demographers expect a leveling-off of human numbers this century.

Every couple of years the United Nations Population Division issues projections of human population growth to 2050. In 2002, UN demographers predicted a somewhat different picture of human population growth to mid-century than what the "population bombers" thought likely a generation ago. World population, growing by 76 million people every year (about 240,000 people per day), will pass 6.4 billion this year. The latest UN mid-range estimate says there will be about 8.9 billion people on Earth by 2050. And, according to this new scenario, total population will begin to shrink over the next hundred years.

These numbers are leading some people to say that the population bomb has been defused. A few nations, such as Italy and Japan, are even worried that birth rates are too low and that their graying populations will be a drain on the economy. (Some studies suggest that China, the world's most populous country, may also "need" more people to help support the hundreds of millions who will retire in coming decades).

We're not out of the woods yet. While the annual rate of population growth has decreased since 1970—from about 2 percent to 1.3 percent today—*the rate is applied to a much larger population* than ever before, meaning that the added yearly increments to the population are also much larger. These numbers show that the largest generation in history has arrived: 1.2 billion people are between 10 and 19. In large measure, it will be their choices—those they have, and those they make—that determine where the global population meter rests by mid-century.

From POPULATION AND ITS DISCONTENTS, September/October 2004, pp. 14–17. Copyright © 2004 by Worldwatch Institute. Reprinted by permission.

Population × Consumption

Potential for catastrophe persists. In many places, population growth is slowly smoldering but could turn into a fast burn. Countries as diverse as Ethiopia, the Democratic Republic of Congo, and Pakistan are poised to more than double their size by 2050 even as supplies of water, forests, and food crops are already showing signs of strain and other species are being squeezed into smaller and smaller ranges. Arid Yemen will likely see its population quadruple to 80 million by 2050. The UN estimates that populations in the world's 48 least-developed countries could triple by 2050. And if the world's women have, on average, a half a child more than the UN predicts, global population could grow to 10.6 billion by mid-century.

But it is a mistake to think that population growth is only a problem for developing countries. While consumption levels need to increase among the 2.8 billion people who now live on less than $2 a day, high rates of population growth combined with high levels of consumption in rich countries are taking a heavy toll on the Earth's natural resources:

- Carbon dioxide levels today are 18 percent higher than in 1960 and an estimated 31 percent higher than they were at the onset of the Industrial Revolution in 1750.
- Half the world's original forest cover is gone and another 30 percent is degraded or fragmented.
- Industrial fleets have fished out at least 90 percent of all large ocean predators—tuna, marlin, swordfish, cod, halibut, skate, and flounder—in just the past 50 years, according to a study in *Nature* in 2003.
- An estimated 10–20 percent of the world's cropland, and more than 70 percent of the world's rangelands, are degraded.

As global consumption of oil, meat, electricity, paper products, and a host of consumer goods rises, the impact of population numbers takes on a new relevance. Although each new person increases total demands on the Earth's resources, the size of each person's "ecological footprint"—the biologically productive area required to support that person—varies hugely from one to another. The largest ecofootprints belong to those in the industrialized world.

Further, new demographic trends can have significant impacts as well. Since 1970, the number of people living together in one household has declined worldwide, as incomes have risen, urbanization has accelerated and families have gotten smaller. With fewer people sharing energy, appliances, and furnishings, consumption actually rises. A one-person household in the United States uses about 17 percent *more* energy per person than a two-person home.

And while some nations are getting nervous about declining birth rates, for most of the world the end of population growth is anything but imminent. Although fertility rates are ratcheting down, this trajectory is not guaranteed. Projections of slower population growth assume that more couples will be able to choose to have smaller families, and that investment in reproductive health keeps pace with rising demand. But along the route to the eventual leveling-off of global population, plateaus are possible. And smaller families are not guaranteed in countries where government resources are strained or where health care, education, and women's rights are low on the list of priorities.

In the West African country of Niger, for example, the availability of family planning and reproductive health services has declined, while birth rates have increased. According to a recent report by the World Bank, the average woman in Niger will give birth to eight children in her lifetime, up from seven in 1998 and more than women in any other nation. Niger is already bulging with young people; 50 percent of the population is under age 15 and 70 percent is under 25.

Biology ≠ Destiny

A series of global conferences in the 1990s—spanning the Rio Earth Summit in 1992, the Cairo population conference (1994), the Beijing women's conference (1995), and the UN's Millennium Summit in 2000—put issues of environment, development, poverty, and women's rights on the global policy table. As a result, discussions of the relationship between growing human numbers and the Earth's ability to provide are increasingly framed by the realities of gender relations. It is now generally agreed that while enabling larger numbers of women and men to use modern methods of family planning is essential, it is not sufficient. Expanding the choices, capacities, and agency of women has become a central thread in the population story. Consumption—what we need and what we want—is, too.

Many studies have shown that women with more education have smaller, healthier families, and that their children have a better chance of making it out of poverty. Likewise, wealthier women and those with the right to make decisions about their lives and bodies also have fewer children. And women who have the choice to delay marriage and childbearing past their teens tend to have fewer children than those women—and there are millions of them still—who marry before they've completed the transition from adolescence. Equalizing relations between women and men is also a social good: not only is it just, but a recent World Bank report found that in developing countries where gender equality lags, efforts to combat poverty and increase economic growth lag, too.

Yet women's rights and voices remain suppressed or muted throughout the world. Over 100 million girls will be married before their 18th birthdays in the next decade, some as young as 8 or 9. Early childbearing is the leading cause of death and disability for women between the ages of 15 and 19 in developing countries. At least 350 million women still lack access to a full range of contraceptive methods, 10 years after the Cairo conference yielded a 20-year plan to balance the world's people with its resources. Demand for services will increase an estimated 40 per cent by 2025.

The assault of HIV/AIDS is also increasingly hurting women: more than 18 million women are living with HIV/AIDS, and in 2003 women's rate of infection for the first time equaled men's. In the region hardest hit, sub-Saharan Africa, 60 percent of adults living with HIV are women. Two-thirds of the world's 876 million illiterates are women and a majority of the 115 million children not attending grade school are girls. In no country in the world are women judged to have political, economic, and social power equal to that of men.

Even in the United States, women's reproductive rights are increasingly constrained by the growing number of restrictions and conditions on choice imposed by state and federal laws. Like the U.S. lifestyle, the current Administration's blinkered view of sexuality has gone global. The United States has withheld $34 million from the UN Population Fund (UNFPA) every year of the Bush Administration due to a dispute over abortion. And the "global gag rule," a relic of the Reagan presidency reimposed by President Bush, binds U.S. population assistance by making taboo any discussion of abortion in reproductive health clinics, even in countries where it is legal.

The impacts reach more deeply than the rhetoric: due to the loss of U.S. population funds, reproductive health services have been scaled back or eliminated in some of the world's poorest countries, precisely where fertility rates are highest and women's access to family planning most tenuous. In Kenya, for instance, the two main providers of reproductive health services refused to sign a pledge to enforce the gag rule, with the result that they lost funds and closed five family planning clinics, eliminating women's access to maternal health care, contraception, and voluntary counseling and testing for HIV/AIDS. In Ethiopia, where only 6 per cent of women use modern methods of contraception, the gag rule has cut a wide swath: clinics have reduced services, laid off staff and curtailed community health programs; many have suffered shortages of contraceptive supplies.

Need ↑ Funds ↓

A recent study by UNFPA and the Alan Guttmacher Institute estimated that meeting women's current unmet need for contraception would prevent each year:

- 23 million unplanned births
- 22 million induced abortions
- 1.4 million infant deaths
- 142,000 pregnancy related-deaths (including 53,000 from unsafe abortions); and
- 505,000 children losing their mothers due to pregnancy-related causes.

The non-medical benefits are not quantified but are considerable: greater self-esteem and decision-making power for women; higher productivity and income; increased health, nutrition, and education expenditures on each child; higher savings and investment rates; and increased equality between women and men. We know this from experience: recent research in the United States, for example, ascribes the large numbers of women entering law, medical, and other professional training programs in the 1970s to the expanded choices afforded by the wide availability of the Pill.

Despite these benedits, vast needs go unmet as the Cairo action plan remains underfunded. The United States is not the only culprit. UNFPA reports that donor funds for a basic package of reproductive health services and population data and policy work totaled about $3.1 billion in 2003–$2.6 billion less than the level agreed to in the ICPD Program. Developing country domestic resources were estimated at $11.7 billion, a major portion of which is spent by just a handful of large countries. A number of countries, particularly the poorest, rely heavily on donor funds to provide services for family planning, reproductive health, and HIV/AIDS, and to build data sets and craft needed policies.

A year from now, donors will be expected to be contributing $6.1 billion annually, $3 billion more than what has already been spent. "A world that spends $800 billion to $1 trillion each year on the military can afford the equivalent of slightly more than one day's military spending to close Cairo's $3 billion external funding gap to save and improve the lives of millions of women and families in developing countries," says UNFPA's executive director, Thoraya Obaid. But as the world's priorities lie in other arenas, it is looking increasingly unlikely that the Cairo targets—despite their modest price tag in a world where the bill for a war can top $100 billion—will be met.

But it isn't only poor people in developing countries who will determine whether the more dire population scenarios pass from speculation to reality. Family size has declined in most wealthy nations, but the U.S. population grew by 32.7 million people (13.1 percent) during the 1990s, the largest number in any 10-year period in U.S. history. At about 280 million people, the United States is now the third most populous nation in the world and its population is expected to reach 400 million by 2050. A recent study suggests that if every person alive today consumed at the rate of an average person in the United States, three more planets would be required to fulfill these demands.

Whether or not birth rates continue to fall, consumption levels and patterns (affluence), coupled with technology, take on new importance. The global consumer class—around 1.7 billion people, or more than a quarter of humanity—is growing rapidly. These people are collectively responsible for the vast majority of meat-eating, paper use, car driving, and energy consumption on the planet, as well as the resulting impact of these activities on its natural resources. As populations surge in developing countries and the world becomes increasingly globalized, more and more people have access to, and the means to acquire, a greater diversity of products and services than ever before.

It is the combined effect of human numbers and human consumption that creates such potent flash-points. Decisions about sexuality and lifestyle are among the most deeply personal and political decisions societies and their citizens can make. The fate of the human presence on the Earth will be shaped in large part by those decisions and how their implications unfold in the coming years. This population story's ending still hasn't been written.

POSTSCRIPT

Are Declining Growth Rates Rather than Rapid Population Growth Today's Major Global Population Problem?

The growth issue can be structured most simply as one of an insurmountable built-in momentum vs. dramatic change in fertility attitudes and behavior. Demographers are correct when they assert that the population explosion of the latter part of the twentieth century had the potential of future fertility disaster because of the high percentage of the population who were either in the middle of or about to enter their reproductive years. They are also correct when they assert that the last decade has witnessed major transitions away from high growth rates, even in many parts of the developing world.

The key word is "potential." Its relevance grows out of the built-in momentum that has caused the developing world's actual population to rise substantially in the last thirty-five years despite a decline in the growth rate. In its 2001 analysis of population patterns, the United Nations suggested that population growth in the first half of the twenty-first century would increase by over 50 percent, or by more than 3 billion (United Nations, *World Population: The 2000 Revision—Highlights,* 2001). This was higher than its projections of just two years earlier, due primarily to higher projected fertility levels in countries that are slow to show signs of fertility decline.

The entire growth of the first half of the twenty-first century would take place in the developing world, according to the study. It acknowledged that population would also grow in the developed world during the first 25 years of the new century, but then it would decline to levels approximating 2000 by mid-century. On the other hand, despite lowering birth rates in the less developed world, the built-in momentum would result in a 65 percent projected growth (from 4.9 billion in 2000 to 8.1 billion in 2050) during the first half of the century.

The UN revised its earlier predictions in 2002, tempering projections for future growth (*World Population Prospects: The 2002 Revision*). The executive director of the UN Population Fund explained it by suggesting that "men and women in larger numbers were making their own decisions on birth spacing and family size, contributing to slower population growth."

It has become increasingly clear that today's youth will not produce at the same level as their parents and grandparents, based on the evidence of the last quarter-century. Two factors are at work here. The demographic transition is evident in those countries of the developing world that are experiencing economic growth. Birth rates have dropped significantly, leading to lowered growth rates. In over one-third of the world's countries, containing 43 percent of the globe's population, women are having no more than two children on average. But while the rates are higher than those for the newly industrializing countries of the third world, growth rates for the remaining developing countries have also dropped in a large number of cases. In the latter situation, policy intervention lies at the heart of such lowered rates. The latter effort has been spearheaded by the United Nations and includes the work of many non-governmental organizations as well. In global conferences held every 10 years (1974, 1984, and 1994), the entire international community has systematically addressed the problem of third-world fertility, admittedly from different perspectives. Yet world population is still growing at a 1.2 percent rate, adding 80 million people annually.

Acknowledging declining birth rates still begs the question: Is there still a built-in momentum that will lead to negative consequences? William P. Butz addresses this question in *The Double Divide: Implosionists and Explosionists Endanger Progress Since Cairo* (Population Reference Bureau, September 2004). The Implosionists argue that the falling birth rate throughout the world is the most important variable. They suggest that the biggest global population challenge in the immediate future will be how to cope with the wide range of challenges confronting countries with declining fertility rates. The Explosionists counter that there will still be substantial population growth, even in those countries that have recently begun to experience low birth rates. To them, world leaders and organizations must not ignore those problems that emerge from populations whose percentage of young are still quite high.

In a sense, both authors are correct, as each acknowledges declining birth rates in the developing world and an eventual leveling-off of growth there. Their disagreement does point out dramatic implications of ever-so-slight variations in both the timing and the degree of fertility reduction among the poorer countries.

The United Nations (www.un.org/popin) serves as an authoritative source on various population data, whether historical, current, or future oriented. One of the UN agencies, the United Nations Population Fund or UNFPA (www.unfpa.org), issues an annual *State of the World Population*, as well as other reports.

Two Washington private organizations, the Population Reference Bureau (PRB) (www.prb.org) and The Population Institute (www.populationinstitute.org), publish a variety of booklets, newsletters, and reports yearly. Admittedly, these sources tend to emphasize the continued urgency rather than the seeds of progress, although recent articles have described the positive aspects of the current population transition. One particularly useful PRB publication is *World Population Beyond Six Billion* (*Population Bulletin*, March 1999). Another important PRB publication is *Global Demographic Divide* by Mary M. Kent and Carl Hub (2005). Still another widely read PRB piece is *Transitions in World Population* (March 2004). Other sources focus on either success stories or the potential for success growing out of recent policy intervention. The Population Council of New York (www.popcouncil.org) falls into the latter category.

For over a decade, until the death of one of the participants, two individuals took center stage in the debate over population growth and its implications. Paul Ehrlich led the call for vigorous action to curb population growth. His co-authored works, *The Population Bomb* (1971) and *The Population Explosion* (1990), advanced the notion that the Earth's resource base could not keep pace with population growth, and thus the survival of the planet was brought into question. The late Julian Simon's *Population Matters: People, Resources, Environment, and Immigration* (1990) and *The Ultimate Resource 2* (1996) challenge Ehrlich's basic thesis. Simon's place in the debate appears to have been assumed by Ronald Bailey, science correspondent for *Reason* magazine. Two important publications from Bailey are *Global Warming and Other Eco-Myths* (Forum, 2002) and *Earth Report 2000* (McGraw-Hill, 2000).

A succinct, centrist, and easily understood analysis of the future of world population can be found in Leon F. Bouvier and Jane T. Bertrand, *World Population: Challenges for the 21st Century* (1999). The annual *State of the World* volume from the Worldwatch Institute typically includes a timely analysis on some aspect of the world population problem.

Numerous web sites can be found under "world population" on www.msn.com (type in "world population web sites").

ISSUE 2

Should the International Community Attempt to Curb Population Growth in the Developing World?

YES: **Robert S. McNamara**, from "The Population Explosion," *The Futurist* (November/December 1992)

NO: **Steven W. Mosher**, from "McNamara's Folly: Bankrolling Family Planning," *PRI Review* (March–April 2003)

ISSUE SUMMARY

YES: Robert McNamara, former president of the World Bank, argues in this piece written during his presidency that the developed countries of the world and international organizations should help the countries of the developing world reduce their population growth rates.

NO: Steven W. Mosher, president of the Population Research Institute, an organization dedicated to debunking the idea that the world is overpopulated, argues that McNamara's World Bank and other international financial lending agencies have served for over a decade as "loan sharks" for those groups and individuals who were pressuring developing countries to adopt fertility reduction programs for self-interest reasons.

The history of the international community's efforts to lower birth rates throughout the developing world goes back to the late 1960s, when the annual growth rate hovered around 2.35 percent. At that time, selected individuals in international governmental organizations, including the United Nations, were persuaded by a number of wealthy national governments as well as by international nongovernmental population agencies that a problem of potentially massive proportions had recently emerged. Quite simply, demographers had observed a pattern of population growth in the poorer regions of the world quite unlike that which had occurred in the richer countries during the previous 150–200 years.

Population growth in the developed countries of the globe had followed a rather persistent pattern during the last two centuries. Prior to the Industrial Revolution, these countries typically experienced both high birth rates and death rates. As industrialization took hold and advances in the quality of life for citizens of these countries occurred, death rates fell, resulting in a period of time when the size of the population rose. Later, birth rates also began to decline, in large part because the newly industrialized societies were better suited to families with fewer children. After awhile, both birth and death rates leveled off at a much lower level than during preindustrial times.

This earlier transition throughout the developed world differed, however, from the newer growth pattern in the poorer regions of the globe observed by demographers in the late 1960s. First, the transition in the developed world occurred over a long period of time, allowing the population to deal more readily with such growth. On the other hand, post-1960s' growth in the developing world had taken off at a much faster pace, far outstripping the capacity of these societies to cope with the changes accompanying such growth.

Second, the earlier growth in the developed world began with a much smaller population base and a much larger resource base than did the developing world, again allowing the richer societies to cope more easily with such growth. The developing world of the 1960s, however, found percentages of increase based on a much higher base. Coping under the latter scenario proved much more difficult.

Finally, industrialization accompanied population change in the developed world, again allowing for those societies to address resultant problems more easily. Today's developing world has no such luxury. New jobs are not available, expanded educational facilities are non-existent, unsatisfactory health services remain unchanged, and modern infrastructures have not been created.

The international community formally placed the population issue—defined primarily as excessive birth rates in the developing world—on the global agenda in 1974 with the first major global conference on population, held in Bucharest, Romania. There was much debate over the motives of both sides. Both rich and poor countries eventually pledged to work together.

Finally, each side bought into the assumption that "the best contraceptive was economic development," but until development was achieved, national family planning programs would help lower growth rates. By 1994 when nations of the world reconvened in Cairo to assess progress, considerable success had been achieved in getting developing countries to accept such programs. As the new millennium occurred, some analysts even called for an ending to such ventures, declaring that the growth problem had diminished.

In the first selection, Robert McNamara argues that high population growth is exacerbating an already dire set of conditions in the developing world and that the industrialized countries of the globe should embark on a massive assistance program to help the "have-not" countries reduce fertility. In the second selection, Steven Mosher views the efforts of organizations such as the United Nations Population Fund much differently. In his view, these organizations have always sought to impose birth-control methods on the developing world in the misguided name of "virtuous and humanitarian motives," while attacking the motives of their opponents as self-serving or worse.

YES

Robert S. McNamara

The Population Explosion

For thousands of years, the world's human population grew at a snail's pace. It took over a million years to reach 1 billion people at the beginning of the last century. But then the pace quickened. The second billion was added in 130 years, the third in 30, and the fourth in 15. The current total is some 5.4 billion people.

Although population growth rates are declining, they are still extraordinarily high. During this decade, about 100 million people per year will be added to the planet. Over 90% of this growth is taking place in the developing world. Where will it end?

The World Bank's latest projection indicates that the plateau level will not be less than 12.4 billion. And Nafis Sadik, director of the United Nations Population Fund, has stated that "the world could be headed toward an eventual total of 14 billion."

What would such population levels mean in terms of alleviating poverty, improving the status of women and children, and attaining sustainable economic development? To what degree are we consuming today the very capital required to achieve decent standards of living for future generations?

More People, Consuming More

To determine whether the world—or a particular country—is on a path of sustainable development, one must relate future population levels and future consumption patterns to their impact on the environment.

Put very simply, environmental stress is a function of three factors: increases in population, increases in consumption per capita, and changes in technology that may tend to reduce environmental stress per unit of consumption.

Were population to rise to the figure referred to by Sadik—14 billion—there would be a 2.6-fold increase in world population. If consumption per capita were to increase at 2% per annum—about two-thirds the rate realized during the past 25 years—it would double in 35 years and quadruple in 70 years. By the end of the next century, consumption per capita would be eight times greater than it is today.

Some may say it is unreasonable to consider such a large increase in the per capita incomes of the peoples in the developing countries. But per capita

From Robert S. McNamara, "The Population Explosion," *The Futurist* (November–December 1992). Copyright © 1992 by The World Future Society. Reprinted by permission of The World Future Society, 7910 Woodmont Avenue, Suite 450, Bethesda, MD 20814.

income in the United States rose at least that much in this century, starting from a much higher base. And today, billions of human beings across the globe are now living in intolerable conditions that can only be relieved by increases in consumption.

A 2.6-fold increase in world population and an eightfold increase in consumption per capita by 2100 would cause the globe's production output to be 20 times greater than today. Likewise, the impact on non-renewable and renewable resources would be 20 times greater, assuming no change in environmental stress per unit of production.

On the assumptions I have made, the question becomes: Can a 20-fold increase in the consumption of physical resources be sustained? The answer is almost certainly "No." If not, can substantial reductions in environmental stress—environmental damage—per unit of production be achieved? Here, the answer is clearly "Yes."

Reducing Environmental Damage

Environmental damage per unit of production can—and will— be cut drastically. There is much evidence that the environment is being stressed today. But there are equally strong indications that we can drastically reduce the resources consumed and waste generated per unit of "human advance."

With each passing year, we are learning more about the environmental damage that is caused by present population levels and present consumption patterns. The superficial signs are clearly visible. Our water and air are being polluted, whether we live in Los Angeles, Mexico City, or Lagos. Disposal of both toxic and nontoxic wastes is a worldwide problem. And the ozone layer, which protects us all against skin cancer, is being destroyed by the concentration of chlorofluorocarbons in the upper atmosphere.

But for each of these problems, there are known remedies—at least for today's population levels and current consumption patterns. The remedies are costly, politically difficult to implement, and require years to become effective, but they can be put in place.

The impact, however, of huge increases in population and consumption on such basic resources and ecosystems as land and water, forests, photosynthesis, and climate is far more difficult to appraise. Changes in complex systems such as these are what the scientists describe as nonlinear and subject to discontinuities. Therefore, they are very difficult to predict.

A Hungrier Planet?

Let's examine the effect of population growth on natural resources in terms of agriculture. Can the world's land and water resources produce the food required to feed 14 billion people at acceptable nutritional levels? To do so would require a four-fold increase in food output.

Modern agricultural techniques have greatly increased crop yields per unit of land and have kept food production ahead of population growth for several decades. But the costs are proving to be high: widespread acceleration of erosion and nutrient depletion of soils, pollution of surface waters, overuse and contamination of groundwater resources, and desertification of overcultivated or overgrazed lands.

The early gains of the Green Revolution have nearly run their course. Since the mid-1980s, increases in worldwide food production have lagged behind population growth. In sub-Saharan Africa and Latin America, per capita food production has been declining for a decade or more.

What, then, of the future? Some authorities are pessimistic, arguing that maximum global food output will support no more than 7.5 billion people. Others are somewhat more optimistic. They conclude that if a variety of actions were taken, beginning with a substantial increase in agricultural research, the world's agricultural system could meet food requirements for at least the next 40–50 years.

However, it seems clear that the actions required to realize that capacity are not now being taken. As a result, there will be severe regional shortfalls (e.g., in sub-Saharan Africa), and as world population continues to increase, the likelihood of meeting global food requirements will become ever more doubtful.

Similar comments could be made in regard to other natural resources and ecosystems. More and more biologists are warning that there are indeed biological limits to the number of people that the globe can support at acceptable standards of living. They say, in effect, "We don't know where those limits are, but they clearly exist."

Sustainability Limits

How much might population grow and production increase without going beyond sustainable levels—levels that are compatible with the globe's capacity for waste disposal and that do not deplete essential resources?

Jim MacNeil, Peter Winsemaus, and Taizo Yakushiji have tried to answer that question in *Beyond Interdependence,* a study prepared recently for the Trilateral Commission. They begin by stating: "Even at present levels of economic activity, there is growing evidence that certain critical global thresholds are being approached, perhaps even passed."

They then estimate that, if "human numbers double, a five- to ten-fold increase in economic activity would be required to enable them to meet [even] their basic needs and minimal aspirations." They ask, "Is there, in fact, any way to multiply economic activity a further five to ten times, without it undermining itself and compromising the future completely?" They clearly believe that the answer is "No."

Similar questions and doubts exist in the minds of many other experts in the field. In July 1991, Nobel laureate and Cal Tech physicist Murray Gell-Mann and his associates initiated a multiyear project to try to understand how "humanity can make the shift to sustainability." They point out that "such a change, if it could be achieved, would require a series of transitions in fields ranging from technology to social and economic organization and ideology."

The implication of their statement is not that we should assume the outlook for sustainable development is hopeless, but rather that each nation individually, and all nations collectively, should begin now to identify and introduce the changes necessary to achieve it if we are to avoid costly—and possibly coercive—action in the future.

One change that would enhance the prospects for sustainable, development across the globe would be a reduction in population growth rates.

Population and Poverty

The developing world has made enormous economic progress over the past three decades. But at the same time, the number of human beings living in "absolute poverty" has risen sharply.

When I coined the term "absolute poverty" in the late 1960s, I did so to distinguish a particular segment of the poor in the developing world from the billions of others who would be classified as poor in Western terms. The "absolute poor" are those living, literally, on the margin of life. Their lives are so characterized by malnutrition, illiteracy, and disease as to be beneath any reasonable definition of human dignity.

Today, their number approaches 1 billion. And the World Bank estimates that it is likely to increase further—by nearly 100 million—in this decade.

A major concern raised by poverty of this magnitude lies in the possibility of so many children's physical and intellectual impairment. Surveys have shown that millions of children in low-income families receive insufficient protein and calories to permit optimal development of their brains, thereby limiting their capacity to learn and to lead fully productive lives. Additional millions die each year, before the age of five, from debilitating disease caused by nutritional deficiencies.

High population growth is not the only factor contributing to these problems; political organization, macroeconomic policies, institutional structures, and economic growth in the industrial nations all affect economic and social advance in developing countries. But intuitively we recognize that the immediate effects of high population growth are adverse.

Our intuition is supported by facts: In Latin America during the 1970s, when the school-age population expanded dramatically, public spending per primary-school student fell by 45% in real terms. In Mexico, life expectancy for the poorest 10% of the population is 20 years less than for the richest 10%.

Based on such analyses, the World Bank has stated: "The evidence points overwhelmingly to the conclusion that population growth at the rates common in most of the developing world slows development. . . . Policies to reduce population growth can make an important contribution to [social advance]."

A Lower Plateau for World Population?

Any one of the adverse consequences of the high population growth rates—environmentally unsustainable development, the worsening of poverty, and the negative impact on the status and welfare of women and children—would be reason enough for developing nations across the globe to move more quickly to reduce fertility rates. Taken together, they make an overwhelming case.

Should not every developing country, therefore, formulate long-term population objectives—objectives that will maximize the welfare of both present and future generations? They should be constrained only by the maximum feasible rate at which the use of contraception could be increased in the particular nation.

If this were done, I estimate that country family-planning goals might lead to national population-stabilization levels that would total 9.7 billion people for the globe. That is an 80% increase over today's population, but it's also 4.3 billion fewer people than the 14 billion toward which we may be heading. At the consumption levels I have assumed, those additional 4.3 billion people could require a production output several times greater than the world's total output today.

Reducing Fertility Rates

Assuming that nations wish to reduce fertility rates to replacement levels at the fastest possible pace, what should be done?

The Bucharest Population Conference in 1974 emphasized that high fertility is in part a function of slow economic and social development. Experience has indeed shown that as economic growth occurs, particularly when it is accompanied by broadly based social advance, birth rates do tend to decline. But it is also generally recognized today that not all economic growth leads to immediate fertility reductions, and in any event, fertility reduction can be accelerated by direct action to increase the use of contraceptives.

It follows, therefore, that any campaign to accelerate reductions in fertility should focus on two components: (1) increasing the pace of economic and social advance, with particular emphasis on enhancing the status of women and on reducing infant mortality, and (2) introducing or expanding comprehensive family-planning programs.

Much has been learned in recent years about how to raise rates of economic and social advance in developing countries. I won't try to summarize those lessons here. I do wish to emphasize, however, the magnitude of the increases required in family planning if individual countries are to hold population growth rates to levels that maximize economic and social advance.

The number of women of childbearing age in developing countries is projected to increase by about 22% from 1990 to 2000. If contraception use were to increase from 50% in 1990 to 65% in 2000, the number of women using contraception must rise by over 200 million.

That appears to be an unattainable objective, considering that the number of women using contraception rose by only 175 million in the past *two* decades, but it is not. The task for certain countries and regions—for example, India, Pakistan, and almost all of sub-Saharan Africa—will indeed be difficult, but other nations have done as much or more. Thailand, Indonesia, Bangladesh, and Mexico all increased use of contraceptives at least as rapidly. The actions they took are known, and their experience can be exported. It is available to all who ask.

Financing Population Programs

A global family-planning program of the size I am proposing for 2000 would cost approximately $8 billion, with $3.5 billion coming from the developed nations (up from $800 million spent in 1990). While the additional funding appears large, it is very, very small in relation to the gross national products and overseas development assistance projected for the industrialized countries.

Clearly, it is within the capabilities of the industrialized nations and the multilateral financial institutions to help developing countries finance expanded family-planning programs. The World Bank has already started on such a path, doubling its financing of population projects in the current year. Others should follow its lead. The funds required are so small, and the benefits to both families and nations so large, that money should not be allowed to stand in the way of reducing fertility rates as rapidly as is desired by the developing countries.

The developed nations should also initiate a discussion of how their citizens, who consume seven times as much per capita as do those of the developing countries, may both adjust their consumption patterns and reduce the environmental impact of each unit of consumption. They can thereby help ensure a sustainable path of economic advance for all the inhabitants of our planet.

Steven W. Mosher

McNamara's Folly: Bankrolling Family Planning

At the same time that Reimert Ravenholt was setting up his "powerful population program," the nations of Western Europe, along with Japan, were being encouraged by the administration of President Lyndon B. Johnson to make family planning a priority of their own aid programs. International organizations, primarily the UN and its affiliated agencies, were also being leveraged on board. Together, they helped to create and maintain the illusion that the international community was solidly behind population control programs. (It wasn't, and isn't, as we shall see.) But it was the World Bank and its billions that was the real prize for the anti-natalists. And they captured it when one of their own, Robert McNamara, was appointed as President in 1968.[1]

McNamara Moves In

McNamara came to the World Bank from the post of Secretary of Defense, where he had unsuccessfully prosecuted the Vietnam War by focusing on "kill ratios" and the "pacification of the natives" instead of victory. A former automobile executive, he was prone to cost-cutting measures which sometimes proved to be false economies, as when he decreed that a new class of ship—the fleet frigate—should have only one screw instead of the customary two. This saved the expense of a second turbine and drive train, but the frigate—known to the Navy as McNamara's Folly—lacked speed, was hard to berth, and had to be retired early.[2] The population policies he was to advocate suffered from similar defects.

When the Boards of the World Bank and the International Monetary Fund convened on October 1 of that year, President Johnson made a surprise appearance.[3] Technology in the underdeveloped nations, he said, had "bought time for family planning policies to become effective. But the fate of development hinges on how vigorously that time is used."

No More People

The stage was now set for McNamara to get up and attack the "population explosion," saying that it was "one of the greatest barriers to the economic

From *PRI Review* by Steven W. Mosher, March/April 2003. Copyright © 2003 by Population Research Institute. Reprinted by permission.

growth and social well-being of our member states." The World Bank would no longer stand idly by in the face of this threat, McNamara said, but would:

> Let the developing nations know the extent to which rapid population growth slows down their potential development, and that, in consequence, the optimum employment of the world's scarce development funds requires attention to this problem. Seek opportunities to finance facilities required by our member countries to carry out family planning programs. Join with others in programs of research to determine the most effective methods of family planning and of national administration of population control programs.[4]

It quickly became evident that "the optimum employment of the world's scarce development funds" meant in practice that the World Bank, the International Monetary Fund (IMF), and its network of regional development banks would act as loan sharks for the anti-natalists, pressuring sovereign nations into accepting family planning programs on pain of forfeiting vital short-term, long-term, and soft loans.[5] This practice is well known in the developing world, as when a Dhaka daily, *The New Nation*, headline read, "WB [World Bank] Conditions Aid to Population Control."[6]

McNamara also began providing loans for population and family planning projects, including those which involved abortion (both surgical and through abortifacient chemicals). By 1976 the National Security Council (NSC) was able to praise the World Bank for being "the principal international financial institution providing population programs."[7] Details are hard to come by, however. The World Bank is one of the most secretive organizations in the world, besides being effectively accountable to no one. It is known that there is a carefully segregated population division, which reportedly employs approximately 500 people. But those who work on conventional development projects are not privy to what goes on in this division, which is off-limits to all but those who work there.[8]

Fewer People, More Money

A rare inside look at the organization's activities in this area is provided by a recent World Bank report, entitled *Improving Reproductive Health: The Role of the World Bank*. Written in a distinctly self-congratulatory tone, the document reveals that the Bank has spent over $2.5 billion over the last twenty-five years to support 130 reproductive health projects in over 70 countries. Indonesia and Lesotho, for example, have been the site of "'information, education and communication' campaigns about sex and reproductive health." India has been the beneficiary of several different programs, which the report claims have "helped bring India two-thirds of the way towards her goal of replacement level fertility." No mention is made of the fact that the Indian campaigns have been notorious for their coercive tactics. Or that McNamara visited India at the height of the compulsory sterilization campaign in 1976 to congratulate the government for its "political will and determination" in the campaign and, one would suspect, to offer new loans.[9]

The World Bank also promotes abortion. *Improving Reproductive Health* openly admits that, since the 1994 Cairo Conference on Population and Development, the first of the World Bank's goals in the area of reproductive health has been "providing access and *choice* in family planning." [italics added] Except for its candor, this promotion of abortion should come as no surprise. In Burkina Faso, for example, we are told that World Bank projects have included "mobilizing public awareness and political support" [that is, lobbying] for abortion and other reproductive health services.

The Bank has long been accused of pressuring nations, such as Nigeria, into legalizing abortion. In 1988, for example, abortion was virtually unthinkable as an official family planning practice in Nigeria. As recently as 1990, the Planned Parenthood Federation of Nigeria was forced to defend itself against allegations that it promoted the sale and use of "contraceptives" that were abortifacient in character. A year later—and two months after approval of a $78 million World Bank population loan—the government announced proposals for allowing abortion under certain circumstances.[10]

Population control loans skyrocketed after the Cairo conference. The Bank reported that, in the two years that followed, it had "lent almost $1 billion in support of population and reproductive health objectives."[11] And the numbers have been climbing since then. But even this is just the tip of the iceberg. As Jacqueline Kasun notes, "Given the conditions which the bank imposes on its lending, the entire $20 billion of its annual disbursements is properly regarded as part of the world population control effort."[12]

No More Reform

Despite his predilection for population control, McNamara never abandoned more conventional aid modalities, roads, dams, power plants, and the like. Not so James Wolfensohn, who became the head of the Bank in 1995. Asked at the 1996 World Food Summit in Rome how the World Bank understood its mission towards the developing world, Wolfensohn replied that there was a "new paradigm" at the Bank. "From now on," Wolfensohn said, "the business of the World Bank will not be primarily economic reform, or governmental reform. The business of the World Bank will primarily be social reform." The Bank has learned, he added, that attempting to reform a nation's economics or government without first reforming the society "usually means failure."

The benefits to nations who are willing to fall into line in the "civil society" will be immediate and intensely attractive. "The World Bank will be willing to look favorably on any reasonable plan for debt reduction—and even debt forgiveness," Wolfensohn told the assembled reporters, "provided that the nation in question is willing to follow a sensible social policy." Wolfensohn went on to tell reporters that population control activities are a *sine qua non* for any social policy to be considered "sensible."[13]

The World Bank is also, according to Wolfensohn, prepared to begin "directly funding—not through loans" certain NGOs in the countries involved, to further ensure that governments adopt "sensible social policies." Thus fueled with money from the World Bank, the heat these favored NGOs will be able to generate on their governments to adopt, say, population control programs, including legalized abortion, will be considerable.[14] Of course, other international organizations, not to mention USAID and European aid agencies, have been using this tactic for many years with great effect. Recalcitrant governments (who may innocently believe that they do not have a population "problem") are thus sandwiched between the demands of international lenders and aid givers on the one hand, and the demands of "local" NGOs—loud, persistent and extremely well-funded—on the other.

Rapid Spread of Programs

With the U.S., international organizations, and an increasing number of developed countries now working in tandem to strong-arm developing countries into compliance, anti-natalist programs spread with startling rapidity. Bernard Berelson, the head of the Population Council, happily reported in 1970 that:

> In 1960 only three countries had anti-natalist population policies (all on paper), only one government was offering assistance [that is, funding population control programs overseas], no international organizations was working on family planning. In 1970 nearly 25 countries on all three developing continents, with 67 percent of the total population, have policies and programs; and another 15 or so, with 12 percent of the population, provide support in the absence of an explicitly formulated policy . . . five to ten governments now offer external support (though only two in any magnitude); and the international assistance system is formally on board (the U.N. Population Division, the UNDP, WHO, UNESCO, FAO, ILO, OECD, the World Bank).[15]

The recklessness with which Ravenholt, McNamara and others forced crude anti-natal programs upon the developing world dismayed many even within the movement. Ronald Freedman, a leading sociologist/demographer, complained in 1975 that, "If reducing the birth rate is that important and urgent, then the results of the expanded research during the 1960s are still *pathetically inadequate. There are serious proposals for social programs on a vast scale to change reproductive institutions and values that have been central to human society for millennia.*"[16] [italics added] This was social engineering with a vengeance, Freedman was saying, and *we don't know what we are doing.*

With even committed controllers saying "Slow down!" one might think that the anti-natalists would hesitate. But their army had already been assembled and its generals had sounded the advance; it could not be halted now. Even Freedman, rhetorically throwing up his hands, conceded that "many people . . . are eager for knowledge that can be used in action programs aimed at accelerating fertility decline," and that the programs would have to proceed by "a process of trial and error." The *trials* of course would be funded by the developed world; while the *errors*, murderous and costly, would be borne by poor women and families in the developed world.

What justification was offered for this massive investment of U.S. prestige and capital in these programs? Stripped of its later accretions—protecting the environment, promoting economic development, advancing the rights of women—at the outset it was mostly blatant self-interest. McNamara, who headed an organization ostensibly devoted to the welfare of the developing countries, had told the World Bank's Board of Governors in 1968 that "population growth slows down their potential development." But he told the *Christian Science Monitor* some years later that continued population growth would lead to "poverty, hunger, stress, crowding, and frustration," which would threaten social, economic and military stability. This would not be "a world that anyone wants," he declared.[17] It was certainly not the world that many in the security establishment wanted, as secret National Security Council deliberations would soon make starkly clear.

Cold War Against Population

As the populations of developing world countries began to grow after World War Two, the U.S. national security establishment—the Pentagon, the Central intelligence Agency, the National Security Agency, and the National Security Council—became concerned. Population was an important element of national power, and countries with growing populations would almost inevitably increase in geopolitical weight. This was obviously a concern in the case of countries opposed to U.S. interests, such as the Soviet Union and China. But even allies might prove less pliable as their populations and economies grew. Most worrisome of all was the possibility that the rapidly multiplying peoples of Asia, Africa and Latin America would turn to communism in their search for independence and economic advancement *unless their birth rate was reduced.* Thus did population control become a weapon in the Cold War. . . .

Earth First (People Last): Environmental Movement Signs On

Every sorcerer deserves an apprentice. Hugh Moore, grand wizard of the population explosion, got his in the person of a young Stanford University entomologist by the name of Paul Ehrlich. In the very first sentence of his very first book Ehrlich proved beyond all doubt that he had already mastered Moore's panic-driven style. "The battle to feed all of humanity is over," he wrote. "In the 1970s the world will undergo famines—hundreds of million of people will starve to death in spite of any crash programs embarked upon now."[18]

In fact, he had gone Moore one better, as overzealous acolytes are prone to do. His book should have been named *The Population Explosion*, instead of *The Population Bomb*, for according to Ehrlich the "bomb" had already gone off and there was nothing to do now but wait for the inevitable human dieback. "Too many people" were chasing "too little food."[19] The most optimistic of Ehrlich's "scenarios" involved the immediate imposition of a harsh regimen of population control and resource conservation around the world, with the goal of reducing the number of people to 1.5 billion (about a fourth of its current level) over the next century or two. Even so, about a fifth of the world's population would still starve to death in the immediate future.

Such a prediction took pluck, for when the book appeared in 1968 there was no hint of massive famine on the horizon. The days of Indian food shortages were past. (We wouldn't learn about China's man-made calamity until a decade later.) The Green Revolution was starting to pay off in increased crop yields. And experts like Dr. Karl Brandt, the Director of the Stanford Food Research Institute, rebuked Ehrlich, saying that "Many nations need more people, not less, to cultivate food products and build a sound agricultural economy . . . every country that makes the effort can produce all the food it needs."[20]

But it wasn't his forecast of a massive human die-off that catapulted Ehrlich into the front rank of environmental prophets. (In a motif that has since become familiar, the book left readers with the impression that this might not be such a bad thing.) Rather it was his startling claim that our reckless breeding had jeopardized earth's ability to support life. All life, not just human life. Our planet was literally dying. Not only were the Children of Earth killing ourselves, we were going to take Mother with us as well.

The Population Bomb

Heavily promoted by the Sierra Club, *The Population Bomb* sold over a million copies. Ehrlich became an instant celebrity, becoming as much of a fixture on the "Tonight Show" as Johnny Carson's sidekick Ed McMahon. He command[ed] hefty lecture fees wherever he went (and he went everywhere), and always drew a crowd. People found it entertaining to hear about the end of the world. Likening the earth to an overloaded spaceship or sinking lifeboat, issuing apocalyptic warnings about the imminent "standing room only" problem, he captured the popular imagination. His prescriptions were always the same: "Join the environmental movement, stop having children, and save the planet."[21]

While Ehrlich fiddled his apocalyptic tunes, Moore burned to commit the growing environmental movement firmly to a policy of population control. His ad campaign, still ongoing, began suggesting that the best kind of environmental protection was population control. "Whatever Your Cause, It's a Lost Cause Unless We Control Population," one ad read. "Warning: The Water You are Drinking May be Polluted," read another, whose text went on to equate more people with more pollution. A third, addressed to "Dear President Nixon," claimed that "We can't lick the environment problem without considering this little fellow." It featured a picture of a newborn baby.

Birth of Earth Day

Moore went all out for the first Earth Day in 1970, printing a third of a million leaflets, folders, and pamphlets for campus distribution. College newspapers received free cartoons highlighting the population crisis and college radio stations a free taped show (featuring Paul Ehrlich). With his genius for marketing, Moore even announced a contest with cash awards for the best slogans relating environmental problems to what he called "popullution" [population pollution]. Students on over 200 campuses participated. The winner, not surprisingly, was "People Pollute."[22]

By 1971 most of the leading environmental groups had signed on to the anti-natal agenda, having been convinced that reducing the human birth rate would greatly benefit the environment. Perhaps it was their interest in "managing" populations of other species—salmon, condors, whales, etc.—that predisposed them to impose technical solutions on their own species. In any event, many of them were population hawks, who believed that simply making abortion, sterilization and contraception widely available was not enough. "Voluntarism is a farce," wrote Richard Bowers of Zero Population Growth as early as 1969. "The private sector effort has failed . . . [even the expenditure] of billions of dollars will not limit growth." Coercive measures were required. He proposed enacting "criminal laws to limit population, if the earth is to survive."[23]

Those who held such views were not content to merely stop people from multiplying, they demanded radical reductions in human numbers. The group Negative Population Growth wanted to cut the-then U.S. population of 200 million by more than half, to 90 million.[24] Celebrated oceanographer Jacques Cousteau told the *UNESCO Courier* in 1991, "In order to stabilize world populations, we must eliminate 350,000 people per day." Garrett Hardin of "The Tragedy of the Commons" fame opined that the "carrying capacity" of the planet was 100 million and that our numbers should be reduced accordingly. (Do we pick the lucky 100 million by lottery?) To carry out these decimations, Malthusian solutions are proposed, as when novelist William T. Vollman stated that, "there are too many people in the world and maybe something like AIDS or something like war may be a good thing on that level."[25] And lest we have compunctions about resorting to such measures, we should bear in mind, as Earth First! Founder Dave Foreman wrote, "We humans have become a disease, the Humanpox."

The Feminist Dilemma

The most radical of the feminists had a different definition of disease. Why should women be "subject to the species gnawing at their vitals," as Simone de Beauvoir so memorably wrote in her feminist classic *The Second Sex*? Why endure pregnancy at all, if contraception, sterilization and, especially, abortion, could be made widely available? With the legalization of abortion in the U.S. in 1973, feminists increasingly looked overseas, eager to extend their newfound rights to "women of color" elsewhere in the world. They had read their Ehrlich as well as their Beauvior, and knew that the world had too many people, or soon would. But family planning, especially abortion, provided a way out. "Let us bestow upon all the women of the world the blessing that we women in the privileged West have received—freedom from fear of pregnancy," the feminists said to themselves. "We will, at the same time and by the same means, solve the problem of too many babies. For surely impoverished Third World women do not actually want all those children they are bearing. Patriarchy has made them into breeding machines, but we will set them free."

Abortion "Needs" Appear

At the time, the population control movement remained ambivalent over the question of abortion. Hugh Moore had long wanted it as a population control measure, but Frank Notestein was still arguing in the early 1970s that the Population Council should "consistently and firmly take the anti-abortion stance and use every occasion to point out that the need for abortions is the proof of program failure in the field of family planning and public health education."[26]

But the women's movement would not be put off with the promise of a perfect contraceptive. They knew, better than anyone (and often from painful personal experience) that contraception, because of the inevitable failures, *always* led to abortion. As Sharon Camp of the Population Crisis Committee wrote "both abortion and contraception are presently on the rise in most developing countries."[27]

Abortion was, in the end, accepted by most controllers because it came to be seen as a necessary part of the anti-natal arsenal. The Rockefeller Commission, established by President Nixon, wrote that "We are impressed that induced abortion has a demographic effect wherever legalized" and on these grounds went on to call for "abortion on demand."[28] The Population Council followed the Commission in endorsing abortion as a means of population control by 1975.

In the end, feminist advocacy of abortion had proven decisive. The feminists had given the population control movement an additional weapon, abortion, to use in its drive to reduce human fecundity, and encouraged its aggressive use.

Third World Women

At the same time, it was soon apparent to many feminists that birth control was not an unmixed blessing for Third World women, who continued to be targets of ever-more aggressive programs in places like Indonesia, India, and Bangladesh. They began to demand further changes in the way programs were carried out, starting with male contraceptives and more vasectomies. Frank Notestein wrote of the feminists that, "As second-generation suffragists they were not at all disposed to allow the brutish male to be in charge of contraception. Women must have their own methods!" But more recent feminists "complain violently that the men are trying to saddle the women with all the contraceptive work. You can't please them if you do, and can't please them if you don't."[29]

Although expressed somewhat crudely, Notestein's comment pointed out the dilemma faced by feminists. On the one hand, they sought to impose a radically pro-abortion agenda on population control programs, whose general purpose—fertility reduction—they applauded. On the other they tried to protect women from the abuses that invariably accompanied such programs. But with the exception of the condom, other methods of contraception all put the burden on women. Vasectomies could easily be performed on men, but it was usually the woman who went under the knife to have her tubes tied. And abortions could only be performed on women. So, as a practical matter, the burden of fertility reduction was placed disproportionately on women. And when programs took a turn towards the coercive, as they were invariably prone to do in the Third World, it was overwhelmingly women who paid the price.

Feminist complaints did lead to some changes, but these were mostly cosmetic. Population controllers did learn, over time, to speak a different language or, rather, several different languages, to disguise the true, anti-natal purpose of their efforts. When Western feminists need to be convinced of the importance of supporting the programs, reproductive rights rhetoric is the order of the day. Thus we hear Nafis Sadik telling Western reporters on the eve of the 1994 U.N. Conference on Population and Development that the heart of the discussion "is the recognition that the low status of women is a root cause of inadequate reproductive health care." Such language would ring strange in the ears of Third World women, who are instead the object of soothing lectures about "child-spacing" and "maternal health." Population control programs were originally unpopular in many Middle Eastern countries and sub-Sarahan countries until they were redesigned, with feminist input, as programs to "help" women. As Peter Donaldson, the head of the Population Reference Bureau, writes, "The idea of limiting the number of births was so culturally unacceptable [in the Middle East and sub-Saharan Africa] that family planning programs were introduced as a means for promoting better maternal and child health by helping women space their births."[30] James Grant of the U.N. Children's Fund (UNICEF), in an address to the World Bank, was even more blunt: "Children and women are to be the Trojan Horse for dramatically slowing population growth."[31]

Corrupted Feminist Movement

The feminists did not imagine, when they signed onto the population control movement, that they would merely be marketing consultants. It is telling that many Third World feminists have refused to endorse population control programs at all, arguing instead that these programs violate the rights of women while ignoring their real needs. It must be painful for Western feminists to contemplate, but their own movement has been used or, to use Betsy Hartmann's term, "co-opted," by another movement for whom humanity as a whole, and women in particular, remain a faceless mass of numbers to be contracepted, sterilized, and aborted. For, despite the feminist rhetoric, the basic character of the programs hasn't changed. They are a numbers-driven, technical solution to the "problem of overpopulation"—which is, in truth, a problem of poverty—and they overwhelmingly target women.

This is, in many respects, an inevitable outcome. To accept the premise that the world is overpopulated and then seek to make the resulting birth control programs "women-friendly," as many feminists have, is a fateful compromise. For it means that concern for the real needs of women is neither the starting point of these programs, nor their ultimate goal, but merely a consideration along the way. Typical of the views of feminists actively involved in the population movement are those of Sharon Camp, who writes, "There is still time to avoid another population doubling, but only if the world community acts very quickly to make family planning universally available and to invest in other social programs, like education for girls, which can help accelerate fertility declines."[32] Here we see the population crisis mentality in an uneasy alliance with programs for women which, however, are justified chiefly because they "help accelerate fertility declines."

The alliance between the feminists and population controllers has been an awkward affair. But the third of the three most anti-natalist movements in history gave the population controllers new resources, new constituencies, new political allies, a new rhetoric, and remains a staunch supporter even today.

Population Firm Funding

Over the past decade the Population Firm has become more powerful than ever. Like a highly organized cartel, working through an alphabet soup of United Nations agencies and "nongovernmental organizations," its tentacles reach into nearly every developing country. It receives sustenance from feeding tubes attached to the legislatures of most developed countries, and further support through the government-financed population research industry, with its hundreds of professors and thousand of students. But unlike any other firm in human history, its purpose is not to produce anything, but rather to destroy—to destroy fertility, to prevent babies from being conceived and born. It diminishes, one might say, the oversupply of people. It does this for the highest of motives—to protect all of us from "popullution." Those who do not subscribe to its ideology it bribes and browbeats, bringing the combined weight of the world's industrial powers to bear on those in countries which are poor.

In 1991, the U.N. estimated that a yearly sum of $4.5 to $5 billion was being directed to population programs in developing countries. This figure, which has grown tremendously in the last 10 years, includes contributions from bilateral donors such as the U.S., the European nations and Japan, from international agencies like those associated with the UN, and from multilateral lending institutions, including the World Bank and the various regional development banks. It includes grants from foundations, like Ted Turner's U.N. Foundation, and wealthy individuals like Warren Buffet.

Moreover, a vast amount of money not explicitly designated as "population" finance is used to further the family planning effort. As Elizabeth Liagin notes, "During the 1980s, the diversion of funds from government non-population budgets to fertility-reduction measures soared, especially in the U.S., where literally hundreds of millions from the Economic Support Funds program, regional development accounts, and other non-population budgets were redirected to "strengthen" population planning abroad."[33]

More Money Spent

An almost unlimited variety of other "development" efforts—health, education, energy, commodity imports, infrastructure, and debt relief, for example—are also used by governments and other international agencies such as the World Bank, to promote population control policies, either through requiring recipient nations to incorporate family planning into another program or by holding funds or loans hostage to the development of a national commitment to tackle the "over-population" problem.

In its insatiable effort to locate additional funds for its insatiable population control programs, USAID has even attempted to redirect "blocked assets"—profits generated by international corporations operating in developing nations that prohibit the transfer of money outside the country—into population control efforts. In September 1992, USAID signed a $36.4 million contract and "statement of work" with the accounting firm Deloitte and Touche to act as a mediator with global corporations and to negotiate deals that would help turn the estimated $200 *billion* in blocked assets into "private" contributions for family planning in host countries. The corporations would in return get to claim a deduction on their U.S. tax return for this "charitable contribution." The Profit initiative, as it is fittingly called, is not limited to applying its funds directly to family planning "services," but is also encouraged to "work for the removal" of "trade barriers for contraceptive commodities" and "assist in the development of a regulatory framework that permits the expansion of private sector family planning services." This reads like a bureaucratic mandate to lobby for the elimination of local laws which in any way interfere with efforts to drive down the birth rate, such as laws restricting abortion or sterilization.[34]

U.S.'s Real Foreign Aid Policy

Throughout the nineties, the idea of the population controllers that people in their numbers were somehow the enemy of all that is good reigned supreme. J. Brian Atwood, who administered the U.S. Agency for International Development in the early days of the Clinton administration, put it this way: "If we aren't able to find and promote ways of curbing population growth, we are going to fail in *all* of our foreign policy initiatives." [italics added] (Atwood also went on to announce that the U.S. "also plans to resume funding in January [1994] to the U.N. Fund for Population Activities (UNFPA).")[35] Secretary of State Warren Christopher offered a similar but even more detailed defense of population programs the following year. "Population and sustainable development are back where they belong in the mainstream of American foreign policy and diplomacy." He went on to say, in a line that comes rights out of U.S. National Security Study Memo 200, that population pressure "ultimately jeopardizes America's security interests." But that's just the beginning. Repeating the now familiar litany, he claimed of population growth that, "It strains resources. It stunts economic growth. It generates disease. It spawns huge refugee flows, and ultimately it threatens our stability. . . . We want to continue working with the other donors to meet the rather ambitious funding goals that were set up in Cairo."[36]

The movement was never more powerful than it was in 2000 in terms of money, other resources, and political clout.

Losing Momentum

Like a wave which crests only seconds before it crashes upon the shore, this appearance of strength may be deceiving. There are signs that the anti-natal movement has peaked, and may before long collapse of its own overreaching.

U.S. spending on coercive population control and abortion overseas have long been banned. In 1998 the U.S. Congress, in response to a flood of reports about human rights abuses, for the first time set limits on what can be done to people in the name of "voluntary family planning."[37] Developing countries are regularly denouncing what they see as foreign interference into their domestic affairs, as the Peruvian Congress did in 2002. Despite strenuous efforts to co-opt them, the opposition of feminists to population control programs (which target women) seems to be growing.[38] Many other groups—libertarians, Catholics, Christians of other denominations, the majority of economists, and those who define themselves as pro-life—have long been opposed.

As population control falls into increasing disrepute worldwide, the controllers are attempting to reinvent themselves, much the same way that the Communists in the old Soviet Union reemerged as "social democrats" following its collapse. Organizations working in this area have found it wise to disguise their agenda by adopting less revealing names. Thus Zero Population Growth in June 2002 became Population Connection, and the Association for Voluntary Surgical Contraception the year before changed its name to Engender Health. Similarly, the U.N., in documents prepared for public consumption, has recently found it expedient to cloak its plans in language about the "empowerment of women," "sustainable development," "safe motherhood," and "reproductive health." Yet the old anti-natal zeal continues to come through in internal discussions, as when Thoraya Obaid affirmed to her new bosses on the U.N. Commission on Population and Development her commitment to "slow and eventually stabilize population growth." "And today I want to make one thing very clear," she went on to say. "The slowdown in population growth does not mean we can slow down efforts for population and reproductive health—quite the contrary. If we want real progress and if we want the projections to come true, we must step up efforts . . . while population growth is slowing, it is still growing by 77 million people every year."[39] And so on.

Such efforts to wear a more pleasing face for public consumption will in the end avail them nothing. For, as we will see, their central idea—the Malthusian notion that you can eliminate poverty, hunger, disease, and pollution by eliminating the poor—is increasingly bankrupt.

Reducing the numbers of babies born has not and will not solve political, societal and economic problems. It is like trying to kill a gnat with a sledgehammer, missing the gnat entirely, and ruining your furniture beyond repair. It is like trying to protect yourself from a hurricane with a bus ticket. Such programs come with massive costs, largely hidden from the view of well-meaning Westerners who have been propagandized into supporting them. And their "benefits" have proven ephemeral or worse. These programs, as in China, have done actual harm to real people in the areas of human rights, health care, democracy, and so forth. And, with falling birth rates everywhere, they are demographic nonsense. Where population control programs are concerned, these costs have been largely ignored (as the cost of doing business) while the benefits to people, the environment, and to the economy, have been greatly exaggerated, as we will see. Women in the developing world are the principal victims.

Notes

1. The World Bank is to a large degree under the control of the United States, which provides the largest amount of funding. This is why the head of the World Bank is always an American. The activities of the Bank are monitored by the National Advisory Council on International Monetary and Financial Policies—called the NAC for short, of the Treasury Department. The 1988 annual report of the NAC states that "the council [NAC] seeks to ensure that . . . the . . . operations [of the World Bank and other international financial institutions] are conducted in a manner consistent with U.S. policies and objectives . . ." International Finance: National Advisory Council on International Monetary and Financial Policies, Annual Report to the President and to the Congress for the Fiscal year 1988, (Wash., D.C., Department of the Treasury), Appendix A, p. 31. Quoted in Jean Guilfoyle, "World Bank Population Policy: Remote Control," *PRI Review* 1:4 (July/August 1991), p. 8.

2. I served on board a ship of this class, the USS Lockwood, from 1974–76. As the Main Propulsion Assistant—the officer in charge of the engine room—I can personally attest that this fleet frigate, as it was called, was anything but fleet. On picket duty, it could not keep up with the big flattops that it was intended to protect from submarine attacks.

3. The 1968 meeting was 23rd joint annual meeting of the Boards of Governors of the World Bank and the International Monetary Fund. The two organizations always hold their annual meetings in tandem, underscoring their collaboration on all matters of importance.

4. McNamara moderated his anti-natal rhetoric on this formal occasion. More often, he sounded like Hugh Moore, as when he wrote: "the greatest single obstacle to the economic and social advancement of the majority of the peoples in the underdeveloped world is rampant population growth. . . . The threat of unmanageable population pressures is very much like the threat of nuclear war. . . . Both threats can and will have catastrophic consequences unless they are dealt with rapidly." *One Hundred Countries, Two Billion People* (London: Pall Mall Press, 1973), pp. 45–46. Quoted in Michael Cromartie, ed., *The 9 Lives of Population Control*, (Washington: Ethics and Public Policy Center, 1995), p. 62. McNamara never expressed any public doubts about the importance of population control, although he did once confide in Bernard Berelson that "many of our friends see family planning as being 'too simple, too narrow, and too coercive.'" As indeed it was—and is. Quote is from Donald Crichtlow, *Intended Consequences: Birth Control, Abortion, and the Federal Government in Modern America*, (Oxford: Oxford University Press, 1999), p. 178.

5. See Fred T. Sai and Lauren A. Chester, "The Role of the World Bank in Shaping Third World Population Policy," in Godfrey Roberst, ed., *Population Policy: Contemporary Issues* (New York: Praeger, 1990). Cited in Jacqueline Kasun, *The War Against Population*, Revised Ed. (San Fransisco: Ignatius Press, 1999), p. 104.

6. 7 September 1994, p. 1. Cited in Kasun, p. 104.

7. U.S. International Population Policy: First Annual report, prepared by the Interagency Task Force on Population Policy, (Wash., D.C., U.S. National Archives, May 1976). Quoted in Jean Guilfoyle, "World Bank Population Policy: Remote Control," *PRI Review* 1:4 (July/ August 1991), p. 8.

8. Personal Communication with the author from a retired World Bank executive who worries that, if his identity is revealed, his pension may be in jeopardy.

9. Peter T. Bauer and Basil S. Yamey, "The Third World and the West: An Economic Perspective," in W. Scott Thompson, ed., *The Third World: Premises of*

U.S. Policy (San Francisco: Institute for Contemporary Studies, 1978), p. 302. Quoted in Kasun, p. 105.

10. See Elizabeth Liagin, "Money for Lies," *PRI Review*, July/October 1998, for the definitive history of the imposition of population control on Nigeria; The Nigerian case is also discussed by Barbara Akapo, "When family planning meets population control," *Gender Review*, June 1994, pp. 8–9.

11. Word Bank, 1995 Annual Report, p. 18. Quoted in *PRI Review* 5:6 (November/December 1995), p. 7.

12. Kasun, p. 277.

13. David Morrison, "Weaving a Wider Net: U.N. Move to Consolidate its Anti-Natalist Gains," *PRI Review* 7:1 (January–February 1997), p. 7.

14. Ibid.

15. Bernard Berelson, "Where Do We Stand," paper prepared for Conference on Technological change and Population Growth, California Institute of Technology, May 1970, p. 1. Quoted in Ronald Freedman, *The Sociology of Human Fertility: An Annotated Bibliography* (New York: Irvington Publishers, 1975), p. 3. It's worth noting that Freedman's book was a subsidized product of the institution Berelson then headed. As Freedman notes in his "Preface," the "staff at the Population Council were very helpful in reading proof, editorial review, and making detailed arrangements for publication. Financial support was provided by the Population Council." (p. 2.)

16. Freedman, p. 4.

17. *Christian Science Monitor*, 5 July 1977. He went on to say that, if present methods of population control "fail, and population pressures become too great, nations will be driven to more coercive methods."

18. Paul R. Ehrlich, *The Population Bomb* (New York: Ballantine books, 1968).

19. The first three sections of Ehrlich's book were called, "Too many people," "Too little Food," "A Dying Planet."

20. *Is there a Population Explosion?*, Daniel Lyons, (Catholic Polls: New York, 1970), p. 5.

21. Ehrlich has continued on the present day, writing one book after another, each one chock full of predictions of imminent disasters that fail to materialize. People wonder why Ehrlich doesn't learn from his experiences? The answer, I think, is that he has learned very well. He has learned that writing about "overpopulation and environmental disaster" sells books, *lots* of books. He has learned that there is no price to pay for being wrong, as long as he doesn't admit his mistakes *in print* and glibly moves on to the next disaster. In one sense, he has far outdone Hugh Moore in this regard. For unlike Moore, who had to spend his own money to publish the original *The Population Bomb*, Ehrlich was able to hype the population scare *and* make money by doing so. He is thus the archetype of a figure familiar to those who follow the anti-natal movement: the population hustler.

22. Lawrence Lader, *Breeding Ourselves to Death*, pp. 79–81.

23. Richard M. Bowers to ZPG members, 30 September 1969, Population Council (unprocessed), RZ. Quoted in Critchlow, p. 156.

24. In later years, as U.S. population continued to grow, NPG has gradually increased its estimate of a "sustainable" U.S. population to 150 million. See Donald Mann, "A No-Growth Steady-State Economy Must Be Our Goal," NPG Position Paper, June 2002.

25. Quoted in David Boaz, "Pro-Life," *Cato Policy Report* (July/August 2002), p. .2.

26. Frank Notestein to Bernard Berelson, 8 February 1971, Rockefeller Brother Fund Papers, Box 210, RA. Quoted in Critchlow, p. 177. These concerns, while

real enough to Notestein, apparently did not cause him to reflect on the fact that he was a major player in a movement that "detracted from the value of human life" by suggesting that there was simply too much of that life, and working for its selective elimination.

27. Population Action International, "Expanding Access to Safe Abortion: Key Policy Issues," Population Policy Information Kit 8 (September 1993). Quoted in Sharon Camp, "The Politics of U.S. Population Assistance," in Mazur, *Beyond the Numbers*, p. 126.

28. Critchlow, p. 165.

29. Frank Notestein to Bernard Berelson, April 27, 1971, Notestein Papers, Box 8, Princeton University.

30. Peter Donaldson and Amy Ong Tsui, "The International Family Planning Movement," in Laurie Ann Mazur, ed., *Beyond the Numbers* (Island Press, Washington, D.C., 1994), p. 118. Donaldson was, at the time, president of the Population Reference Bureau, and Tsui was deputy director of the Carolina Population Center.

31. World Bank 1993 International Development Conference, Washington, D.C., 11 January 1993, p. 3. Also quoted in *PRI Review* (September–October 1994), p. 9.

32. Sharon Camp, "Politics of U.S. Population Assistance," in *Beyond the Numbers*, pp. 130–1. Camp for many years worked at the Population Crisis Committee, founded by Hugh Moore in the early sixties. It has recently, perhaps in recognition of falling fertility rates worldwide, renamed itself Population Action International.

33. Quoted from Elizabeth Liagin, "Profit or Loss: Cooking the Books at USAID," *PRI Review* 6:3 (November/December 1996), p. 1.

34. Ibid., p. 11.

35. John M. Goshko, "Planned Parenthood gets AID grant . . . ," *Washington Post*, 23 November 1993, A12-13.

36. Reuters, "Christopher defends U.S. population programs," Washington, D.C, 19 December 1994.

37. The Tiahrt Amendment.

38. See Betsy Hartmann, *Reproductive Rights and Wrongs* (Boston: South End Press, 1995).

39. Thoraya Ahmed Obaid, "Reproductive Health and Reproductive Rights With Special Reference to HIV/AIDS," Statement to the U.N. Commission on Population and Development, 1 April 2002.

POSTSCRIPT

Should the International Community Attempt to Curb Population Growth in the Developing World?

There are at least two basic dimensions to this issue. First, should the international community involve itself in reducing fertility throughout the developing world? That is, is it a violation of either national sovereignty (a country should be free from extreme outside influence) or human rights (an individual has the right to make fertility decisions unencumbered by outside pressure, particularly those from another culture)? And second, if so, what should its motives be? To put it another way, is advocacy of population control really a form of ethnic or national genocide?

McNamara answers these questions with the same set of arguments. His belief that the international community has an obligation to get involved is based on his assessment of the resultant damage to both the developing world where such high levels of growth are found and the rest of the world that must compete with the poorer countries for increasingly scare resources. For McNamara, a 2.6-fold increase in population and an 8-fold increase in consumption per capita (by the end of the twenty-first century) would result in a 20-fold impact on nonrenewable and renewable resources. He cites the projected agricultural needs to demonstrate that the Earth cannot sustain such consumption levels. Additionally, McNamara suggests that other consequences include environmentally unsustainable development, worsening of poverty, and negative impacts on the welfare of women and children. They appear to make an overwhelming case for fertility reduction. And the cost of policy intervention is small, in his judgment, compared to both gross national products and overseas general development assistance. And finally, given the projections, the developing world should embrace such policy intervention for its own good.

Those who oppose such intervention point to several different reasons. The first, originally articulated at the 1974 Bucharest conference, argues that economic development is the best contraceptive. The demographic transition worked in the developed world, and there was no reason to assume that it will not work in the developing world. This was the dominant view of the developing countries at Bucharest, who feared international policy intervention, and at the same time wanted and needed external capital to develop. They soon came to realize that characteristics particular to the developing world in the latter part of the twentieth century meant that foreign aid alone would not be enough. It had to be accompanied by fertility reduction programs, wherever the origin.

Second, some who oppose intervention do not see the extreme negative environmental consequences of expanding populations. For them, the pronatalists (those favoring fertility reduction programs) engage in inflammatory discourse, exaggerated arguments, and scare tactics devoid of much scientific evidence.

Mosher goes further in the second selection, accusing McNamara, who served as U.S. Secretary of Defense during the Vietnam era, of using his position as one of the most important leaders in the international financial lending world to help prevent continuing population growth in the developing world because of its potential threat to the national security of countries like the United States. He also concludes that the interventionists are engaged in social engineering, not economic development.

And finally, cultural constraints can counteract any attempt to impose fertility reduction values from the outside. Children are valued in most societies. They, particularly male heirs, serve as the social security system for most families. They become producers at an early age, contributing to the family income.

McNamara is correct if one assumes that no changes—technological, social, and organizational—accompany predicted population growth. But others would argue that society has always found a way to use technology to address newly emerging problems successfully. The question remains, though: Are the current problems and the world in which they function too complex to allow for a simple solution?

The bottom line is that each has a valid point. Environmental stress is a fact of life, and increased consumption does play a role. If the latter originates because of more people, than fertility reduction is a viable solution. Left to its own devices, the developing world is unable to provide both the rationale for such action and the resources to accomplish it. The barriers are too great. While the international community must guard against behavior that is or appears to be either genocidal in nature or violates human rights, it must nonetheless move forward.

An excellent account of the 1994 Cairo population conference's answer to how the international community ought to respond is found in Lori S. Ashford's *New Perspectives on Population: Lessons from Cairo* (1995). Other sources that address the question of the need for international action include William Hollingworth's *Ending the Explosion: Population Policies and Ethics for a Humane Future* (1996), Elizabeth Liagin's *Excessive Force: Power, Politics, and Population* (1996), and Julian Simon's *The Ultimate Resource 2* (1996).

There are both United Nations and external analyses of the Cairo conference and assessments of progress made in implementing its Plan of Action. The United Nations Population Fund's Web site (www.unfpa.org) has a section entitled *ICPD and Key Actions*. There you will find numerous in-house analyses of both the conference and post-1994 action. In the broader UN Web site (www.un.org/popin), one finds a major study, *Progress Made in Achieving the Goals and Objectives of the Program of Action of the International Conference on Population and Development.* An external assessment is the Population Institute's report entitled *The Hague Forum: Measuring ICPD Progress Since Cairo* (1999). See its Web site at www.populationinstitute.org.

ISSUE 3

Is Global Aging in the Developed World a Major Problem?

YES: The Center for Strategic and International Studies (CSIS), from *Meeting the Challenge of Global Aging: A Report to World Leaders from the CSIS Commission on Global Aging* (CSIS Press, 2002)

NO: Rand Corporation, from "Population Implosion?" Research Brief, Rand Europe (2005)

ISSUE SUMMARY

YES: The CSIS Report, *Meeting the Challenge of Global Aging: A Report to World Leaders from the CSIS Commission on Global Aging,* suggests that the wide range of changes brought on by global aging poses significant challenges in the ability of countries to address problems associated with the elderly directly and to the national economy as a whole.

NO: This Rand Corporation study suggests that because of declining fertility, European populations are either growing more slowly or have actually begun to decline. Although these trends "portend difficult times ahead," European governments should be able to confront these challenges successfully.

T he developed world is now faced with an aging population brought on by declining birth rates and an increasing life expectancy. The phenomenon first appeared during the last quarter of the previous century with the demographic transition from high birth and death rates to lower rates. The drop in death rates in these countries was a function of two basic factors: (1) the dramatic decline in both infant mortality (within the first year of birth) and child mortality (within the first five years) due to women being healthier during pregnancy and nursing periods, and to the virtually universal inoculation of children against principal childhood diseases; and (2) once people reach adulthood, they are living longer, in large part because of medical advances against key adult illnesses, such as cancer and heart disease.

Declining mortality rates yield an aging population in need of a variety of services—heath care, housing, and guards against inflation, for example—provided, in large part, by the tax dollars of the younger, producing sector of society. As the "gray" numbers of society grow, the labor force is increasingly called upon to provide more help for this class.

Declining birth and death rates mean that significantly more services will be needed to provide for the aging populations of the industrialized world, while at the same time, fewer individuals will be joining the workforce to provide the resources to pay for these services. However, some experts say that the new workforce will be able to take advantage of the skills of the more aged, unlike previous eras. In order for national economies to grow in the information age, an expanding workforce may not be as important a prerequisite as it once was. Expanding minds, not bodies, may be the key to expanding economies and increased abilities to provide public services.

However, the elderly and the young are not randomly distributed throughout society, which is likely to create a growing set of regional problems. In the United States, for example, the educated young are likely to leave the "gray belt" of the north for the Sun Belt of the south, southwest, and west. Who will be left in the older, established sectors of the country that were originally at the forefront of the industrial age to care for the disproportionately elderly population? Peter G. Peterson introduces the phrase, "the Floridization of the developed world," where retirees continue to flock in unprecedented numbers, in order to capture the essence of the problems associated with the changing age composition in industrial societies. What will happen 30 and 40 years from now, when the respective sizes of the young and the elderly populations throughout the developed world will yield a much larger population at the twilight of their existence? Although the trend is most evident in the richer part of the globe, people are also living longer in the developing world, primarily because of the diffusion of modern medical practices. But unless society can accommodate their skills of later years, they may become an even bigger burden for their national governments in the future.

A recent report, *Preparing for an Aging World: The Case for Cross-National Research* (National Academy of Sciences), identified a number of areas in which policymakers need a better understanding of the consequences of aging and resultant appropriate policy responses. Unless national governments of the developed world can effectively respond to these issues, the economic and social consequences can have a significant negative impact in both the aging population cohort as well as throughout the entire society.

In the first selection, The Center for Strategic and International Studies (CSIS) Commission on Global Aging suggests that the inevitable changing demographics of a much higher percentage of the population falling into the elderly category will create "significant challenges" to society to maintain and improve the quality of life for seniors as well as for the entire population of the developed world. In the second selection, the Rand Corporation study suggests that because of declining fertility, European populations are either growing more slowly or have actually begun to decline. Although these trends "portend difficult times ahead," European governments should be able to confront these challenges successfully.

Foreword

The advent of modern old age represents one of the great achievements of the democratic, market economies of the developed world. Advances in medicine and public health have led to a substantial increase in life spans, producing nearly a threefold increase in over-65 populations among the OECD countries since 1950. Meanwhile, pension income from public and private sources now provides the kind of income security in old age that was only a dream a century ago.

Although these advances have greatly improved the quality of life for the elderly, demographic developments beginning later this decade and continuing for the next 50 years will create significant challenges to this achievement that must be addressed in a timely manner. The surge of baby-boomer retirements and further gains in longevity, combined with continued below-replacement fertility rates, will lead to a significant increase in old-age dependency ratios in all developed nations. The portion of elderly in the populations of developed countries is expected to double from current levels, even as working populations grow more slowly or decline. This, in turn, will bring substantial fiscal pressures and, absent offsetting trends, is likely to constrain economic growth. In sum, a demographic transition is occurring in all the developed nations, albeit at varying degrees from one nation to another, making it a global aging phenomenon—one that requires the attention of world leaders, the academic and business communities, electorates, and individuals.

This challenge prompted the Center for Strategic and International Studies in 1999 to launch its two-year Global Aging Initiative and convene the Commission on Global Aging. The commission is a panel of 85 leading voices in politics, government, business, academia, and the nongovernmental sector from three continents.

Following more than a year of intensive research and discussion, CSIS surveyed the commission members during the spring of 2001 to determine their views on the fiscal, economic, financial, and international political challenges presented by global aging. Many were contacted by telephone and otherwise for more extensive input. The responses of 73 of these commissioners are presented and analyzed in this report.

Summary and Highlights

With a high degree of consensus, the commission found that the changes wrought by global aging are fundamental and unprecedented. They pose significant challenges to the ability of nations to sustain current benefits for the elderly as well as to sustain economic growth rates and the recent historic rise in living standards.

The commission found that pessimism is not warranted if the nations most directly affected act promptly to manage their aging transitions. The commission also found, however, that postponing reforms will lead to greater sacrifice later on. Rising dependency ratios will require industrial nations to examine carefully their pay-as-you-go old-age guarantees. In some cases these guarantees are not sustainable in their present form at current tax rates and will require significant reforms this decade. In other cases, the challenges facing pay-as-you-go pensions are less daunting if addressed in a timely manner.

The commission called on countries to undertake a multifaceted program of policy reforms to social protection schemes, private pensions, labor law, financial services, family policy, immigration, civil society, and international diplomacy in order to manage the aging transition and ensure sustainable old-age pension and health care systems. Although there are many ways that nations can prepare for the aging transition, no single policy reform by itself will be sufficient.

The growing interdependence of the major economies with one another, and with the larger world economy, means that the interests at stake transcend national boundaries. Although retirement policies must reflect unique national issues such as demographics and culture, the commission recommended that nations develop a framework of international consultation and monitoring. This will enable nations to share best practices and, where appropriate, coordinate policies.

Several major themes emerged from the commission's deliberations:

- **The fiscal challenge.** The commission was unanimous in its finding that demographic trends throughout the developed world pose a significant challenge to the sustainability of current social insurance pension and health guarantees for the aged. This remains true in spite of more than a decade of reforms around the world. The commission found that this fiscal challenge could make it difficult for nations to afford other important spending priorities, including infrastructure, defense, and education. Further, it found that indebted nations with large social insurance guarantees would be vulnerable to economic shocks that would reduce revenue and lead to large, potentially destabilizing budget deficits. Such instability could adversely affect the global economy.
- **The economic challenge.** The commission found that a rising portion of the retired populations in the developed world combined with slow growing—or, in several cases, sharply declining—numbers of labor force entrants will lead to sluggish economic growth and slower gains in standards of living. These forces may be compounded as total populations decline in Japan and most European Union member states. To

the extent that nations can find ways to boost capital and labor productivity, they will be able to partially offset this effect. This can be done by, among other things, labor law reform and economic and sectoral restructuring.

- **Aged-society industry as a boost for economy.** Population aging will result in large-scale sectoral shifts that could provide a stimulus to economic growth. The commission found that although some economic sectors, such as real estate, may decline owing to population aging, the expansion of aged-society-industry sectors will provide substantial growth opportunities. These sectors include not only medical and nursing care but also supply of the products and services to meet the needs of the aged, including barrier-free and universal-design products, housework assistance, and leisure activities.

- **Prefunding of public pensions.** The commission called on nations to undertake a gradual transition to market financing of public pension systems, provided that such reforms retain a public guarantee of an adequate retirement income beginning at a specified eligibility age. Such financing could take the form of the investment of government trust funds in equities and other private assets or the creation of individual accounts. The course any country takes would depend on its unique political and technical circumstances. Because funded systems are not directly affected by changes in the old-age dependency ratio, commissioners view funded systems as more sustainable in the long term. Funded systems also can help to insulate retirement security from adverse national economic trends through cross-border diversification. However, funded systems are not completely immune to demographic vulnerabilities. In particular, the commission found that the trend to dissaving after 2020 could reduce equity return in some countries. Funded pension systems also are vulnerable to fluctuations in economic performance.

- **Funded supplement to pay-as-you-go systems.** In some nations, pay-as-you-go systems are generous enough to sustain preretirement living standards without the need for saving. The commission recommended gradual reductions in benefit levels in these nations and in other nations where it is deemed appropriate owing to fiscal pressures and high and rising government debt. To make up for those benefit reductions, the commissioners also recommend that nations adopt supplemental funded pension systems designed to attract widespread participation. The combination of pay-as-you go and funded pensions should reduce risk by diversifying retiree income sources.

- **Older populations are becoming fitter.** The commissioners found that advances in medicine have brought—and will continue to bring— longer, healthier, more productive lives with declining rates of disability for the elderly. Further, the commission found that some countries may be significantly underestimating future gains in longevity. To the extent that people who have reached pension eligibility age are willing to continue working, current and future medical advances could enable more of the elderly to work past normal retirement age. Increasing full-time and part-time opportunities for older people who wish to work offers one of the best potential prescriptions to the challenges presented by population aging. The commissioners also unanimously agreed that increased labor participation by the elderly could produce significant social benefit.

- **Lengthening work lives.** To alleviate what is expected to be a severe tightening of labor markets in coming years, the commission recommended that nations enact reforms designed to reduce programmatic incentives for early retirement and to make it possible for those who wish to work after the statutory retirement age to do so. The commission also endorses raising statutory retirement ages under social security systems in correspondence with increases in longevity. Some commissioners, particularly in the United States where the retirement age is already relatively high, strongly believe that raising the eligibility age may not be necessary, particularly if prudent actions are taken soon to ensure long-term sustainability of social security systems. Thus, although the commission endorsed raising eligibility ages, the application of this measure would need to take into account the unique circumstances of each country. The private sector should likewise prepare for a world where employers will need older workers.

- **Opportunities for women.** Nations where the labor force participation rates of women are low may want to pursue policies that allow more women who wish to do so to participate in the labor force. Employers are encouraged to provide more flexible work opportunities that make it easier for women to both work and have a family. The commission strongly endorses the concept of phased retirement together with rules that forbid age and gender discrimination.

- **Financial markets.** The commission found that a rise in the old-age dependency ratio may depress saving rates and the value of equities. Demand for public sector borrowing to meet pension obligations could increase at the same time that large retired populations are spending down their retirement savings and also when defined pension plans are paying out more in benefits than they are receiving in employer contributions. The combined effect may put upward pressure on interest rates and downward pressure on equity and other financial asset values. To help ensure adequate global savings and stable financial markets after 2020, the commission recommended that nations without employer-sponsored pension schemes establish them. It also recommended that developed nations encourage the establishment of funded social insurance schemes in both the developed and developing worlds. The commission recommended that funded pensions be managed in accord with prudent-expert fiduciary standards and a minimum of cross-border investment restrictions and that financial markets be well regulated and transparent.

- **The role of globalization.** The commission found that economic integration with the "youthful" developing world could produce several benefits for developed and developing societies. It could enable the developed nations to restructure their economies around high-value-added activities, thereby raising labor productivity. It also could allow pension fund managers and individuals to invest a portion of their retirement assets in the developing world, both enhancing investment returns and spurring global growth. Finally, immigration from the developing world could increase the labor supply and support economic specialization. But it should also reflect the needs of the developing countries. In support of this objective, the commission recommended that the developed nations take steps to promote tolerance of nonnative races or cultures, and to make it easier for nonnative residents to achieve citizenship or permanent residency.

- **Mechanisms for international consultation.** The commission recommended that the challenges of global aging be a permanent item on the agenda of the Group of Eight nations. They also recommend that a mechanism be created for senior-level consultations among officials and experts around the world on an ongoing basis to better develop and coordinate beneficial cooperative strategies. These consultations should include the emerging market economies of the developing world.

In conclusion, we wish to thank our fellow commissioners and the Global Aging Initiative staff of the Center for Strategic and International Studies for their many contributions to the analyses and ideas reflected in this broad and far-reaching report. Like all important research, our work raised as many questions as it has answered. Inevitably, these uncertainties gave rise to disagreement among the commissioners. Yet the breadth and depth of our consensus reflects the reality that global aging will usher in a profoundly new era in which everything from the way we plan our lives to the role of nations in the global system will change.

We commend this report to the world's leaders and the international public in the hope that you will join us in meeting the fundamental challenge of global aging.

Ryutaro Hashimoto
Former Prime Minister of Japan

Walter F. Mondale
Former Vice President of the United States

Karl Otto Pöhl
Former President of Deutsche Bundesbank

January 7, 2002

Rand Corporation

Population Implosion?

Across Europe, birth rates are falling and family sizes are shrinking. The total fertility rate is now less than two children per woman in every member nation in the European Union. As a result, European populations are either growing very slowly or beginning to decrease.

At the same time, low fertility is accelerating the ageing of European populations. As a region, Europe in 2000 had the highest percentage of people age 65 or older—15 percent. According to data from the U.S. Bureau of the Census, this percentage is expected to nearly double by 2050.[1]

These demographic trends portend difficult times ahead for European economies. For example, a shrinking workforce can reduce productivity. At the same time, the growing proportion of elderly individuals threatens the solvency of pension and social insurance systems. As household sizes decrease, the ability to care for the elderly diminishes. Meanwhile, elderly people face growing health care needs and costs. Taken together, these developments could pose significant barriers to achieving the European Union (EU) goals of full employment, economic growth, and social cohesion.

Concern over these trends has sparked intense debate over the most effective policies to reverse them or mitigate their impact. The policy debate has focused on three approaches: (1) promoting increased immigration of working-age people; (2) encouraging more childbearing, especially among younger couples; and (3) reforming social policy to manage the negative consequences of these trends—including measures that could raise the retirement age or encourage more women to join the workforce. To date, this debate has produced more heat than light, and solid research-based evidence to inform the debate remains sketchy. Many aspects of the relationship between national policies and demographic trends are either disputed or not well understood, and it remains difficult to disentangle the effects of specific policy initiatives from the effects of broader social, political, and economic conditions.

To help inform EU policy deliberations, analysts from RAND Europe and RAND U.S. examined the relationships between policy and demographic change. The RAND team analysed the interrelationships between European government policies and demographic trends and behaviour, and assessed which policies can prevent or mitigate the adverse consequences of current low fertility and population ageing. The monograph, *Low Fertility and Population Ageing: Causes, Consequences, and Policy Options*, documents the study's findings.

The study carried out three tasks: It analysed European demographic trends; it examined the relationship between national-level policies on the one hand and demographic trends and household behaviours on the other hand; and it conducted case studies of five countries—France, Germany, Poland, Spain, and Sweden—which represent a mix of original and new member states, with varying levels of fertility and net immigration, and with different policy approaches.

The study found that:

- Immigration is not a feasible way of reversing population ageing or its consequences.
- National policies can slow fertility declines under the right circumstances.
- However, no single policy intervention necessarily works.
- And, what works in one country may not work in another. Social, economic, and political contexts influence the effect of policies.
- Policies designed to improve broader social and economic conditions may affect fertility, indirectly.
- Population policies take a long time to pay dividends—increases in fertility taking a generation to translate into an increased number of workers—making such policies politically unattractive.

Increased Immigration Will Not Reverse Population Ageing

Allowing large numbers of working-age immigrants to enter EU countries is not a feasible solution to the problem of population ageing. The sheer numbers of immigrants needed to offset population ageing in the EU states would be unacceptable in Europe's current sociopolitical climate. Furthermore, over the longer term, these immigrants would themselves age. The study concluded that the debate should focus on using immigration as a potential tool for slowing—as opposed to overcoming—population ageing.

Government Policies Can Slow Fertility Declines

Government policies can have an impact on fertility. For example, Spain and France present an instructive contrast. Currently, Spain has the second-lowest rate of fertility among the original 15 EU member states. However, a generation ago (in 1971), Spain's fertility was among the highest in Europe. The dramatic decline in fertility since then is associated with a shift from the pronatalist Franco regime—which banned contraception and encouraged large families—to a democratic regime that has no explicit population policy.

In contrast, France, which was the first European nation to experience a decline in its fertility rate and which has had an aggressive set of pronatalist policies in place for many decades, now has the second-highest fertility rate in Europe (behind Ireland). The fertility rate in France has not declined as much as that in other countries, and it actually increased between 1993 and 2002.

Reversing Fertility Decline Has Involved a Mix of Policies

No single policy intervention has worked to reverse low fertility. Historically, governments have attempted to boost fertility through a mix of policies and programmes. For example, France in recent decades has employed a suite of policies intended to achieve two goals: reconciling family life with work and reversing declining fertility. To accomplish the first goal, for example, France instituted generous child-care subsidies. To accomplish the second, families have been rewarded for having at least three children.

Sweden, by contrast, reversed the fertility declines it experienced in the 1970s through a different mix of policies, none of which specifically had the objective of raising fertility. Its parental work policies during the 1980s allowed many women to raise children while remaining in the workforce. The mechanisms for doing so were flexible work schedules, quality child care, and extensive parental leave on reasonable economic terms.

Political, Economic, and Social Contexts Influence Policy Impacts

Designing successful interventions is complicated by the fact that policies that work in one country may not work in another. Different interventions have varying effects because of the diverse, complex, and shifting political, economic, and social contexts in which they are implemented. The impact of these contexts appears in some of the sweeping political transitions Europe has witnessed over the past two decades.

For example, fertility declines in the former East Germany after unification appear to owe more to a shifting social environment than to policy change. Women who faced the unification with concerns about their economic situations were less likely to have children in the following year or two. The contrast between the former East Germany and West Germany is instructive. After reunification, the former West Germany's fertility rate remained relatively stable, whereas the former East Germany saw a precipitous decline over the next three years. Similarly, the transition to a free-market economy in Poland changed the economic environment and incentives for childbearing. Since 1989, Poland has experienced a sharp decrease in fertility.

Sweden provides a different kind of example. Although Sweden did not undergo political or economic transformation, its economic conditions nonetheless have affected fertility. Unlike in most other countries, fertility rates in Sweden are positively related to the earnings of women—likely because women's earnings in Sweden constitute a substantial proportion of dual-earner household income. Since parental-leave benefits are proportional to earnings, improvements in economic conditions lead to higher parental benefits, which can help promote increased fertility. Part of the decrease in fertility in Sweden during the 1990s is likely related to the decline in economic conditions.

Population Change Is Slow

Government policies intended to reverse fertility declines typically have a long-term focus and require many years to bear fruit. A few population policies may have an immediate effect (for example, those restricting/allowing abortion), but they are exceptions, and their effects tend to be on the timing of births rather than on completed fertility. Population policies to increase fertility take at least one generation before they ultimately increase the number of new entrants to the labour force.

As a result, there is a disconnect between electoral cycles (typically, 4–5 years) and the longer cycle of population policy. Politicians have limited incentives to advocate such policies, especially when doing so might entail the expenditure of political capital in entering a contentious policy domain. Therefore, population policies tend to lack both political appeal and political champions.

Instead, politicians tend to focus on policies for mitigating the effects of population ageing, which have shorter time horizons. One policy for doing so is encouraging participation in the workforce. This can mean promoting a longer working life and encouraging new entrants, such as women, into the workforce. Related to this are policies that seek to enhance the productivity of older workers.

Conclusion

This study showed that, under certain conditions, European governments can successfully confront the looming economic threats of declining fertility rates and ageing populations. Policies that remove workplace and career impediments to childrearing are a critical part of any solution. However, reversing long-term ageing and low fertility remains problematic, given that policies for doing so may not pay dividends until the next generation reaches working age. Prior to that time, millions of baby boomers will have retired. Hence, a solution will require long-term vision and political courage.

Note

1. Kevin Kinsella and Victoria A. Velkoff, *An Aging World: 2001*, Washington, D.C.: U.S. Census Bureau, Series P95/01-1, 2001, p. 9.

POSTSCRIPT

Is Global Aging in the Developed World a Major Problem?

The issue of the changing age composition in the developed world was fore-seen a few decades ago but its heightened visibility is relatively recent. This visibility culminated in a UN-sponsored conference on aging in Madrid in April 2002. Its plan of action commits governments to address the problem of aging and provided them with a set of 117 specific recommendations covering three basic areas: older individuals and development, advancing health and well-being into old age, and ensuring enabling and supportive environments.

With the successful demographic transition in the industrial world, the percentage of those above the age of 60 is on the rise, while the labor force percentage is decreasing. In 1998, 19 percent of the first world fell into the post-60 category (10 percent worldwide). Children under age 15 also make up 19 percent of the developed world's population while the labor force is at 62 percent. With birth rates hovering around 1 percent or less, and life expectancy increasing, the percentages will likely continue to grow toward the upper end of the scale.

Paul Peterson has argued that the costs of global aging will not only outweigh the benefits, the capacity of the developed world to pay for these costs is questionable at best. The economic burden on the labor force will be "unprecedented," he suggests, and he offers a number of solutions ("Gray Dawn: The Global Aging Crisis" in *Foreign Affairs* (January/February 1999).

A particularly outspoken opponent of the "gloom" viewpoint is Phil Mullan. His book, *The Imaginary Time Bomb; Why an Ageing Population Is Not a Social Problem* (I. B. Tauris, New York, 2000), criticizes how the idea of an aging developed world has become "a kind of mantra for opponents of the welfare state and for a collection of alarmists." Mullan is joined by Phillip Longman, author of *The Empty Cradle* (Basic Books, 2004), who suggests in a *Foreign Affairs* article ("The Global Baby Bust," May/June 2004) that the coming "baby bust" will yield a variety of positive consequences as well as negative ones.

An excellent overview of the issues to be addressed in an era of global aging is *Global Aging: The Challenge of Success* by Kevin Kinsella and David R. Phillips (Population Reference Bureau, *Population Bulletin*, vol. 60, no 1, 2005).

A good introduction to global aging is found in "Aging of Population" by Leonid A. Gavrilov and Patrick Heuveline (*The Encyclopedia of Population*, Macmillan Reference, 2003). The most succinct presentation of the effects of global aging can be found in John Hawksworth's "Seven Key Effects of Global Aging" (PricewaterhouseCoopers' web site, www.pwcglobal.com). This is "must reading" for those who want a concise objective description of the potential consequence of the changing demographics associated with age distribution. His presentation focuses on the effects on economic growth, pensions systems, working lives, equity and bond markets, international capital flows, migration, and business strategies.

A good source for the effect of declining population in Europe is "Population Policy Dilemmas in Europe at the Dawn of the Twenty-First Century" (*Population and Development Review*, The Population Council, Inc., March 2003).

Some, such as Leon F. Bouvier and Jane T. Bertrand (*World Population: Challenges for the 21st Century*, Seven Locks Press, 1999), there seems to be a potential silver lining on the horizon. Although future increases in immigration will counterbalance the decline of the indigenous population, they assert, the real advance will be the decoupling of productivity expansion and work-force increases. The information age is knowledge-intensive, and becoming more so, not labor-intensive.

One author who accepts Bouvier and Bertrand's thesis is the noted scholar of management, Peter Drucker. In "The Future That Has Already Happened", *The Futurist* (November 1998), Drucker predicts that retirement age in the developed world will soon rise to 75, primarily because the greatest skill of this age group—knowledge—will become even more of an asset. He maintains that knowledge resources will become the most important commodity.

An important book on the fiscal problems facing the developed world because of aging can be found in Robert Stowe England's *The Fiscal Challenge of an Aging Industrial World* (CSIS *Significant Issues* Series, November 2001). An earlier report from CSIS is *Global Aging: The Challenge of the New Millennium* (CSIS and Watson Wyatt Worldwide). This document's presentation of the raw data is particularly useful.

The Center for Strategic and International Studies (CSIS) in Washington is at the forefront of research and policy advocacy on the issue of global aging in the developed world. Its Global Aging Initiative (GAI) explores the international economic, financial, political, and security implications of aging and depopulation. One can find numerous speeches and other short papers on the issue on its web site, including the most extensive bibliography on the effects of global aging in the developed world (www.csis.org).

Other web sites include LinkAge 2000: Policy Implications of Global Aging (library.thinkquest.org/), The International Center on Global Aging (www.globalaging.org/resources), and The Environmental Literacy Council (www.enviroliteracy.org).

Finally, the Second World Assembly on Aging in Madrid in April 2002 produced a huge number of documents. Its Plan of Action and other reports can be found at www.un.org/ageing/coverage.

ISSUE 4

Does Global Urbanization Lead Primarily to Undesirable Consequences?

YES: Divya Abhat, Shauna Dineen, Tamsyn Jones, Jim Motavalli, Rebecca Sanborn, and Kate Slomkowski, from "Today's 'Mega-Cities' Are Overcrowded and Environmentally Stressed," http://www.emagazine.com (September/October 2005)

NO: Robert McDonald, from "A World of the City, by the City, for the City," ZNet/Activism, http://www.zmag.org (December 20, 2005)

ISSUE SUMMARY

YES: Jim Motavalli, editor of *E/The Environmental Magazine*, suggests that the world's major cities suffer from a catalog of environmental ills, among them pollution, poverty, fresh water shortages, and disease.

NO: Robert McDonald, a postdoctoral fellow at Harvard University, suggests that global urbanization presents a great opportunity for the world to achieve international peace. It creates new possibilities for democracy and a sharing of common interests across national boundaries.

In 1950, 55 percent of the population of the developed world resided in urban areas, compared to only 18 percent in the developing world. By 2000, 76 percent of those in the developed world were urbanized, and it is expected, according to UN projections, to reach 82 percent by 2025. But because there will be low population growth throughout the developed world in the coming decades, the impact will not be substantial.

The story is different in the developing world. In 2000, the level of urbanization had risen to almost 40 percent, and is projected to be 54 percent in 2025. The percentages tell only part of the story, however, as they are not based on a stable national population level, but will occur in the context of substantial increases in the national population level. To illustrate the dual implication of urban growth as the consequence of both migration to the cities and increased births to those already living in urban areas, the UN projects that the urban population in the developing world will nearly double in size between 2000 and 2025, from just under 2 billion to more than 3.5 billion. And urban growth will not stop in 2025.

The UN reported in *UN Chronicle* (no. 3, 2002) a number of other conclusions relating to urban growth:

- Almost all population increase in the next three decades will occur in the urban areas of the developed world.
- This growth in the developed world's urban centers will continue at a rate of at least 3 percent.
- There are marked differences in the level and pace of urbanization among major geographical areas of the developing world.
- The proportion of the world's population living in megacities is small, as only 3.7 percent reside in cities of at least 10 million inhabitants.

There are two ways to examine rapid urbanization. One is to study the ability of society to provide services to the urban population. A second approach is to examine the adverse impacts of the urbanized area on the environment. The best place to begin a discussion of urbanization's effects is found in "An Urbanizing World" (Martin P. Brockerhoff, *Population Bulletin*, Population Reference Bureau, 2000). To Brockerhoff, increasing urbanization in the poor countries can be seen "as a welcome or as an alarming trend." He suggests that cities have been the "engines of economic development and the centers of industry and commerce." The diffusion of ideas is best found in cities around the world. And Brockerhoff observes the governmental cost savings of delivering goods and services to those in more densely populated environments.

The current problem is not that cities of the developing world are growing, but that they are expanding at a rapid pace. This calls into question the ability of both government and the private sector to determine what is necessary for urbanites to not only survive but to thrive. Many researchers believe that poverty and health problems (both physical and mental) are consequences of urbanization. Brockerhoff also alludes to the potential for greater harm to residents of cities from natural disasters and environmental hazards.

There is a more recent concern emerging among researchers about urbanization's impact—this time on biodiversity. One source has coined the phrase "heavy ecological footprints" ("Impact on the Environment," *Population Reports*, 2002) to describe the adverse effects. One study concludes, for example, that urban sprawl in the United States endangers more species than any other human behavior (Michael L. McKinney, "Urbanization, Biodiversity, and Conservation," *Bioscience*, 2002).

The two selections for this issue address the question of the consequences of urbanization. In the first selection, Jim Motavalli argues that most urban dwellers in the developing world are already confronted by severe environmental problems that will only increase in nature as population continues to grow. In the second selection, Robert McDonald suggests that global urbanization presents a great opportunity for the world to achieve international peace. It creates new possibilities for democracy and a sharing of common interests across national boundaries.

Cities of the Future: Today's "Mega-Cities" Are Overcrowded and Environmentally Stressed

We take big cities for granted today, but they are a relatively recent phenomenon. Most of human history concerns rural people making a living from the land. But the world is rapidly urbanizing, and it's not at all clear that our planet has the resources to cope with this relentless trend. And, unfortunately, most of the growth is occurring in urban centers ill-equipped for the pace of change. You've heard of the "birth dearth"? It's bypassing Dhaka, Mumbai, Mexico City and Lagos, cities that are adding population as many of their western counterparts contract.

The world's first cities grew up in what is now Iraq, on the plains of Mesopotamia near the banks of the Tigris and Euphrates Rivers. The first city in the world to have more than one million people was Rome at the height of its Empire in 5 A.D. At that time, world population was only 170 million. But Rome was something new in the world. It had developed its own sophisticated sanitation and traffic management systems, as well as aqueducts, multi-story lowincome housing and even suburbs, but after it fell in 410 A.D. it would be 17 centuries before any metropolitan area had that many people.

The first large city in the modern era was Beijing, which surpassed one million population around 1800, followed soon after by New York and London. But at that time city life was the exception; only three percent of the world's population lived in urban areas in 1800.

The rise of manufacturing spurred relocation to urban centers from the 19th through the early 20th century. The cities had the jobs, and new arrivals from the countryside provided the factories with cheap, plentiful labor. But the cities were also unhealthy places to live because of crowded conditions, poor sanitation and the rapid transmission of infectious disease. As the Population Reference Bureau reports, deaths exceeded births in many large European cities until the middle of the 19th century. Populations grew, then, by continuing waves of migration from the countryside and from abroad.

From First World to Third

In the first half of the 20th century, the fastest urban growth was in western cities. New York, London and other First World capitals were magnets for immigration and job opportunity. In 1950, New York, London, Tokyo and Paris boasted of having the world's largest metropolitan populations. (Also in the top 10 were Moscow, Chicago and the German city of Essen.) By then, New York had already become the first "mega-city," with more than 10 million people. It would not hold on to such exclusivity for long.

In the postwar period, many large American cities lost population as manufacturing fled overseas and returning soldiers taking advantage of the GI Bill fueled the process of suburbanization. Crime was also a factor. As an example, riot-torn Detroit lost 800,000 people between 1950 and 1996, and its population declined 33.9 percent between 1970 and 1996. Midwestern cities were particularly hard-hit. St. Louis, for instance, lost more than half its population in the same period, as did Pittsburgh. Cleveland precipitously declined, as did Buffalo, Cincinnati, Minneapolis and many other large cities, emerging as regional players rather than world leaders.

Meanwhile, while many American cities shrank, population around the world was growing dramatically. In the 20th century, world population increased from 1.65 billion to six billion. The highest rate of growth was in the late 1960s, when 80 million people were added every year.

According to the "World Population Data Sheet," global population will rise 46 percent between now and 2050 to about nine billion. While developed countries are losing population because of falling birth rates and carefully controlled immigration rates (only the U.S. reverses this trend, with 45 percent growth to 422 million predicted by 2050), population is exploding in the developing world.

India's population will likely grow 52 percent to 1.6 billion by 2050, when it will surpass China as the world's most populous country. The population in neighboring Pakistan will grow to 349 million, up 134 percent in 2050. Triple-digit growth rates also are forecast for Iraq, Afghanistan and Nepal.

Africa could double in population to 1.9 billion by 2050. These growth rates hold despite the world's highest rates of AIDS infection, and despite civil wars, famines and other factors. Despite strife in the Congo, it could triple to 181 million by 2050, while Nigeria doubles to 307 million.

Big Cities Get Bigger—and Poorer

According to a 1994 UN report, 1.7 billion of the world's 2.5 billion urban dwellers were then living in less-developed nations, which were also home to two thirds of the world's mega-cities. The trend is rapidly accelerating. *People and the Planet* reports that by 2007, 3.2 billion people—a number larger than the entire global population of 1967—will live in cities. Developing countries will absorb nearly all of the world's population increases between today and 2030. The estimated urban growth rate of 1.8 percent for the period between

2000 and 2030 will double the number of city dwellers. Meanwhile, rural populations are growing scarcely at all.

Also by 2030, more than half of all Asians and Africans will live in urban areas. Latin America and the Caribbean will at that time be 84 percent urban, a level comparable to the U.S. As urban population grows, rural populations will shrink. Asia is projected to lose 26 million rural dwellers between 2000 and 2030.

For many internal migrants, cities offer more hope of a job and better health care and educational opportunities. In many cases, they are home to an overwhelming percentage of a country's wealth. (Mexico City, for example, produces about 30 percent of Mexico's total Gross Domestic Product.) Marina Lupina, a Manila, Philippines resident, told *People and the Planet* that she and her two children endure the conditions of city living (inhabiting a shack made from discarded wood and cardboard next to a fetid, refuse-choked canal) because she can earn $2 to $3 a day selling recycled cloth, compared to 50 cents as a farm laborer in the rural areas. "My girls will have a better life than I had," she says. "That's the main reason I came to Manila. We will stay no matter what."

Movement like this will lead to rapidly changing population levels in the world's cities, and emerging giants whose future preeminence can now only be guessed. "By 2050, an estimated two-thirds of the world's population will live in urban areas, imposing even more pressure on the space infrastructure and resources of cities, leading to social disintegration and horrific urban poverty," says Werner Fornos, president of the Washington-based Population Institute.

Today, the most populous city is Tokyo (26.5 million people in 2001), followed by Sao Paulo (18.3 million), Mexico City (18.3 million), New York (16.8 million) and Bombay/Mumbai (16.5 million). But by 2015 this list will change, with Tokyo remaining the largest city (then with 27.2 million), followed by Dhaka (Bangladesh), Mumbai, Sao Paulo, New Delhi and Mexico City (each with more than 20 million). New York will have moved down to seventh place, followed by Jakarta, Calcutta, Karachi and Lagos (all with more than 16 million).

The speed by which some mega-cities are growing has slowed. Thirty years ago, for instance, the UN projected Mexico City's population would grow beyond 30 million by 2000, but the actual figures are much lower. Other cities not growing as much as earlier seen are Rio de Janeiro, Calcutta, Cairo and Seoul, Korea. But against this development is the very rapid growth of many other cities (in some cases, tenfold in 40 years) such as Amman (Jordan), Dar es Salaam (Tanzania), Lagos and Nairobi.

The rise of mega-cities, comments the *Washington Post*, "poses formidable challenges in health care and the environment, in both the developed and developing world. The urban poor in developing countries live in squalor unlike anything they left behind . . . In Caracas, more than half the total housing stock is squatter housing. In Bangkok, the regional economy is 2.1 percent smaller than it otherwise would be because of time lost in traffic jams. The mega-cities of the future pose huge problems for waste management, water use and climate change."

In Cairo, Egypt, the rooftops of countless buildings are crowded with makeshift tents, shacks and mud shelters. It's not uncommon to see a family cooking their breakfast over an open fire while businesspeople work in their cubicles below. The city's housing shortage is so severe that thousands of Egyptians have moved into the massive historic cemetery known as the City of the Dead, where they hang clotheslines between tombs and sleep in mausoleums.

By 2015, there will be 33 mega-cities, 27 of them in the developing world. Although cities themselves occupy only two percent of the world's land, they have a major environmental impact on a much wider area. London, for example, requires roughly 60 times its own area to supply its nine million inhabitants with food and forest products. Mega-cities are likely to be a drain on the Earth's dwindling resources, while contributing mightily to environmental degradation themselves.

The Mega-City Environment

Mega-cities suffer from a catalog of environmental ills. A World Health Organization (WHO)/United Nations Environment Programme (UNEP) study found that seven of the cities—Mexico City, Beijing, Cairo, Jakarta, Los Angeles, Sao Paulo and Moscow—had three or more pollutants that exceeded the WHO health protection guidelines. All 20 of the cities studied by WHO/UNEP had at least one major pollutant that exceeded established health limits.

According to the World Resources Institute, "Millions of children living in the world's largest cities, particularly in developing countries, are exposed to life-threatening air pollution two to eight times above the maximum WHO guidelines. Indeed, more than 80 percent of all deaths in developing countries attributable to air pollution-induced lung infections are among children under five." In the big Asian mega-cities such as New Delhi, Beijing and Jakarta, approximately 20 to 30 percent of all respiratory disease stems from air pollution.

Almost all of the mega-cities face major fresh water challenges. Johannesburg, South Africa is forced to draw water from highlands 370 miles away. In Bangkok, saltwater is making incursions into aquifers. Mexico City has a serious sinking problem because of excessive groundwater withdrawal.

More than a billion people, 20 percent of the world's population, live without regular access to clean running water. While poor people are forced to pay exorbitant fees for private water, many cities squander their resources through leakages and illegal drainage. "With the population of cities expected to increase to five billion by 2025," says Klaus Toepfer, executive director of the UNEP, "the urban demand for water is set to increase exponentially. This means that any solution to the water crisis is closely linked to the governance of cities."

Mega-city residents, crowded into unsanitary slums, are also subject to serious disease outbreaks. Lima, Peru (with population estimated at 9.4 million by 2015) suffered a cholera outbreak in the late 1990s partly because, as the *New York Times* reported, "Rural people new to Lima . . . live in houses without running water and use the outhouses that dot the hillsides above." Consumption of unsafe food and water subjects these people to life-threatening diarrhea and dehydration.

It's worth looking at some of these emerging mega-cities in detail, because daily life there is likely to be the pattern for a majority of the world's population. Most are already experiencing severe environmental problems that will only be exacerbated by rapid population increases. Our space-compromised list leaves out the largest European and American cities. These urban centers obviously face different challenges, among them high immigration rates.

Jakarta, Indonesia

A Yale University graduate student, who served as a college intern at the U.S. Embassy in Jakarta, brought back this account: "Directly adjacent to the Embassy's high-rise office building was a muddy, trash-filled canal that children bathed in every morning. The view from the top floors was unforgettable: a layer of brown sky rising up to meet the blue—a veritable pollution horizon. In the distance the tips of skyscrapers stretched up out of the atmospheric cesspool below, like giant corporate snorkels. Without fresh air to breathe, my days were characterized by nausea and constant low-grade headaches. I went to Indonesia wanting a career in government, and left determined to start a career working with the environment."

Jakarta is one of the world's fastest-growing cities. United Nations estimates put the city's 1995 population at 11.5 million, a dramatic increase from only 530,000 in 1930. Mohammad Dannisworo of the Bandung Institute of Technology (ITB) says 8.5 million people live within the city's boundaries at night and an additional 5.5 million migrate via 2.5 million private cars, 3.8 million motorcycles and 255,000 public transportation vehicles into the city during the day. This daily parade of combustion engines clogs the city streets and thickens the air, making Jakarta the world's third-most-polluted city after Bangkok and Mexico City.

Rapid growth has become one of the capital city's greatest challenges, as migrants continue to pour into Jakarta from the surrounding countryside in search of higher-paying jobs. An estimated 200,000 people come to the city looking for employment every year. In the face of such growth, the city has been unable to provide adequate housing, despite repeated attempts to launch urban improvement programs. The Kampung Improvement Program (KIP), established in the 1980s, was initially highly successful in boosting living conditions for more than 3.5 million established migrants, but it has been unable to accommodate the persistent migrant influx. There is an acute housing shortage, with a demand for 200,000 new units a year unfulfilled.

As Encarta describes it, "In the 1970s, efforts failed to control growth by prohibiting the entry of unemployed migrants. The current strategy emphasizes family planning, dispersing the population throughout the greater [metropolitan] region, and promoting transmigration (the voluntary movement of families to Indonesia's less-populated islands). Jakarta is a magnet for migrants . . .

[During the late 1980s] most were between the ages of 15 and 39 years, many with six years of education or less."

The UN reports that the city's drinking water system is ineffective, leading 80 percent of Jakarta inhabitants to use underground water, which has become steadily depleted. In lowlying North Jakarta, groundwater depletion has caused serious land subsidence, making the area more vulnerable to flooding and allowing seawater from the Java Sea to seep into the coastal aquifers. According to Suyono Dikun, Deputy Minister for Infrastructure at the National Development Planning Board, more than 100 million people in Indonesia are living without proper access to clean water.

Jakarta's environment has been deteriorating rapidly, with serious air pollution and the lack of a waterborne sewer. Jakarta officials have only recently begun to acknowledge the source of over half of the city's air pollution, and have begun to take action against automobile congestion. The Blue Skies Program, founded in 1996, is dedicated to updating the city's public and private transportation technology. The project's successes to date include an increase in the percentage of vehicles meeting pollution standards, a near-complete phasing out of leaded gasoline, and an increase in the number of natural gas-fueled vehicles to 3,000 taxis, 500 passenger cars and 50 public buses.

The Blue Skies Project is pushing Jakarta toward a complete natural gas conversion and is working towards the installation of dedicated filling stations, establishing a fleet of natural gasfueled passenger busses, supplying conversion kits for gasoline-fueled cars, and creating adequate inspection and maintenance facilities.

Jakarta has acknowledged its traffic problems and undertaken both small and large scale projects to alleviate the stresses of pollution and congestion. The city has launched a "three-in-one" policy to encourage carpooling, demanding that every car on major thruways carry at least three passengers when passing through special zones from 4:30 p.m. to 7:30 p.m. The city has also undertaken the construction of a nearly 17-mile monorail system.

But if Jakarta really wants to alleviate its infrastructure problems, it has to work from within, says Gordon Feller of the California-based Urban Age Institute. "The mayor needs to create a partnership between the three sectors—the government, the local communities and the nongovernmental agencies. The job of the mayor is to empower the independent innovators, not to co-opt or block them."

Dhaka, Bangladesh

Dhaka had only 3.5 million people in 1951; now it has more than 13 million. The city has been gaining population at a rate of nearly seven percent a year since 1975, and it will be the world's second-largest city (after Tokyo) by 2015. According to a recent Japanese environmental report, "Dhaka city is beset with a number of socio-environmental problems. Traffic congestion, flooding, solid waste disposal, black smoke from vehicular and industrial emissions, air and noise pollution, and pollution of water bodies by industrial discharge. . . .

Black smoke coming out from the discharge is intolerable to breathe, burning eyes and throats. The city dwellers are being slowly poisoned by lead concentration in the city air 10 times higher than the government safety limit."

Because of a heavy concentration of cars burning leaded gasoline, Dhaka's children have one of the highest blood lead levels in the world. Almost 90 percent of primary school children tested had levels high enough to impair their developmental and learning abilities, according to a scientific study.

Water pollution is already rampant. According to the Japanese report, "The river Buriganga flows by the side of the densely populated area of the old city. Dumping of waste to the river by . . . industries is rather indiscriminate. . . . The indiscriminate discharge of domestic sewage, industrial effluents and open dumping of solid wastes are becoming a great concern from the point of water-environment degradation."

Nearly half of all Bangladeshis live below the poverty line, able only to glance at the gleaming new malls built in Dhaka. Urbanization and the pressures of poverty are severely stressing the country's once-abundant natural resources. According to U.S. Aid for International Development (USAID), "Pressures on Bangladesh's biological resources are intense and growing."

They include:

- Poor management of aquatic and terrestrial resources;
- Population growth;
- Overuse of resources;
- Unplanned building projects; and
- Expansion of agriculture onto less-productive lands, creating erosion and runoff, among other by-products.

Bangladesh's expanding population destroys critical habitats, reports USAID, causing a decrease in biodiversity. Most of Bangladesh's tropical forests and almost all of the freshwater floodplains have been negatively affected by human activities.

But despite all the negatives, there is a growing environmental movement in Bangladesh that is working to save Dhaka's natural resources. The Bangladesh Environmental Network (BEN), for instance, works on reducing the high level of arsenic in Bangladesh's water supply (more than 500 percent higher than World Health Organization standards), combats the country's severe flooding problem and tries to defeat India's River Linking Project, which could divert an estimated 10 to 20 percent of Bangladesh's water flow. Bangladesh Poribesh Andolon holds demonstrations and international action days to increase citizen awareness of endangered rivers.

International development projects are also addressing some of the country's environmental woes, including a $44 million arsenic mitigation project launched in 1998 and jointly financed by the World Bank and the Swiss Development and Cooperation Agency. The project is installing deep wells, installing hardware to capture rainwater, building sanitation plants and expanding distribution systems. A $177 million World Bank project works with the government of Bangladesh to improve urban transportation in

Dhaka. Private companies from Bangladesh and Pakistan recently announced a joint venture to construct a waste management plant that could handle 3,200 metric tons of solid waste per day, turning it into organic fertilizer.

Mexico City

Mexico City is like an anxious teenager, growing up faster than it probably should. That phenomenon manifests itself in awkward contrasts: Sports cars zipping down crowded streets, choked with air pollution; a Wal-Mart rising against a skyline of the ancient ruins of Teotihuacan; and trendy designer knock-off bags lining the walls of a grungy street stall.

The locale has long been a cultural hub—the ancient Aztec capital of Tenochtitlán, where Mexico City now stands, was the largest city in the Americas in the 14th century with a population of about 300,000. When the Spanish razed Tenochtitlán they erected Mexico City in its place, though a smallpox epidemic knocked the population back to 30,000. Mexico City served as the center of Spain's colonial empire during the 1500s, but the modern-day metropolis only began to materialize in the late 1930s when a combination of rapid economic growth, population growth, and a considerable rural migration filled the city with people.

The larger metropolitan area now engulfs once-distinct villages and population estimates range from 16 million to 30 million, depending on how the city's boundaries are drawn. Regardless, Mexico City is now widely considered the world's third-largest city, and still growing; birth rates are high and 1,100 new residents migrate to the capital each day.

With so many people crammed into a closed mountain valley, many environmental and social problems are bound to arise. Mexico City's air was ranked by WHO as the most contaminated in the world in 1992. By 1998, the Mexican capital had added the distinction of being "the world's most dangerous city for children." Twenty percent of the city's population lives in utter poverty, the Mega-Cities Project reports, 40 percent of the population lives in "informal settlements," and wealth is concentrated in very few hands.

A combination of population, geography and geology render air pollution one of the city's greatest problems. WHO studies have reported that it is unhealthy to breathe air with over 120 parts per billion of ozone contaminants more than one day a year, but residents breathe it more than 300 days a year. More than one million of the city's more than 18 million people suffer from permanent breathing problems.

According to the U.S. Energy Information Administration, "Exhaust fumes from Mexico City's approximately three million cars are the main source of air pollutants. Problems resulting from the high levels of exhaust are exacerbated by the fact that Mexico City is situated in a basin. The geography prevents winds from blowing away the pollution, trapping it above the city."

The International Development Research Center has observed that "despite more than a decade of stringent pollution control measures, a haze hangs over Mexico City most days, obscuring the surrounding snow-capped mountains and endangering the health of its inhabitants. Many factors have contributed to this situation: industrial growth, a population boom and the proliferation of vehicles." More than 30 percent of the city's vehicles are more than 20 years old.

Solid waste creates another major problem, and officials estimate that, of the 10,000 tons of waste generated each day, at least one quarter is dumped illegally. The city also lacks an effective sanitation and water distribution system. According to the United Nations, "Urbanization has had a serious negative effect on the ecosystem of Mexico City. Although 80 percent of the population has piped inside plumbing, residents in the peripheral areas cannot access the sewage network and a great percentage of wastewater remains untreated as it passes to the north for use as irrigation water."

Perhaps three million residents at the edge of the city do not have access to sewers, says the Mega-Cities Project. Untreated waste from these locations is discharged directly into water bodies or into the ground, where it can contaminate ground water. Only 50 percent of residents in squatter settlements have access to plumbing, and these residents are more likely to suffer from health effects linked to inadequate sanitation. Furthermore, Mexico City is now relying on water pumped from lower elevations to quench an ever-deepening thirst; as the city continues to grow, the need for water and the politics surrounding that need are likely only to intensify.

Mexican industry is centered within the city and is primarily responsible for many of the city's environmental problems as well as for the prosperity that certain areas have achieved. Mexico City houses 80 percent of all the firms in the country, and 2.6 million cars and buses bring people to work and shop in them. Sandwiched in between slums and sewers are glitzy, luxurious neighborhoods and shopping centers, as chic as any in New York or Los Angeles.

The streets of the Zócalo, a central city plaza modeled after Spanish cities, serve as Mexico City's cultural hub. Unwittingly, the plaza has become one of the economic centers as well. Most job growth in Mexico occurs in the underground sector—in street stalls that cover every square inch of sidewalk space, women flipping tortillas curbside, and kids hawking phone cards or pirated CDs to passersby. Despite efforts to clean up activities that are illegal or considered eyesores, street vendors make up an enormous part of Mexico's job force and, according to the *Los Angeles Times,* are primarily responsible for keeping the official unemployment rate below that of the United States. . . .

Robert McDonald  **NO**

A World of the City, by the
City, for the City

Rip Van Winkle took his famous nap on the outskirts of Palenville in the Catskill Mountains. In Washington Irving's original story, Rip slept for only 20 years, managing to miss the entire American Revolution in the process. Let us imagine that Rip, being incredibly long-lived due to his many hours of restorative sleep, is still wandering around the Hudson Valley. How different it must look to him! The New York City megalopolis alone now holds more Americans than the Empire State and all of New England did in 1900. Rip has just been the witness to one of the most dramatic transformations of the last century, the shift from a rural to urban existence for the vast majority of Americans. In 1900, 60% of the U.S. population was rural. Today, less than 25% of the population is. It was a transformation that changed the very character of life for Americans, and drove a series of political and cultural changes that continue today.

I've heard that Rip Van Winkle has grown tired again. This time, however, in his quest to find a quiet place to rest his head, he's ventured to a calm spot along the China coast. What can Rip expect to see when he awakens, another 50 or 100 years hence?

The United Nations Population Division is making plans to celebrate the day, sometime in the next couple of years, when, for the first time in the history of the world, a majority of humanity will live in cities. Most of this growth will occur in developing countries, particularly India and China. In all likelihood, Rip will awaken from another long slumber to find a landscape far more modified than that of upstate New York. He will be witness to a momentous, qualitative change in the lived experience of humanity, one that poses a grave challenge to the goal of sustainable development, and yet offers hope of achieving the dream of international peace.

Sustainable Urbanization

This massive urbanization occurring worldwide creates new possibilities for environmental problems. An increase in urban population necessarily requires an increase in urban land area, on a grand scale: by one estimate, there will be more housing units built in the 21st century than in the rest of recorded human history. And as urban land increases, there will be a host of impacts on the environment. Many of the negative impacts will be clustered in poor neighborhoods in fast growing cities in developing countries: the fast growing *favelas* of Rio de Janeiro, the shantytowns of Mumbai. Understanding and managing this array of environmental impacts is thus crucial not just for environmentalists, but for anyone interested in reducing poverty and human misery.

Foremost among the problems that the megacities of the 21st century will face is the pollution of drinking water by industrial waster. The Chinese city of Harbin, for example, just had to shut off its public water supply after a petrochemical plant upstream exploded. Its three million residents had to survive without easily available water for several days. Even more common in cities in poorer countries that lack sewer systems are frequent outbreaks of cholera and other water borne diseases. The UN estimates that more than a billion people live in slums, the vast majority without clean drinking water, and this number will likely continue to grow unless there are substantial increases in infrastructure investment.

If Harbin is a warning of the water crises that may become common in the near future, then Lanshou is an example of the worst kind of air pollution that can occur with poorly planned development. This West China city has arguably the world's worst levels of total suspended particulates, which can cause bronchitis and asthma when inhaled, some 8 times more than recommended levels. This kind of air pollution problem is now common to virtually all big cities in developing countries. However, there's some evidence that as countries become wealthier they can clean up their air. A whole field of economics has sprung up investigating the relationship between a country's development and the emissions of different pollutants. For some pollutants such as suspended particulates, any increase in per capita income beyond a certain point is correlated with reductions in pollution levels. In effect, these pollutants can be reduced or eliminated—it just takes money and expertise that many rapidly growing cities in developing countries do not have. Increases in literacy levels and political rights are further correlated with reduction in these pollutants, suggesting that further democratization in the developing world will be good for the environment.

One of the most common mistakes environmentalists and progressives from developed countries often make is to assume that urbanization is always bad for the environment. This viewpoint focuses on urbanization without having a proper baseline—what would happen if all these people didn't move to cities? What would the world look like? People would be more dispersed across the landscape, multiplying the impact on the environment many times. One example is energy use, which is markedly more efficient in cities than in rural areas in many countries. Most major uses of energy, like heating and transportation, are more efficient at higher densities. For instance, 53% of New York City residents take public transit to work, far above the United States overall average of 5%, a gain that is possible in large part because of the increased population density of that city. This substantially reduces the energy required to transport people and goods: on average, European cities are 5 times denser than American cities and consume 3.6 times less transport energy per capita. In this sense, the massive migration to urban regions in the developing world provides an opportunity for energy savings, if planned properly. In effect, the world faces the largest urban planning challenge in the history of the human race. The question is, are we ready for it?

Convergence of Needs

If urbanization is a great challenge for sustainable development, it is also a great opportunity for progressives to build a more peaceful world. Urbanization involves a move away from rural areas, which often are isolated, without adequate education or political representation, toward an urban lifestyle. Despite the many problems of the slums, its residents often enjoy better education than their rural counterparts, and the sheer proximity of poor citizens to one another can greatly facilitate political organizing. The slum dwellers in El Alto in Bolivia, for example, through a series of protests, succeeded in forcing the state to take back control of water and sewer service from the private consortium that had failed to provide the slum dwellers with affordable service. In short, urbanization creates new potentialities for democracy.

City dwellers have much in common with one another, even if they are in different regions or nations. The slum dwellers in El Alto in Bolivia have more in common with the slum dwellers in Khan in Mumbai, India, than they do with those living in rural regions of Bolivia. While this commonality is often obscured by cultural differences and nationalist sentiment, it is a powerful political force that will be ascendant in the 21st century. Urban dwellers have a convergence of needs and desires that makes them a class with shared interests.

The increased mobility of many city dwellers heightens this sense of commonality among urban populations. It is worthwhile to note, for example, that much *international* migration is not rural to urban but interurban. In the developed world, the European Union provides a good example of what the future probably looks like: younger workers willing to move to find jobs will bounce between cities frequently. One acquaintance of mine, for example, moved from Barcelona to Paris to London in the space of 3 years, all in a quest of a better, more stable job. This kind of cross-cultural movement reduces the sense that this or that city seems foreign. Indeed, what seems foreign to many EU youth is a more rural lifestyle, even if it is in their native country.

This is not to sound sanguine about the political effects of this convergence of needs caused by the ongoing global urbanization. Significant resistance to a sense of commonality remains a political fact. Internally, cultural ties may bind urban dwellers to a rural culture long after that rural culture has ceased to exist. The United States political scene is a prime example of this, with the family farmer being seen as decent and the urban dweller seen as corrupt—never mind the fact that the net flow of tax dollars is from productive cities to stagnant rural areas, or that almost 80% of Americans are urbanites. That isn't to say that the good rural traits—hard work, honesty, simplicity—should disappear, but rather that politically we must re-envision these ideals in an urban context.

Among countries, nationalism may prevent an acknowledgement of how similar the urban realities are. Still, progressives must be aware of the great opportunity this convergence of needs presents, and seize the opening to push for real improvements in the lives of urban dwellers. Progressives must also avoid any blanket condemnation of urbanization in the developing world, recognizing that any such condemnation is pointless and highly offensive to many Third World residents.

POSTSCRIPT

Does Global Urbanization Lead Primarily to Undesirable Consequences?

It appears self-evident that rapidly growing urbanized areas, particularly in the developing world, create special circumstances. Our visual image of such places accents this fact. First-world travelers, particularly to the developing world, are likely to take away from that experience a litany of pictures that paint a bleak image of life there. The critical question, though, is whether such environments really do create major problems for those who live within such areas and policy makers who must provide goods and services. Conventional wisdom that such problems must exist is found everywhere throughout the urbanization literature. We easily could have selected any one of a dozen or more articles that asserted this situation with much conviction.

Perhaps, however, those problems observed by urban visitors might, in fact, be a consequence of some other situation unrelated to urbanization. Moreover, others allude to the advantages of urbanization for modernization. These individuals tend to bring a historical perspective to their analysis, but therein lies a potentially fatal flaw. Urbanization occurred first in the current developed world, where cities grew more slowly and where governments had the capacity and the inclination to provide a better quality of life for urban dwellers.

The most definitive source of information about urbanization is found in the United Nations' *2005 Revision of World Urbanization Prospects* (U.N., 2006). Another important work is the National Research Council's *Cities Transformed: Demographic Change and Its Implications in the Developing World* (National Academies Press, 2003). The Population Reference Bureau has produced a comprehensive monogram called "An Urbanizing World" by Martin P. Brockerhoff (*Population Bulletin*, September 2004). An interesting approach to the subject is found in Mike Davis' "Planet of Slums" (*Harper's*, June 2004). A sophisticated analysis is found in a 2003 article in *Sociological Perspectives* by John M. Shandra, Bruce London, and John B. Williamson ("Environmental Degradation, Environmental Sustainability, and Overurbanization in the Developing World: A Quantitative, Cross-National Analysis," 2003). Klaus Toepler, executive director of the United Nations Environment Programme at the time, provides a thoughtful discussion of challenges facing megacities of the world.

Two articles with differing viewpoints are useful sources. Barbara Boyle Torrey, a member of the Population Reference Bureau's Board of Trustees and a writer/consultant, argues that extremely high urban growth rates are resulting in and will continue to create a range of negative environmental problems ("Urbanization: An Environmental Force to Be Reckoned With," Population Reference Bureau, April 2004). Gordon McGranahan and David Satterthwaite, members of London's International Institute for Environment and Development, suggest that there is little research about urban centers and their ability to provide sustainable development ("Urban Centers: An Assessment of Sustainability," *Annual Review of Energy and the Environment*, vol. 28, 2003). McGranahan and Satterthwaite do acknowledge that ecologically more sustainable patterns of urban development are needed. While they admit the existence of certain environmental impacts of urbanization, they are more concerned about how cities will handle these impacts in the future rather than concentrating on present deficiencies.

On the Internet . . .

UNEP World Conservation Monitoring Centre

The United Nations Environment Programme's World Conservation Monitoring Centre Web site contains information on conservation and sustainable use of the globe's natural resources. The center provides information to policymakers concerning global trends in conservation, biodiversity, loss of species and habitats, and more. This site includes a list of publications and environmental links.

http://www.unep-wcmc.org

The International Institute for Sustainable Development

This nonprofit organization based in Canada provides a number of reporting services on a range of environmental issues, with special emphasis on policy initiatives associated with sustainable development.

http://www.iisd.ca

The Hunger Project

The Hunger Project is a nonprofit organization that seeks to end global hunger. This organization asserts that society-wide actions are needed to eliminate hunger and that global security depends on ensuring that everyone's basic needs are fulfilled. Included on this site is an outline of principles that guide the organization, information on why ending hunger is so important, and a list of programs sponsored by the Hunger Project in 11 developing countries across South Asia, Latin America, and Africa.

http://www.thp.org

Global Warming Central

The Global Warming Central Web site provides information on the global warming debate. Links to the best global warming debate sites as well as key documents and reports are included. Explore the recent news section to find the latest articles on the subject.

http://www.law.pace.edu/env/energy/globalwarming.html

International Association for Environmental Hydrology

The International Association for Environmental Hydrology (IAEH) is a worldwide association of environmental hydrologists dedicated to the protection and cleanup of freshwater resources. The IAEH's mission is to provide a place to share technical information and exchange ideas and to provide a source of inexpensive tools for the environmental hydrologist, especially hydrologists and water resource engineers in developing countries.

http://www.hydroweb.com

United Nations Environment Programme (UNEP)

UNEP's general Web site provides a variety of information and links to other sources.

http://www.ourplanet.com

PART 2

Global Resources and the Environment

*T*he availability of resources and the manner in which the planet's inhabitants use them characterize another major component of the global agenda. Many believe that environmentalists overstate their case because of ideology, not science. Many others state that renewable resources are being consumed at a pace that is too fast to allow for replenishment, while non-renewable resources are being consumed at a pace that is faster than our ability to find suitable replacements.

The production, distribution, and consumption of these resources also leave their marks on the planet. A basic set of issues relates to whether these impacts are permanent, too degrading to the planet, too damaging to one's quality of life, or simply beyond a threshold of acceptability.

- Do Environmentalists Overstate Their Case?

- Should the World Continue to Rely on Oil as a Major Source of Energy?

- Will the World Be Able to Feed Itself in the Foreseeable Future?

- Is the Threat of Global Warming Real?

- Is the Threat of a Global Water Shortage Real?

ISSUE 5

Do Environmentalists Overstate Their Case?

YES: Ronald Bailey, from "Debunking Green Myths," *Reason* (February 2002)

NO: David Pimentel, from "Skeptical of the Skeptical Environmentalist," *Skeptic* (vol. 9, no. 2, 2002)

ISSUE SUMMARY

YES: Environmental journalist Ronald Bailey in his review of the Bjørn Lomborg controversial book, *The Skeptical Environmentalist: Measuring the Real State of the World* (Cambridge University Press, 2001), argues in the subtitle of his critique that "An environmentalist gets it right," suggests that finally someone has taken the environmental doomsdayers to task for their shoddy use of science.

NO: Bioscientist David Pimentel takes to task Bjørn Lomborg's findings, accusing him of selective use of data to support his conclusions.

$\mathbf{F}$or a few decades, those skeptics of the claims of many environmentalists that the world was in danger of ecological collapse and in the not so distant future, looked to Julian Simon for guidance. And Simon did not disappoint, as he constantly questioned these researchers' motives and methodology—their models, data, and data analysis techniques. Two seminal works, *The Ultimate Resource* and *The Ultimate Resource 2*, in particular, attempted to demonstrate that much research was really bad science. Simon's popularity reached its height when he took on the leading spokesperson of pending environmental catastrophe, Paul Ehrlich, in the late 1970s. Ehrlich, a professor at Stanford, along with his wife Anne (also a Stanford professor), had been echoing "the sky is falling message" since the late 1960s. Simon challenged Ehrlich to a "forecasting duel," betting him $10,000 that the cost of five non-government raw materials (to be selected by Ehrlich) would fall by the end of the next decade (1990). Ehrlich won the bet. With Simon's death in 1998, the critics of environmental doomsdayers lost their most effective voice and their central rallying cry.

Bjørn Lomborg, a young Danish political scientist changed all of that with the 2001 publication of *The Skeptical Environmentalist: Measuring the Real State of the World* (a take-off on the annual State of the World Series produced by Lester Brown and the Worldwatch Institute). Lomborg's central thesis is that statistical analyses of principal environmental indicators reveal that environmental problems have been overstated by most leading figures in the environmental movement.

What distinguished Simon's body of work that earned him the unofficial title of "doomslayer" from Lomborg's book was that the latter received much greater and more widespread attention, both by the popular media and by those in academic and scientific circles. In effect, it has become the most popular anti-environmental book ever, prompting a huge backlash by those vested in the scientific community. Because the popular press appeared to accept Lomberg's assertions with an uncritical eye, the scientific community began a comprehensive counter-attack against *The Skeptical Environmentalist. Scientific American*, in January 2002, published almost a dozen pages of critiques of the book by four experts and concluded that the book's purpose of showing the real state of the world was a failure.

The attention paid Lomborg's book by *Scientific American* was typical of the responses found in every far corner of the scientific community. Not only was Lomberg's analyses attacked, but his credentials were as well. Researchers scurried to discredit both him and his work, with a passion unseen heretofore in the debate over the potential for global environmental catastrophe. The Danish Committees on Scientific Dishonesty were called upon to investigate the work. The Danish Ministry of Science, Technology, and Innovation found serious flaws in Lomborg's critique. One reviewer concluded by observing that he wished he could find that the book had some scientific merit but he could not. The British Broadcasting Company (BBC) devoted a three-part series to Lomborg's claims. One critique was titled "No Friend of the Earth."

These examples illustrate the debate put forth in this issue, namely, do environmentalists overstate the case for environmental decay and potential catastrophe? Ronald Bailey, probably the unofficial successor to Julian Simon's role as principal critic of environmentalist ideology, provides one of the few positive critiques of *The Skeptical Environmentalist.* His initial statement places the genesis of modern environmentalism in the radical movements of the 1960s, suggesting that their aim is to demonstrate that "the world is going to hell in a handbasket." Calling environmentalism an ideology, Bailey argues that like Marxists, environmentalists "have had to force the facts to fit their ideology." In sum, he suggests that the book deals a major blow to environmentalist ideology "by superbly documenting a response to environmental doomsaying." David Pimentel, a professor of insect ecology and agricultural sciences, argues that those who contend that the environment is not threatened are using data selectively while ignoring much evidence to the contrary.

YES

Ronald Bailey

Debunking Green Myths:
An Environmentalist Gets It Right

Modern environmentalism, born of the radical movements of the 1960s, has often made recourse to science to press its claims that the world is going to hell in a handbasket. But this environmentalism has never really been a matter of objectively describing the world and calling for the particular social policies that the description implies.

Environmentalism is an ideology, very much like Marxism, which pretended to base its social critique on a "scientific" theory of economic relations. Like Marxists, environmentalists have had to force the facts to fit their theory. Environmentalism is an ideology in crisis: The massive, accumulating contradictions between its pretensions and the actual state of the world can no longer be easily explained away.

The publication of *The Skeptical Environmentalist,* a magnificent and important book by a former member of Greenpeace, deals a major blow to that ideology by superbly documenting a response to environmental doomsaying. The author, Bjorn Lomborg, is an associate professor of statistics at the University of Aarhus in Denmark. On a trip to the United States a few years ago, Lomborg picked up a copy of Wired that included an article about the late "doomslayer" Julian Simon.

Simon, a professor of business administration at the University of Maryland, claimed that by most measures, the lot of humanity is improving and the world's natural environment was not critically imperiled. Lomborg, thinking it would be an amusing and instructive exercise to debunk a "right-wing" anti-environmentalist American, assigned his students the project of finding the "real" data that would contradict Simon's outrageous claims.

Lomborg and his students discovered that Simon was essentially right, and that the most famous environmental alarmists (Stanford biologist Paul Ehrlich, Worldwatch Institute founder Lester Brown, former Vice President Al Gore, Silent Spring author Rachel Carson) and the leading environmentalist lobbying groups (Greenpeace, the World Wildlife Fund, Friends of the Earth) were wrong. It turns out that the natural environment is in good shape, and the prospects of humanity are actually quite good.

Lomborg begins with "the Litany" of environmentalist doom, writing: "We are all familiar with the Litany. . . . Our resources are running out. The

From *Reason* by Ronald Bailey, February 2002, pp. 396–403, 406–409, 416–420. Copyright © 2002 by Reason Foundation. Reprinted by permission.

population is ever growing, leaving less and less to eat. The air and water are becoming ever more polluted. The planet's species are becoming extinct in vast numbers. . . . The world's ecosystem is breaking down. . . . We all know the Litany and have heard it so often that yet another repetition is, well, almost reassuring." Lomborg notes that there is just one problem with the Litany: "It does not seem to be backed up by the available evidence."

Lomborg then proceeds to demolish the Litany. He shows how, time and again, ideological environmentalists misuse, distort, and ignore the vast reams of data that contradict their dour visions. In the course of The Skeptical Environmentalist, Lomborg demonstrates that the environmentalist lobby is just that, a collection of interest groups that must hype doom in order to survive monetarily and politically.

Lomborg notes, "As the industry and farming organizations have an obvious interest in portraying the environment as just-fine and no-need-to-do-anything, the environmental organizations also have a clear interest in telling us that the environment is in a bad state, and that we need to act now. And the worse they can make this state appear, the easier it is for them to convince us we need to spend more money on the environment rather than on hospitals, kindergartens, etc. Of course, if we were equally skeptical of both sorts of organization there would be less of a problem. But since we tend to treat environmental organizations with much less skepticism, this might cause a grave bias in our understanding of the state of the world." Lomborg's book amply shows that our understanding of the state of the world is indeed biased.

So what is the real state of humanity and the planet?

Human life expectancy in the developing world has more than doubled in the past century, from 31 years to 65. Since 1960, the average amount of food per person in the developing countries has increased by 38 percent, and although world population has doubled, the percentage of malnourished poor people has fallen globally from 35 percent to 18 percent, and will likely fall further over the next decade, to 12 percent. In real terms, food costs a third of what it did in the 1960s. Lomborg points out that increasing food production trends show no sign of slackening in the future.

What about air pollution? Completely uncontroversial data show that concentrations of sulfur dioxide are down 80 percent in the U.S. since 1962, carbon monoxide levels are down 75 percent since 1970, nitrogen oxides are down 38 percent since 1975, and ground level ozone is down 30 percent since 1977. These trends are mirrored in all developed countries.

Lomborg shows that claims of rapid deforestation are vastly exaggerated. One United Nations Food and Agriculture survey found that globally, forest cover has been reduced by a minuscule 0.44 percent since 1961. The World Wildlife Fund claims that two-thirds of the world's forests have been lost since the dawn of agriculture; the reality is that the world still has 80 percent of its forests. What about the Brazilian rainforests? Eighty-six percent remain uncut, and the rate of clearing is falling. Lomborg also debunks the widely circulated claim that the world will soon lose up to half of its species. In fact, the best evidence indicates that 0.7 percent of species might be lost in the next 50 years if nothing is done. And of course, it is unlikely that nothing will be done.

Finally, Lomborg shows that global warming caused by burning fossil fuels is unlikely to be a catastrophe. Why? First, because actual measured temperatures aren't increasing nearly as fast as the computer climate models say they should be—in fact, any increase is likely to be at the low end of the predictions, and no one thinks that would be a disaster. Second, even in the unlikely event that temperatures were to increase substantially, it will be far less costly and more environmentally sound to adapt to the changes rather than institute draconian cuts in fossil fuel use.

The best calculations show that adapting to global warming would cost $5 trillion over the next century. By comparison, substantially cutting back on fossil fuel emissions in the manner suggested by the Kyoto Protocol would cost between $107 and $274 trillion over the same period. (Keep in mind that the current yearly U.S. gross domestic product is $10 trillion.) Such costs would mean that people living in developing countries would lose over 75 percent of their expected increases in income over the next century. That would be not only a human tragedy, but an environmental one as well, since poor people generally have little time for environmental concerns.

Where does Lomborg fall short? He clearly understands that increasing prosperity is the key to improving human and environmental health, but he often takes for granted the institutions of property and markets that make progress and prosperity possible. His analysis, as good as it is, fails to identify the chief cause of most environmental problems. In most cases, imperiled resources such as fisheries and airsheds are in open-access commons where the incentive is for people to take as much as possible of the resource before someone else beats them to it. Since they don't own the resource, they have no incentive to protect and conserve it.

Clearly, regulation has worked to improve the state of many open-access commons in developed countries such as the U.S. Our air and streams are much cleaner than they were 30 years ago, in large part due to things like installing catalytic converters on automobiles and building more municipal sewage treatment plants. Yet there is good evidence that assigning private property rights to these resources would have resulted in a faster and cheaper cleanup. Lomborg's analysis would have been even stronger had he more directly taken on ideological environmentalism's bias against markets. But perhaps that is asking for too much in an already superb book.

"Things are better now," writes Lomborg, "but they are still not good enough." He's right. Only continued economic growth will enable the 800 million people who are still malnourished to get the food they need; only continued economic growth will let the 1.2 billion who don't have access to clean water and sanitation obtain those amenities. It turns out that ideological environmentalism, with its hostility to economic growth and technological progress, is the biggest threat to the natural environment and to the hopes of the poorest people in the world for achieving better lives.

"The very message of the book," Lomborg concludes, is that "children born today—in both the industrialized world and the developing countries—will live longer and be healthier, they will get more food, a better education, a higher standard of living, more leisure time and far more possibilities—without the global environment being destroyed. And that is a beautiful world."

David Pimentel  **NO**

Skeptical of the Skeptical Environmentalist

Bjørn Lomborg discusses a wide range of topics in his book and implies, through his title, that he will inform readers exactly what the real state of world is. In this effort, he criticizes countless world economists, agriculturists, water specialists, and environmentalists, and furthermore, accuses them of misquoting and/or organizing published data to mislead the public concerning the status of world population, food supplies, malnutrition, disease, and pollution. Lomborg bases his optimistic opinion on his selective use of data. Some of Lomborg's assertions will be examined in this review, and where differing information is presented, extensive documentation will be provided.

Lomborg reports that "we now have more food per person than we used to."[1] In contrast, the Food and Agricultural Organization (FAO) of the United Nations reports that food per capita has been declining since 1984, based on available cereal grains.[2] Cereal grains make up about 80% of the world's food. Although grain yields per hectare (abbreviated ha) in both developed and developing countries are still increasing, these increases are slowing while the world population continues to escalate.[3] Specifically from 1950 to 1980, U.S. grains yields increased at about 3% per year, but after 1980 the rate of increase for corn and other grains has declined to only about 1%.

Obviously fertile cropland is an essential resource for the production of foods but Lomborg has chosen not to address this subject directly. Currently, the U.S. has available nearly 0.5 ha of prime cropland per capita, but it will not have this much land if the population continues to grow at its current rapid rate.[4] Worldwide the average cropland available for food production is only 0.25 ha per person.[5] Each person added to the U.S. population requires nearly 0.4 ha (1 acre) of land for urbanization and transportation.[6] One example of the impact of population growth and development is occurring in California where an average of 156,000 ha of agricultural land is being lost each year.[7] At this rate it will not be long before California ceases to be the number one state in U.S. agricultural production.

In addition to the quantity of agricultural land, soil quality and fertility is vital for food production. The productivity of the soil is reduced when it is eroded by rainfall and wind.[8] Soil erosion is not a problem, according to Lomborg, especially in the U.S. where soil erosion has declined during the past decade. Yes, as Lomborg states, instead of losing an average of 17 metric tons per hectare per year on

From *Skeptic* by David Pimentel, vol. 9 no. 2, 2002, pp. 90–93. Copyright © 2002 by David Pimentel. Reprinted by permission of the author.

cropland, the U.S. cropland is now losing an average of 13 t/ha/yr.[9] However, this average loss is 13 times the sustainability rate of soil replacement.[10] Exceptions occur, as during the 1995–96 winter in Kansas, when it was relatively dry and windy, and some agricultural lands lost as much as 65 t/ha of productive soil. This loss is 65 times the natural soil replacement in agriculture.[11]

Worldwide soil erosion is more damaging than in the United States. For instance, in India soil is being lost at 30 to 40 times its sustainability.[12] Rate of soil loss in Africa is increasing not only because of livestock overgrazing but also because of the burning of crop residues due to the shortages of wood fuel.[13] During the summer of 2000, NASA published a satellite image of a cloud of soil from Africa being blown across the Atlantic Ocean, further attesting to the massive soil erosion problem in Africa. Worldwide evidence concerning soil loss is substantiated and it is difficult to ignore its effect on sustainable agricultural production.

Contrary to Lomborg's belief, crop yields cannot continue to increase in response to the increased applications of more fertilizers and pesticides. In fact, field tests have demonstrated that applying excessive amounts of nitrogen fertilizer stresses the crop plants, resulting in declining yields.[14] The optimum amount of nitrogen for corn, one of the crops that require heavy use of nitrogen, is approximately 120 kg/ha.[15]

Although U.S. farmers frequently apply significantly more nitrogen fertilizer than 120 kg/ha, the extra is a waste and pollutant. The corn crop can only utilize about one-third of the nitrogen applied, while the remainder leaches either into the ground or surface waters.[16] This pollution of aquatic ecosystems in agricultural areas results in the high levels of nitrogen and pesticides occurring in many U.S. water bodies.[17] For example, nitrogen fertilizer has found its way into 97% of the well-water supplies in some regions, like North Carolina.[18] The concentrations of nitrate are above the U.S. Environmental Protection Agency drinking-water standard of 10 milligrams per liter (nitrogen) and are a toxic threat to young children and young livestock.[19] In the last 30 years, the nitrate content has tripled in the Gulf of Mexico,[20] where it is reducing the Gulf fishery.[21]

In an undocumented statement Lomborg reports that pesticides cause very little cancer.[22] Further, he provides no explanation as to why human and other nontarget species are not exposed to pesticides when crops are treated. There is abundant medical and scientific evidence that confirms that pesticides cause significant numbers of cancers in the U.S. and throughout the world.[23] Lomborg also neglects to report that some herbicides stimulate the production of toxic chemicals in some plants, and that these toxicants can cause cancer.[24]

In keeping with Lomborg's view that agriculture and the food supply are improving, he states that "fewer people are starving."[25] Lomborg criticizes the validity of the two World Health Organization reports that confirm more than 3 billion people are malnourished.[26] This is the largest number and proportion of malnourished people ever in history! Apparently Lomborg rejects the WHO data because they do not support his basic thesis. Instead, Lomborg argues that only people who suffer from calorie shortages are malnourished, and ignores the fact that humans die from deficiencies of protein, iron, iodine, and vitamin A, B, C, and D.[27]

Further confirming a decline in food supply, the FAO reports that there has been a three-fold decline in the consumption of fish in the human diet during the past seven years.[28] This decline in fish per capita is caused by overfishing, pollution, and the impact of a rapidly growing world population that must share the diminishing fish supply.

In discussing the status of water supply and sanitation services, Lomborg is correct in stating that these services were improved in the developed world during the 19th century, but he ignores the available scientific data when he suggests that these trends have been "replicated in the developing world" during the 20th century. Countless reports confirm that developing countries discharge most of their untreated urban sewage directly into surface waters.[29] For example, of India's 3,119 towns and cities, only eight have full waste water treatment facilities.[30] Furthermore, 114 Indian cities dump untreated sewage and partially cremated bodies directly into the sacred Ganges River. Downstream the untreated water is used for drinking, bathing, and washing.[31] In view of the poor sanitation, it is no wonder that water borne infectious diseases account for 80% of all infections worldwide and 90% of all infections in developing countries.[32]

Contrary to Lomborg's view, most infectious diseases are increasing worldwide.[33] The increase is due not only to population growth but also because of increasing environmental pollution.[34] Food-borne infections are increasing rapidly worldwide and in the United States. For example, during 2000 in the U.S. there were 76 million human food-borne infections with 5,000 associated deaths.[35] Many of these infections are associated with the increasing contamination of food and water by livestock wastes in the United States.[36]

In addition, a large number of malnourished people are highly susceptible to infectious diseases, like tuberculosis (TB), malaria, schistosomiasis, and AIDS.[37] For example, the number of people infected with tuberculosis in the U.S. and the world is escalating, in part because medicine has not kept up with the new forms of TB. Currently, according to the World Health Organization,[38] more than 2 billion people in the world are infected with TB,[39] with nearly 2 million people dying each year from it.[40]

Consistent with Lomborg's thesis that world natural resources are abundant, he reports that the U.S. Energy Information Agency for the period 2000 to 2020 projects an almost steady oil price over the next two decades at about $22 per barrel. This optimistic projection was crossed late in 2000 when oil rose to $30 or more per barrel in the United States and the world.[41] The best estimates today project that world oil reserves will last approximately 50 years, based on current production rates.[42]

Lomborg takes the World Wildlife Fund (WWF) to task for their estimates on the loss of world forests during the past decade and their emphasis on resulting ecological impacts and loss of biodiversity. Whether the loss of forests is slow, as Lomborg suggests, or rapid as WWF reports, there is no question that forests are disappearing worldwide. Forests not only provide valuable products but they harbor a vast diversity of species of plants, animals and microbes. Progress in medicine, agriculture, genetic engineering, and environmental quality depend on maintaining the species diversity in the world.[43]

This reviewer takes issue with Lomborg's underlying thesis that the size and growth of the human population is not a major problem. The difference between Lomborg's figure that 76 million humans were added to the world population in 2000, or the 80 million reported by the Population Reference Bureau,[44] is not the issue, though the magnitude of both projections is of serious concern. Lomborg neglects to explain that the major problem with world population growth is the young age structure that now exists. Even if the world adopted a policy of only two children per couple tomorrow, the world population would continue to increase for more than 70 years before stabilizing at more than 12 billion people.[45] As an agricultural scientist and ecologist, I wish I could share Lomborg's optimistic views, but my investigations and those of countless scientists lead me to a more conservative outlook. The supply of basic resources, like fertile cropland, water, energy, and an unpolluted atmosphere that support human life is declining rapidly, as nearly a quarter million people are daily added to the Earth. We all desire a high standard of living for each person on Earth, but with every person added, the supply of resources must be divided and shared. Current losses and degradation of natural resources suggest concern and a need for planning for future generations of humans. Based on our current understanding of the real state of the world and environment, there is need for conservation and protection of vital world resources.

References

1. Lomborg, B. 2001. *The Skeptical Environmentalist.* Cambridge University Press, 61.

2. FAO, 1961–1999. *Quarterly Bulletin of Statistics.* Food and Agriculture Organization of the United Nations.

3. Ibid.; PRB 2000. *World Population Data Sheet.* Washington, DC: Population Reference Bureau.

4. USBC, 2000. *Statistical Abstract of the United States 2000.* Washington, DC: U.S. Bureau of the Census, U.S. Government Printing Office.

5. PRB, 2000; WRI 1994. *World Resources 1994-95.* Washington, DC: World Resources Institute.

6. Helmlich, R. 2001. Economic Research Service, USDA, Washington, DC, personal communication.

7. UCBC, 2000, op. cit.

8. Lal, R., and B. A. Stewart, 1990. *Soil Degradation.* New York: Springer-Verlag: Troeh, F. R., Hobbs, J. A., & Donahue, R. L. 1991. *Soil and Water Conservation* (2nd ed.). Englewood Cliffs, NJ: Prentice Hall.

9. USDA, 1994. *Summary Report 1992 National Resources Inventory.* Washington, DC: Soil Conservation Service, U.S. Department of Agriculture.

10. Pimentel, D., and N. Kounang, 1998. "Ecology of Soil Erosion in Ecosystems," *Ecosystems,* 1, 416–426.

11. Lal and Stewart, 1990; Troeh et al., 1991, op. cit.

12. Khoshoo, T. N. & Tejwani, K. G. 1993. "Soil Erosion and Conservation in India (status and policies)." In Pimentel, D. (ed.) *World Soil Erosion and Conservation.* pp. 109–146. Cambridge: Cambridge University Press.

13. Tolba, M. K. 1989. "Our Biological Heritage Under Siege." *BioScience,* 39: 725–728.

14. Romanova, A. K., Kuznetsova, L. G., Golovina, E. V., Novichkova, N. S., Karpilova, I. F., & Ivanov, B. N. 1987. *Proceedings of the Indian National Science Academy, B (Biological Sciences),* 53(5–6): 505–512.

15. Troeh, F. R., & Thompson, L. M. 1993. *Soils and Soil Fertility* (5th ed.). New York: Oxford University Press.

16. Robertson, G. P. 2000. "Dinitrification." In *Handbook of Soil Science.* M. E. Summer (Ed). pp. C181–190. Boca Raton, FL: CRC Press.

17. Ibid.; Mapp, H. P. 1999. "Impact of Production Changes on Income and Environmental Risk in the Southern High Plains." *Journal of Agricultural and Applied Economics,* 31(2): 263–273; Gentry, L. E., David, M. B., Smith-Starks, K. M., and Kovacics, 2000. "Nitrogen Fertilizer and Herbicide Transport from Tile Drained Fields." *Journal of Environmental Quality,* 29(1): 232–240.

18. Smith, V. H., Tilman, G. D. and Nekola, J. C. 1999. "Eutrophication: Impacts of Excess Nutrient Inputs on Freshwater, Marine, and Terrestrial Ecosystems." *Environment and Pollution,* 100(1/3): 179–196.

19. Ibid.

20. Goolsby, D. A., Battaglin, W. A., Aulenbach, B. T. and Hooper, R. P. 2000. "Nitrogen Flux and Sources in the Mississippi River Basin." *Science and the Total Environment,* 248(2–3): 75–86.

21. NAS, 2000. *Clean Coastal Waters: Understanding and Reducing the Effects of Nutrient Pollution.* Washington, DC: National Academy of Sciences Press.

22. Lomborg, 2001, op. cit., 10.

23. WHO, 1992. *Our Planet, Our Health: Report of the WHO Commission on Health and Environment.* Geneva: World Health Organization: Ferguson, L. R. 1999. "Natural and Man-Made Mutagens and Carcinogens in the Human Diet." *Mutation Research, Genetic Toxicology and Environmental Mutagenesis,* 443(1/2): 1–10; NAS, 2000. *The Future Role of Pesticides in Agriculture.* Washington, DC: National Academy of Sciences Press.

24. Culliney, T. W., Pimentel, D., & Pimentel, M. H. 1992. "Pesticides and Natural Toxicants in Foods." *Agriculture, Eco-systems and Environment,* 41, 297–320.

25. Lomborg, 2001, op. cit., 328.

26. WHO, 1996. *Micronutrient Malnutrition—Half of the World's Population Affected* (Pages 1–4 No. Press Release WHO No. 78). World Health Organization; WHO, 2000a. *Malnutrition Worldwide* http://www.who.int/nut/malnutrition_worldwide. htm, July 27, 2000.

27. Sommer, A. and K. P. West, 1996. *Vitamin A Deficiency: Health, Survival and Vision.* New York: Oxford University Press; Tomashek, K. M., Woodruff, B. A., Gotway, C. A., Bloand, P. & Mbaruku, G. 2001. "Randomized Intervention Study Comparing Several Regimens for the Treatment of Moderate Anemia Refugee Children in Kigoma Region, Tanzania." *American Journal of Tropical Medicine and Hygiene,* 64(3/4): 164–171.

28. FAO, 1991. *Food Balance Sheets.* Rome: Food and Agriculture Organization of the United Nations; FAO, 1998. *Food Balance Sheets.* http://armanncorn:98ivysub@ faostat.fao.org/lim . . . ap.pl?

29. WHO, 1993. "Global Health Situation." *Weekly Epidemiological Record,* World Health Organization 68 (12 February): 43–44; Wouters, A. V. 1993. "Health Care Utilization Patterns in Developing Countries: Role of the Technology Environment in 'Deriving' the Demand for Health Care." *Boletin de la Oficina Sanitaria Panamericana,* 115(2): 128–139; Biswas; M. R. 1999. "Nutrition, Food, and Water Security." *Food and Nutrition Bulletin,* 20(4): 454–457.

30.   WHO, 1992, op. cit.

31.   NGS, 1995, *Water: A Story of Hope*. Washington, DC: National Geographic Society.

32.   WHO, 1992, op. cit.

33.   Ibid.

34.   Pimentel, D., Tort, M., D'Anna, L., Krawic, A., Berger, J., Rossman, J., Mugo, F., Doon, N., Shriberg, M., Howard, E. S., Lee, S., & Talbot, J. 1998. "Ecology of Increasing Disease: Population Growth and Environmental Degradation." *BioScience*, 48, 817–826.

35.   Taylor, M. R. & Hoffman, S. A. 2001. "Redesigning Food Safety: Using Risk Analysis to Build a Better Food Safety System." *Resources*. Summer, 144: 13–16.

36.   DeWaal, C. S., Alderton, L., and Jacobson, M. J. 2000. *Outbreak Alert! Closing the Gaps in Our Federal Food-Safety Net*. Washington, DC: Center for Science in the Public Interest.

37.   Chandra, R. K. 1979. "Nutritional Deficiency and Susceptibility to Infection." *Bulletin of the World Health Organization*, 57(2): 167–177; Stephenson, L. S., Latham, M. C. & Ottesen, E. A. 2000a. "Global Malnutrition." *Parasitology*. 121: S5–S22; Stephenson, L. S., Latham, M. C. & Ottesen, E. A. 2000b. "Malnutrition and Parasitic Helminth Infections." *Parasitology*. S23–S38.

38.   WHO, 2001. "World Health Organization. Global Tuberculosis Control." *WHO Report 2001*. Geneva, Switzerland, WHO/CDS/TB/2001. 287 (May 30, 2001).

39.   WHO, 2000b. "World Health Organization. Tuberculosis." *WHO Fact Sheet 2000 No104*. Geneva, Switzerland, www.who.int/gtb (May 30, 2001).

40.   WHO, 2001, op. cit.

41.   BP, 2000. *British Petroleum Statistical Review of World Energy*. London: British Petroleum Corporate Communications Services; Duncan, R. C. 2001. "World Energy Production, Population Growth, and the Road to the Olduvai Gorge." *Population and Environment*, 22(5), 503–522.

42.   Youngquist, W. 1997. *Geodestinies: The Inevitable Control of Earth Resources Over Nations and Individuals*. Portland, OR: National Book Company; Duncan, 2001, op. cit.

43.   Myers, N. 1996. "The World's Forests and Their Ecosystem Services." In G. C. Dailey (Ed.), *Ecosystem Services: Their Nature and Value* (pp. 1–19 in press). Washington, DC: Island Press.

44.   PRB, 2000, op. cit.

45.   Population Action International, 1993. *Challenging the Planet: Connections Between Population and the Environment*. Washington, DC: Population Action International.

POSTSCRIPT

Do Environmentalists Overstate Their Case?

The issue of whether science or ideology is at the heart of the environmental debate is a vexing one. The issue is framed by the juxtaposition of three groups. The first are those individuals, commonly called political or environmental activists, who emerged in the late 1960s and early 1970s following the success of the early civil rights movement. Taking its inspiration from the 1962 publication of Rachel Carson's *The Silent Spring*, which exposed the dangers of the pesticide DDT, many politically active individuals found a new cause. When the book received legitimacy because of President John Kennedy's order that his Science Advisory Committee address the issues raised therein, the environmental movement was under way. The second group, government policy makers were then a part of the mix and the third group, scientists, were soon to come on board. The first global environmental conference sponsored by the United Nations was held in Stockholm in 1972 to address atmospheric pollution on Scandinavian lakes. Emerging from the conference was a commitment of the international policy making community to put environmental issues on the new global agenda. Environmentalism was now globalized.

Since the early 1970s, through a variety of forums and arenas, the issue has been on the forefront of this global agenda. As with any issue where debates focus not only on how to address problems but whether, in fact, the problems really exist in the first place, many disparate formal and informal interest groups have become involved in all aspects of the debate—from trying to make the case that a problem exists and will ultimately have dire consequences if left unsolved, to specific prescriptions for solving the issue. The intersection of science, public policy, and political activism then becomes like the center ring at a boxing match, where contenders vie for success. Objectivity clashes with passion as well intentioned and not so well intentioned individuals attempt to influence the debate and the ultimate outcome. In many cases, the doomsdayers gain the upper hand as their commitment to change seems greater than those who urge caution until all the scientific evidence is in.

The reaction to Lomborg illustrates this point perfectly. He has become the arch villain to environmentalists. One such website proclaims in headline, "Something is Rotten in Denmark" and then proceeds to "fight fire with fire" in attacking him (www.gristmagazine.com). *Grist Magazine* devoted a special issue (December 12, 2001) to the debate where experts in specific environmental fields took issue with Lomborg's conclusions.

Another source provides a variety of links to the debate fueled by *The Skeptical Scientist*. The journal *Scientific American* launched an extreme attack against Lomborg, while *The Economist* came to his defense. Google.com shows 247,000 references to the young Danish political scientist at last count. Amazon.com provides an array of related books that fall into the same genre.

In sum, one is struck by both the forcefulness with which Lomborg makes his case and the even greater passion with which the scientific community responds. While the latter may be more accurate with respect to the true state of the world, to paraphrase the essence of the debate, Lomborg does provide a valuable service by reminding us that at the heart of any meaningful prescription for effective public policy is an accurate assessment of the nature of the problem. Science, not ideology, provides the instruments for such an assessment.

One principal source that consistently sounds the alarm on environment issues is the Worldwatch Institute. Its web site, worldwatch@worldwatch.org, yields an extraordinary amount of resources on environmental issues.

In the first few decades after environment was placed on the global agenda, Julian Simon was one of the few, and certainly the most read, critic of environmentalists for their ideological approach to environmental problems, causing them, in Simon's view, to ignore science when science yielded an answer different from the one sought by the environmentalists. His *The Ultimate Resource* and *The Ultimate Resource 2* represented two harshly critical books that sought to show how science had taken a back seat to ideology. Since his death in 1998, his role as principal vocal critic of extremists in the environmental movement has been assumed by Ronald Bailey, science correspondent for the monthly magazine *Reason*. His *Global Warming and Other Eco-Myths* (Forum, 2002) charges the environmentalists with using "False Science to Scare Us to Death" (part of the book's subtitle). The titles of earlier books also suggest his basic message: *Earth Report 2000: Revisiting the True State of the Planet* (McGraw-Hill, 1999); *ECOSCAM: The False Prophets of Ecological Apocalypse* (St. Martin's Press, 1993); and *The True State of the Planet* (The Free Press, 1995).

Perhaps the most helpful website for gathering information about the debate as it relates to Lomborg is www.anti-lomborg.com. The websites name is misleading as it provides a list of pro-Lomborg sources in addition to those that attack him.

ISSUE 6

Should the World Continue to Rely on Oil as a Major Source of Energy?

YES: Red Cavaney, from "Global Oil Production about to Peak? A Recurring Myth," *Worldwatch* (January/February 2006)

NO: James Howard Kunstler, from *The Long Emergency* (Grove/Atlantic, 2005)

ISSUE SUMMARY

YES: Red Cavaney, president and chief executive officer of the American Petroleum Institute, argues that recent revolutionary advances in technology will yield sufficient quantities of available oil for the foreseeable future.

NO: James Howard Kunstler, author of *The Long Emergency* (2005), suggests that simply passing the all-time production peak of oil and heading toward its steady depletion will result in a global energy predicament that will substantially change our lives.

T he new millennium witnessed an oil crisis almost immediately, the third major crisis in the last 30 years (1972–73 and 1979 were the dates of earlier problems). The crisis of 2000 manifested itself in the United States via much higher gasoline prices and in Europe via both rising prices and shortages at the pump. Both were caused by the inability of national distribution systems to adjust to the Organization of Petroleum Exporting Countries' (OPEC's) changing production levels. The 2000 panic eventually subsided but reappeared in 2005 in the wake of the uncertainty surrounding the Iraq war and the war on terrorism. Four major crises in 34 years thus characterize the oil issue.

These four major fuel crises are discrete episodes in a much larger problem facing the human race, particularly the industrial world. That is, oil, the earth's current principal source of energy, is a finite resource that ultimately will be totally exhausted. And unlike earlier energy transitions, where a more attractive source invited a change (such as from wood to coal and from coal to oil), the next energy transition is being forced upon the human race in the absence of an attractive alternative. In short, we are being pushed out of our almost total reliance on oil toward a new system with a host of unknowns.

What will the new fuel be? Will it be from a single source or some combination? Will it be a more attractive source? Will the source be readily available at a reasonable price, or will a new cartel emerge that controls much of the supply? Will its production and consumption lead to major new environmental consequences? Will it require major changes to our lifestyles and standards of living? When will we be forced to jump into using this new source?

Before considering new sources of fuel, other questions need to be asked. Are the calls for a viable alternative to oil premature? Are we simply running scared without cause? Did we learn the wrong lessons from the earlier energy crises? More specifically, were these crises artificially created or a consequence of the actual physical unavailability of the energy source? Have these crises really been about running out of oil globally, or were they due to other phenomena at work, such as poor distribution planning by oil companies or the use of oil as a political weapon by oil-exporting countries?

For well over half a century now, Western oil-consuming countries have been predicting the end of oil. Using a model known as Hubbert's Curve (named after a U.S. geologist who designed it in the 1930s), policymakers have predicted that the world would run out of oil at various times; the most recent prediction is that oil will run out a couple of decades from now. Simply put, the model visualizes all known available resources and the patterns of consumption on a time line until the wells run dry. Despite such prognostication, it was not until the crisis of the early 1970s that national governments began to consider ways of both prolonging the oil system and finding a suitable replacement. Prior to that time, governments, as well as the private sector, encouraged energy consumption. "The more, the merrier" was an oft-heard refrain. Increases in energy consumption were associated with economic growth. After Europe recovered from the devastation of World War II, for example, every 1 percent increase in energy consumption brought a similar growth in economic output. To the extent that governments engaged in energy policymaking, it was designed solely to encourage increased production and consumption. Prices were kept low, and the energy was readily available. Policies making energy distribution systems more efficient and consumption patterns both more efficient and lowered were almost non-existent.

But today the search for an alternative to oil still continues. Nuclear energy, once thought to be the answer, may play a future role, but at a reduced level. Both water power and wind power remain possibilities, as do biomass, geothermal, and solar energy. Many also believe that the developed world is about to enter the hydrogen age in order to meet future energy needs. The question before us, therefore, is whether the international community has the luxury of some time before all deposits of oil are exhausted.

The two selections for this issue suggest different answers to this last question. Red Cavaney argues that oil should and will define the energy future for some time to come. He argues that despite forecasts of gloom, we have found more oil nearly every year than we have used and reserves continue to grow. The answer is technology. James Howard Kunstler suggests that we are facing the end of cheap fossil fuels as we have passed the all-time production peak of oil and are witnessing its steady depletion.

YES

Red Cavaney

Global Oil Production about to Peak? A Recurring Myth

Once again, we are hearing that world oil production is "peaking," and that we will face a steadily diminishing oil supply to fuel the global economy. These concerns have been expressed periodically over the years, but have always been at odds with energy and economic realities. Such is the case today.

Let's look at some history: In 1874, the chief geologist of Pennsylvania predicted we would run out of oil in four years—just using it for kerosene. Thirty years ago, groups such as the Club of Rome predicted an end of oil long before the current day. These forecasts were wrong because, nearly every year, we have found more oil than we have used, and oil reserves have continued to grow.

The world consumes approximately 80 million barrels of oil a day. By 2030, world oil demand is estimated to grow about 50 percent, to 121 million barrels a day, even allowing for significant improvements in energy efficiency. The International Energy Agency says there are sufficient oil resources to meet demand for at least the next 30 years.

The key factor here is technology. Revolutionary advances in technology in recent years have dramatically increased the ability of companies to find and extract oil—and, of particular importance, recover more oil from existing reservoirs. Rather than production peaking, existing fields are yielding markedly more oil than in the past. Advances in technology include the following:

Directional Drilling. It used to be that wellbores were basically vertical holes. This made it necessary to drill virtually on top of a potential oil deposit. However, the advent of miniaturized computers and advanced sensors that can be attached to the drill bit now allows companies to drill directional holes with great accuracy because they can get real-time information on the subsurface location throughout the drilling process.

Horizontal Drilling. Horizontal drilling is similar to directional drilling, but the well is designed to cut horizontally through the middle of the oil or natural gas deposit. Early horizontal wells penetrated only 500 to 800 feet of reservoir laterally, but technology advances recently allowed a North Slope operator to penetrate 8,000 feet of reservoir horizontally. Moreover, horizontal wells can operate up to 10 times more productively than conventional wells.

3-D Seismic Technology.   Substantial enhancements in computing power during the past two decades have allowed the industry to gain a much clearer picture of what lies beneath the surface. The ability to process huge amounts of data to produce three-dimensional seismic images has significantly improved the drilling success rate of the industry.

Primarily due to these advances, the U.S. Geological Survey (USGS), in its 2000 *World Petroleum Assessment,* increased by 20 percent its estimate of undiscovered, technically recoverable oil. USGS noted that, since oil became a major energy source about 100 years ago, 539 billion barrels of oil have been produced outside the United States. USGS estimates there are 649 billion barrels of undiscovered, technically recoverable oil outside the United States. But, importantly, USGS also estimates that there will be an *additional* 612 billion barrels from "reserve growth"—nearly equaling the undiscovered resources. Reserve growth results from a variety of sources, including technological advancement in exploration and production, increases over initially conservative estimates of reserves, and economic changes.

The USGS estimates reflected several factors:

- As drilling and production within discovered fields progresses, new pools or reservoirs are found that were not previously known.
- Advances in exploration technology make it possible to identify new targets within existing fields.
- Advances in drilling technology make it possible to recover oil and gas not previously considered recoverable in the initial reserve estimates.
- Enhanced oil recovery techniques increase the recovery factor for oil and thereby increase the reserves within existing fields.

Here in the United States, rather than "running out of oil," potentially vast oil and natural gas reserves remain to be developed. According to the latest published government estimates, there are more than 131 billion barrels of oil and more than 1,000 trillion cubic feet of natural gas remaining to be discovered in the United States. However, 78 percent of this oil and 62 percent of this gas are expected to be found beneath federal lands—much of which are non-park and non-wilderness lands—and coastal waters. While there is plenty of oil in the ground, oil companies need to be allowed to make major investments to find and produce it.

The U.S. Energy Information Administration has projected that fossil fuels will continue to dominate U.S. energy consumption, with oil and natural gas providing almost two-thirds of that consumption in the year 2025, even though energy efficiency and renewables will grow faster than their historical rates. However, renewables in particular start from a very small base; and the major shares provided by oil, natural gas, and coal in 2025 are projected to be nearly identical to those in 2003.

Those who block oil and natural gas development here in the United States and elsewhere only make it much more difficult to meet the demand for oil, natural gas, and petroleum products. Indeed, it is not surprising that some of the end-of-oil advocates are the same people who oppose oil and natural gas development everywhere.

Failure to develop the potentially vast oil and natural gas resources that remain in the world will have a high economic cost. We must recognize that we live in a global economy, and that there is a strong link between energy and economic growth. If we are to continue to grow economically, here in the United States, in Europe, and the developing world, we must be cost-competitive in our use of energy. We need *all* sources of energy. We do not have the luxury of limiting ourselves to one source to the exclusion of others. Nor can we afford to write off our leading source of energy before we have found cost-competitive and readily available alternatives.

Consider how oil enhances our quality of life—fueling growth and jobs in industry and commerce, cooling and warming our homes, and getting us where we need to go. Here in the United States, oil provides about 97 percent of transportation fuels, which power nearly all of the cars and trucks traveling on our nation's highways. And plastics, medicines, fertilizers, and countless other products that extend and enhance our quality of life are derived from oil.

In considering our future energy needs, we also need to understand that gasoline-powered automobiles have been the dominant mode of transport for the past century—and the overwhelming preference of hundreds of millions of people throughout the world. Regardless of fuel, the automobile—likely to be configured far differently from today—will remain the consumer's choice for personal transport for decades to come. The freedom of mobility and the independence it affords consumers is highly valued.

The United States—and the world—cannot afford to leave the Age of Oil before realistic substitutes are fully in place. It is important to remember that man left the Stone Age not because he ran out of stones—and we will not leave the Age of Oil because we will run out. Yes, someday oil will be replaced, but clearly not until substitutes are found—substitutes that are proven more reliable, more versatile, and more cost-competitive than oil. We can rely on the energy marketplace to determine what the most efficient substitutes will be.

As we plan for our energy future, we also cannot afford to ignore the lessons of recent history. In the early 1970s, many energy policymakers were sure that oil and natural gas would soon be exhausted, and government policy was explicitly aimed at "guiding" the market in a smooth transition away from these fuels to new, more sustainable alternatives. Price controls, allocation schemes, limitations on natural gas, massive subsidies to synthetic fuels, and other measures were funded heavily and implemented.

Unfortunately, the key premises on which these programs were based, namely that oil was nearing exhaustion and that government guidance was desirable to safely transition to new energy sources, are now recognized as having been clearly wrong—and to have resulted in enormously expensive mistakes.

Looking into the distant future, there will be a day when oil is no longer the world's dominant energy source. We can only speculate as to when and how that day will come about. For example, there is an even bigger hydrocarbon resource that can be developed to provide nearly endless amounts of energy: methane hydrates (methane frozen in ice crystals). The deposits of methane hydrates are so vast that when we develop the technology to bring them to market, we will have clean-burning energy for 2,000 years. It's just one of the exciting scenarios we may see in the far-off future. But we won't be getting there anytime soon, and until we do, the Age of Oil will continue.

 **NO**

The Long Emergency

A few weeks ago, the price of oil ratcheted above fifty-five dollars a barrel, which is about twenty dollars a barrel more than a year ago. The next day, the oil story was buried on page six of the *New York Times* business section. Apparently, the price of oil is not considered significant news, even when it goes up five bucks a barrel in the span of ten days. That same day, the stock market shot up more than a hundred points because, CNN said, government data showed no signs of inflation. Note to clueless nation: Call planet Earth.

Carl Jung, one of the fathers of psychology, famously remarked that "people cannot stand too much reality." What you're about to read may challenge your assumptions about the kind of world we live in, and especially the kind of world into which events are propelling us. We are in for a rough ride through uncharted territory.

It has been very hard for Americans—lost in dark raptures of nonstop infotainment, recreational shopping and compulsive motoring—to make sense of the gathering forces that will fundamentally alter the terms of everyday life in our technological society. Even after the terrorist attacks of 9/11, America is still sleepwalking into the future. I call this coming time the Long Emergency.

Most immediately we face the end of the cheap-fossil-fuel era. It is no exaggeration to state that reliable supplies of cheap oil and natural gas underlie everything we identify as the necessities of modern life—not to mention all of its comforts and luxuries: central heating, air conditioning, cars, airplanes, electric lights, inexpensive clothing, recorded music, movies, hip-replacement surgery, national defense—you name it.

The few Americans who are even aware that there is a gathering global-energy predicament usually misunderstand the core of the argument. That argument states that we don't have to run out of oil to start having severe problems with industrial civilization and its dependent systems. We only have to slip over the all-time production peak and begin a slide down the arc of steady depletion.

The term "global oil-production peak" means that a turning point will come when the world produces the most oil it will ever produce in a given year and, after that, yearly production will inexorably decline. It is usually represented graphically in a bell curve. The peak is the top of the curve, the halfway point of the world's all-time total endowment, meaning half the world's oil will be left. That seems like a lot of oil, and it is, but there's a big catch: It's the half that is much more difficult to extract, far more costly to get, of much poorer quality and located mostly in places where the people hate us. A substantial amount of it will never be extracted.

The United States passed its own oil peak—about 11 million barrels a day—in 1970, and since then production has dropped steadily. In 2004 it ran just above 5 million barrels a day (we get a tad more from natural-gas condensates). Yet we consume roughly 20 million barrels a day now. That means we have to import about two-thirds of our oil, and the ratio will continue to worsen.

The U.S. peak in 1970 brought on a portentous change in geoeconomic power. Within a few years, foreign producers, chiefly OPEC, were setting the price of oil, and this in turn led to the oil crises of the 1970s. In response, frantic development of non-OPEC oil, especially the North Sea fields of England and Norway, essentially saved the West's ass for about two decades. Since 1999, these fields have entered depletion. Meanwhile, worldwide discovery of new oil has steadily declined to insignificant levels in 2003 and 2004.

Some "cornucopians" claim that the Earth has something like a creamy nougat center of "abiotic" oil that will naturally replenish the great oil fields of the world. The facts speak differently. There has been no replacement whatsoever of oil already extracted from the fields of America or any other place.

Now we are faced with the global oil-production peak. The best estimates of when this will actually happen have been somewhere between now and 2010. In 2004, however, after demand from burgeoning China and India shot up, and revelations that Shell Oil wildly misstated its reserves, and Saudi Arabia proved incapable of goosing up its production despite promises to do so, the most knowledgeable experts revised their predictions and now concur that 2005 is apt to be the year of all-time global peak production.

It will change everything about how we live.

To aggravate matters, American natural-gas production is also declining, at five percent a year, despite frenetic new drilling, and with the potential of much steeper declines ahead. Because of the oil crises of the 1970s, the nuclear-plant disasters at Three Mile Island and Chernobyl and the acid-rain problem, the U.S. chose to make gas its first choice for electric-power generation. The result was that just about every power plant built after 1980 has to run on gas. Half the homes in America are heated with gas. To further complicate matters, gas isn't easy to import. Here in North America, it is distributed through a vast pipeline network. Gas imported from overseas would have to be compressed at minus-260 degrees Fahrenheit in pressurized tanker ships and unloaded (re-gasified) at special terminals, of which few exist in America. Moreover, the first attempts to site new terminals have met furious opposition because they are such ripe targets for terrorism.

Some other things about the global energy predicament are poorly understood by the public and even our leaders. This is going to be a permanent energy crisis, and these energy problems will synergize with the disruptions of climate change, epidemic disease and population overshoot to produce higher orders of trouble.

We will have to accommodate ourselves to fundamentally changed conditions.

No combination of alternative fuels will allow us to run American life the way we have been used to running it, or even a substantial fraction of it. The wonders of steady technological progress achieved through the reign of cheap oil have lulled us into a kind of Jiminy Cricket syndrome, leading many Americans to believe that anything we wish for hard enough will come true. These days, even people who ought to know better are wishing ardently for a seamless transition from fossil fuels to their putative replacements.

The widely touted "hydrogen economy" is a particularly cruel hoax. We are not going to replace the U.S. automobile and truck fleet with vehicles run on fuel cells. For one thing, the current generation of fuel cells is largely designed to run on hydrogen obtained from natural gas. The other way to get hydrogen in the quantities wished for would be electrolysis of water using power from hundreds of nuclear plants. Apart from the dim prospect of our building that many nuclear plants soon enough, there are also numerous severe problems with hydrogen's nature as an element that present forbidding obstacles to its use as a replacement for oil and gas, especially in storage and transport.

Wishful notions about rescuing our way of life with "renewables" are also unrealistic. Solar-electric systems and wind turbines face not only the enormous problem of scale but the fact that the components require substantial amounts of energy to manufacture and the probability that they can't be manufactured at all without the underlying support platform of a fossil-fuel economy. We will surely use solar and wind technology to generate some electricity for a period ahead but probably at a very local and small scale.

Virtually all "biomass" schemes for using plants to create liquid fuels cannot be scaled up to even a fraction of the level at which things are currently run. What's more, these schemes are predicated on using oil and gas "inputs" (fertilizers, weed-killers) to grow the biomass crops that would be converted into ethanol or bio-diesel fuels. This is a net energy loser—you might as well just burn the inputs and not bother with the biomass products. Proposals to distill trash and waste into oil by means of thermal depolymerization depend on the huge waste stream produced by a cheap oil and gas economy in the first place.

Coal is far less versatile than oil and gas, extant in less abundant supplies than many people assume and fraught with huge ecological drawbacks—as a contributor to greenhouse "global warming" gases and many health and toxicity issues ranging from widespread mercury poisoning to acid rain. You can make synthetic oil from coal, but the only time this was tried on a large scale was by the Nazis under wartime conditions, using impressive amounts of slave labor.

If we wish to keep the lights on in America after 2020, we may indeed have to resort to nuclear power, with all its practical problems and eco-conundrums. Under optimal conditions, it could take ten years to get a new generation of nuclear power plants into operation, and the price may be beyond our means. Uranium is also a resource in finite supply. We are no closer to the more difficult project of atomic fusion, by the way, than we were in the 1970s.

The Long Emergency is going to be a tremendous trauma for the human race. We will not believe that this is happening to us, that 200 years of modernity can be brought to its knees by a world-wide power shortage. The survivors will have to cultivate a religion of hope—that is, a deep and comprehensive belief that humanity is worth carrying on. If there is any positive side to stark changes coming our way, it may be in the benefits of close communal relations, of having to really work intimately (and physically) with our neighbors, to be part of an enterprise that really matters and to be fully engaged in meaningful social enactments instead of being merely entertained to avoid boredom. Years from now, when we hear singing at all, we will hear ourselves, and we will sing with our whole hearts.

POSTSCRIPT

Should the World Continue to Rely on Oil as a Major Source of Energy?

The twenty-first century ushered in another in a series of energy crises that have plagued the developed world since 1972. Gas prices jumped to record heights, and then rose even higher in 2006, and the prospects of a return to $2.00-a-gallon levels seem increasingly remote. Once again, cries for eliminating dependence on foreign oil and for developing alternatives to the twentieth century's principal energy source were heard.

Yet when one reads the UN assessment of foreseeable world energy supplies (Hisham Khatib et al., *World Energy Assessment: Energy and the Challenge of Sustainability,* United Nations Development Programme, 2002), a sobering message appears. Don't panic just yet. The study reveals no serious energy shortage during the first half of the twenty-first century. In fact, the report suggests that oil supply conditions have actually improved since the crises of the 1970s and early 1980s. The report goes further in its assessment, concluding that fossil fuel reserves are "sufficient to cover global requirements throughout this century, even with a high-growth scenario."

Francis R. Stabler argues in "The Pump Will Never Run Dry," (*The Futurist,* November 1998) that technology and free enterprise will combine to allow the human race to continue its reliance on oil far into the future. For Stabler, the title of his article tells the reader everything. The pump will not run dry! He may not be a neutral observer, as he has worked as a systems engineer in the automotive and defense industries for 30 years, and is currently a technology planner for General Motors. The latter has a vested interest in the internal combustion engine with its fuel requirements of gasoline. Nonetheless, his article lays out a systematic argument for his position, with an appropriate level of evidence in support of his conclusions.

To be sure, his view of the future availability of gas is a minority one. One supporter is Julian L. Simon who argues in his *The Ultimate Resource 2* (1996) that even God may not know exactly how much oil and gas are "out there." Chapter 11 of Simon's book is entitled "When Will We Run Out of Oil? Never!." Simon takes the reader through a twelve-step process to demonstrate that the doomsayers are wrong. Another Stabler supporter is Bjørn Lomborg in *The Skeptical Environment: Measuring the Real State of the World* (Cambridge University Press, 2001). Arguing that the world seems to find more fossil energy than it consumes, he concludes that "we have oil for at least 40 years at present consumption, at least 60 years' worth of gas, and 230 years' worth of coal."

Simon and Lomborg are joined by Michael C. Lynch in a published article on the Web under global oil supply (msn.com) entitled "Crying Wolf: Warnings about Oil Supply."

Seth Dunn, on the other hand, follows conventional wisdom in his article. That is, because oil is a finite resource, its supply will end some day, and that day will be sooner rather than later. In fact, he suggests that new renewable energy sources are in the same position as oil a century ago, that is, "gaining footholds" in the energy market. Dunn has argued elsewhere (Christopher Flavin and Seth Dunn, "Reinventing the Energy System," *State of the World 1999,* Worldwatch Institute,1999) that the global economy has been built on the rapid depletion of non-renewable resources, and such consumption levels cannot possibly be maintained throughout the twenty-first century, as they were the previous century. Although Flavin and Dunn's arguments probably have received a receptive audience among most scholars who are concerned with the increasing scarcity of nonrenewable resources, they require the reader to accept a set of assumptions about the acceleration of future energy consumption. But one can easily be seduced by the logic of their argument, because it "just seems to make sense."

An excellent report is the Worldwatch Institute's "Energy for Development: The Potential Role of Renewable Energy in Meeting the Millennium Goals" (September 15, 2005). The report identifies those renewable energy options currently in wide use somewhere in the world. Another Worldwatch Institute report, "Biofuels for Transportation: Global Potential and Implications for Sustainable Agriculture and Energy in the 21st Century" (June 7, 2006), addresses one particular alternative to oil.

Lester R. Brown, et al. in *Beyond Malthus* (1999) suggest that most writers point to between the years 2010 and 2025 as the time when world oil production will peak. The consequence, if that is accurate, is a need for alternative sources. The student of energy politics, however, must be careful not to ignore how advances in energy source exploration and extraction have tended to expand known reserves. Is the future lesson that the tide has finally turned and no significant reserves remain to be discovered? Or is the lesson that history will repeat itself and modern science will yield more oil and gas deposits, as well as make their extraction cost effective?

Finally, James J. MacKensie has provided a comprehensive yet succinct article on the peaking of oil in "Oil as a Finite Resource: When Is Global Production Likely to Peak?" (World Resources Institute). Seth Dunn of the Worldwatch Institute in *State of the World 2001* suggests that a new energy system is fast approaching because of a series of revolutionary new technologies and approaches.

Finally, David R. Francis provides a balanced assessment of the peaking debate in "Has Global Oil Production Peaked?" (*Christian Science Monitor,* January 29, 2004).

The msn.com web site provides numerous citations of articles on both sides of the issue.

ISSUE 7

Will the World Be Able to Feed Itself in the Foreseeable Future?

YES: Sylvie Brunel, from *The Geopolitics of Hunger, 2000-2001: Hunger and Power* (Lynne Rienner, 2001)

NO: Janet Raloff, from "Global Food Trends," *Science News Online* (May 31, 2003)

ISSUE SUMMARY

YES: Sylvie Brunel, former president of Action Against Hunger, argues that "there is no doubt that world food production is enough to meet the needs of" all the world's peoples.

NO: Janet Raloff, a writer for *Science News,* looks at a number of factors—declining per capita grain harvests, the world's growing appetite for meat, the declining availability of fish for the developing world, and continuing individual poverty.

Visualize two pictures. The first snapshot, typical of photographs that have graced the covers of the world's magazines, reveals a group of people in Africa, including a significant number of small children, who show dramatic signs of advanced malnutrition and even starvation. The second picture (really several in sequence) shows an apparently wealthy couple finishing a meal at a rather expensive restaurant. The waiter removes their plates still half-full of food, an untouched loaf of French bread, and assorted other morsels from the table, and deposits them in the kitchen garbage can. These scenarios once highlighted a popular film about world hunger. The implication was quite clear. If only the wealthy would share their food with the poor, no one would go hungry. Today the simplicity of this image is obvious.

This issue addresses the question of whether or not the world will be able to feed itself by the middle of the twenty-first century. A prior question, of course, is whether or not enough food is grown throughout the world today to handle current nutritional and caloric needs of all the planet's citizens. News accounts of chronic food shortages somewhere in the world seem to have been appearing with regularly consistency for about 30 years. This time period has witnessed graphic accounts in news specials about the consequences of insufficient food, usually somewhere in sub-Saharan Africa. Also, several national and international studies have been commissioned to address world hunger. An American study organized by President Carter, for example, concluded that the root cause of hunger was poverty.

One might deduce from all of this activity that population growth has outpaced food production and that the planet's agricultural capabilities are no longer sufficient. Yet, the ability of most countries to grow enough food has not yet been challenged. During the 1970–2000 period, only one region of the globe, sub-Saharan Africa, was unable to have its own food production keep pace with population growth. All other regions of the world experienced food increases greater than human growth.

This is instructive because, beginning in the early 1970s, a number of factors conspired to lessen the likelihood that all humans would go to bed each night adequately nourished. Weather in major food-producing countries turned bad; a number of countries, most notably Japan and the Soviet Union, entered the world grain importing business with a vengeance; the cost of energy used to enhance agricultural output rose dramatically; and less capital was available to poorer countries as loans or grants for purchasing agricultural inputs or the finished product (food) itself. Yet the world has had little difficulty growing sufficient food, enough to provide every person with two loaves of bread per day as well as other commodities. Major food-producing countries even cut back the amount of acreage devoted to agriculture.

Why then did famine and other food-related maladies appear with increasing frequency? The simple answer is that food is treated as a commodity, not a nutrient. Those who can afford to buy food or grow their own do not go hungry. However, the world's poor became increasingly unable to afford either to create their own successful agricultural ventures or to buy enough food.

The problem for the next half-century, then, has several facets to it. First, can the planet physically sustain increases in food production equal to or greater than the ability of the human race to reproduce itself? This question can only be answered by examining both factors in the comparison—likely future food production levels and future fertility scenarios. A second question relates to the economic dimension associated with an efficient global food distribution system. Will those poorer countries of the globe that are unable to grow their own food have sufficient assets to purchase it, or will the international community create a global distribution network that ignores a country's ability to pay? And third, will countries that want to grow their own food be given the opportunity to do so?

The selections for this issue address the specific question of the planet's continuing ability to grow sufficient food to feed its growing population. Syvie Brunel contends that if world food production were distributed among all the world's peoples, there would be plenty of food. In the second selection, Janet Raloff looks at a number of dangerous world food trends—declining per capita grain harvests, the world's growing appetite for meat, the declining availability of fish for the developing world, and continuing individual poverty.

YES

Sylvie Brunel

Increasing Productive Capacity:
A Global Imperative

Can the earth feed its inhabitants? Despite the most alarmist predictions—those of Lester Brown in his State of the World published each year by the World Watch Institute of Washington; those of Paul Ehrlich, author of *The Population Bomb;* or those of the Club of Rome—there is no doubt that world food production, if equally distributed among all the world's peoples, is enough to meet the needs of them all. It is true that the increase in world agricultural production has slowed in recent years, a situation that has led to an immediate flood of alarmist predictions. The world, however, is not heading toward famine. And this is for a number of reasons.

The Increase in World Agricultural Production Continues to Outpace Population Growth

Only persons who are ill informed or of bad faith can argue that the trend is toward a decline in food production. They may even succeed in proving that claim. It is enough for them to select as the base year one in which harvests were particularly good and a second year in which they declined steeply in order to show a "disturbing" trend. Hervé Kempf demonstrated, for example, how a comparison of 1984 and 1991 would show an increase in cereal production of only 0.7 percent per year, which would be "disturbing," since it is far below the 1.7% annual rate of population increase. By selecting the preceding years, one can show, on the contrary, that world agriculture has never been more productive: A comparison between 1983 and 1990 shows an increase in cereal production of 2.7 percent per year. These two statistics are clearly equally deceptive, and Joseph Klatzmann, in a refreshing little book, repeatedly warned against the danger of blindly trusting statistical data taken out of context.

If we examine world agricultural production over a long period, it becomes clear that the production curve exceeds the population growth curve. While world population did indeed double in one generation, grain production increased more than threefold, from 600 million to approximately 1,900 million tons per year. Each human being has available in theory 20 percent more food than in the early 1970s, or 2,700 calories per person per day, which is far more than a person's estimated need of between 2,000 and 2,200 calories, depending on the sources. However, half of the current grain production does not directly benefit people: Approximately 20 percent is used to feed cattle, 5 percent is kept for seeds, and the remaining 25 percent is quite simply lost as a result of poor storage or destruction by rodents, insects, and so on, especially in developing countries. It is therefore not the impossibility of increasing agricultural production that threatens mankind, but rather the way in which this increase is achieved and for the benefit of whom.

Indeed, it is in fact not in the countries of the so-called Third World but rather in the developed countries that agricultural production has slowed, in other words precisely where the problems of hunger have been overcome. (At least they have been overcome in quantitative terms; in qualitative terms, obesity, on the one hand, and malnutrition caused by the economic and social marginalization of certain categories of persons, on the other, have become real societal problems.) The developed countries have chosen to voluntarily limit their agricultural production in order to adapt it to the level of demand at which production would be profitable, in other words, to the consumer market. The fact that there are some 800 million people suffering from malnutrition in the world in no way changes this calculation, since those persons are too poor to buy food.

The reduction or slowdown in the rate of increase in world food production is thus attributable mainly to the developed countries, for reasons that have nothing to do with ecological limitations. Pierre Le Roy estimates at 20 million tons the reduction in supply that results from Europe's policy of limitation of production (land left fallow), an amount that represents twice the total of all food imports by sub-Saharan Africa.

Food imports by the Third World are indeed increasing, rising from 20 million tons in 1960, or 2 percent of consumption, to 120 million tons in the mid 1990s, or 20 percent of consumption. Economic forecasts suggest that this dependency is likely to increase even further in the decades ahead and to rise to 160 million tons within two decades. The reasons for this growing dependence, which will create problems without precedent for the economies of poor countries that will face increasingly onerous food import bills, are both negative and positive.

The negative factor of continuing population growth and spreading urbanization in the countries of the South, where nearly half the population now lives in cities, explains why more and more people are consuming food that their farmers are incapable of providing. The positive factor of the increase in average living standards in the developing countries and the emergence of a middle class that consumes more meat and dairy products places increasing pressure on the demand for cereals, in particular secondary cereals for stock feed.

Two-Speed Agricultural Policies

Why cannot the Third World feed itself, even though self-sufficiency in food was the grand slogan of the 1970s and 1980s?

The "technical" impossibility of increasing agricultural production in the South is not the problem: The earth is far from reaching its maximum agricultural potential, and the Food and Agriculture Organization (FAO) has pointed out that the useful agricultural surface in developing countries (700 million hectares) could be doubled without encroaching on protected areas such as forests or areas in which people live. Latin America and Africa hold the greatest potential in this regard. In addition, the potential for increased production through more intensive farming methods remains considerable. Only 11 kilograms of fertilizer are used per hectare in Africa, compared with 66 kilograms in Latin America and 139 kilograms in Asia, and only 5 percent of land is irrigated in Africa (most of this in countries that are unable to take advantage of it, such as Sudan and Madagascar), compared to 37 percent in Asia and 14 percent in Latin America. This situation offers tremendous potential for growth.

But the political and economic choices made by the countries of the Third World have thus far been detrimental to agriculture, and in particular to small peasant farming. Investments in agriculture have been concentrated in regions in which purchasing power is greatest and are characterized by a concern to protect the income of farmers, which has been steadily declining. As a result, these investments are moving in the direction of a two-tiered world that is becoming increasingly unequal in terms of access to food.

On one hand, the developed countries enjoy rapid growth, and despite the fact that farmers represent on average no more than 3 percent of the active population, their food supply is abundant and diversified, prices are low, and import levels are low as a result of the massive support given to the agricultural sector (in the mid-1990s, Organization for Economic Cooperation and Development (OECD) countries each year spent more than two hundred billion dollars to support their agricultural sectors). In that part of the world, the concern is no longer the fear of shortage, but rather the quality of the food consumed. The agrofood industry, now powerful after being forced to steadily increase its output over the past decades in order to keep up with the steadily rising demand, is today facing another challenge, namely, shifting to production methods that focus less on quantity and more on the quality of the inputs used and on the quality of the final product. Producers are also concerned about the methods used to satisfy the demand of consumers, who now want food that is not only abundant but also varied and, above all, healthy. The successive scandals of mad-cow disease, salmonella poisoning in chickens, hormone-treated beef cattle; the rejection by consumer groups of genetically modified plants; animal feed that includes mud from cleaning stations; and the questions raised about the production of eggs by battery hens are indicative of a new era in which insistence on quality is now a greater challenge than the demand for quantity.

On the other side are the poor and vulnerable countries, where malnutrition is endemic and where a majority of the population still depends on agriculture. The food supply remains insufficient, however, because of the poor yields that result from the low level of technology used, the absence of incentives to produce because of economic policies that discourage agriculture, and unfavorable exchange rates that make the importation of agricultural imports expensive. It is therefore precisely in those countries that agricultural production needs to be increased. First, increased production would reduce the cost of food, particularly in the large urban centers, and thereby make it accessible to this large sector of the population that is too poor to eat properly. Second, increased production would reduce the food import bills of countries that are increasingly dependent on imports, mainly from the rich countries.

The Food Supply Is a Regional, Not a Global Problem

At the global level, the food supply is increasing for reasons that are both positive and negative. On the positive side, the agricultural sector in Eastern Europe, which needed to be restructured following the collapse of the Iron Curtain, is now on the road to recovery. On the negative side, the economic difficulties of East Asia have led to a decline in food imports by that region.

At the regional level, the structural overproduction that the world has been experiencing for the past twenty years hardly prevents sub-Saharan Africa and South Asia from experiencing hunger. Of the approximately 800 million malnourished people in the world, more than 200 million live in Africa (nearly 40 percent of the population) and 530 million in South Asia (or one person out of five).

What is therefore responsible for this disastrous and paradoxical situation at a time when, in order to reduce the supply of food, rich countries are destroying mountains of surplus food each year and forcing their farmers to leave a portion of their land uncultivated, through subsidies for land that is left fallow?

One answer is wars and conflicts, particularly in Africa, that disrupt agricultural production. A second, even more important answer is mass poverty, which prevents an entire sector of the world's population (one out of every five inhabitants of the Third World) from obtaining adequate food. That sector is incapable of producing enough food to meet its needs and lacks the means to purchase it, even when the food is available and can be bought. Worldwide, some 1.5 billion people live below the poverty level. Mass poverty is all the more serious, as it is always combined with ignorance: It is always the poorest classes that commit the most harmful errors of nutrition, since they lack the advantage of basic education. The errors of nutrition committed by pregnant women and children, who make up the primary groups at risk of hunger, should be the focus of particular attention, since these errors have disastrous consequences for the future of the entire society.

According to the United Nations Children's Fund (UNICEF), half of the world's malnourished children live in Asia, which has 100 million of the 200 million total, including 70 million in India. That country alone has two and a half times more malnourished children than all of sub-Saharan Africa.

Writing about South Asia . . . , Gilbert Etienne remarked on the extent to which the problem of hunger remains unresolved because of mass poverty and the slowdown in investments in agriculture. This is despite the notable progress achieved on the Asian continent.

The case of Africa gives cause for even greater concern:

1. Unlike the situation in other continents, the rate of malnutrition is not declining. Quite the opposite, in fact, because chronic malnutrition still affects nearly 40 percent of the region's population.
2. The high proportion of young people in the population, a sign of a vigorous population still characterized by high birthrates, has led to a high proportion of unemployed in relation to the number of those who are in a position to contribute to production. The burden on the economies of African countries is therefore particularly heavy, especially since most of these countries suffer from an acute lack of financial resources.
3. The continent's dependency on food from foreign countries is due to the low productivity of its own agriculture and the growing number of Africans now living in urban areas (more than one in three compared with one in ten a generation ago). This dependency is effectively addressed neither by imports (9 to 10 million tons per year), because of the lack of adequate financial resources, nor by food aid (approximately 2 million tons), which has been falling drastically for some years now. Consequently, the food needs of Africans are not being satisfied, since the widespread poverty of a sector of the urban population does not permit that sector to obtain food at market prices. At the same time many rural dwellers are unable to provide for themselves during the period between harvests, on account of the low productivity levels and inadequate access to food.
4. A large number of people in Africa are affected by war or internal conflict. Even in countries in which the populations could, in theory, be properly fed, there is an adverse impact on the food situation of the population because of the insecurity of the economic actors, the weakness of the State, and the destruction or confiscation of crops. In this regard, Africa is by far the continent most affected by conflicts, which also result in massive populations of refugees and displaced persons who depend on international aid for their survival.
5. Poverty and the pressure on land and resources of the high population growth rate are not matched by corresponding investments in agriculture that would bring about increases in yields. Africa is thus the continent in which the problems of deforestation, desertification, and soil erosion are most acute. It is also the continent in which access to drinking water and irrigation is still very limited.

The situation is not desperate, however: Despite the lack of investment in small peasant farming, agricultural production in Africa has risen by 2 percent per year since the early 1960s. Grain production has more than doubled, from 30 million to 66 million tons. This rate of increase is insufficient to meet the needs of a growing population (3 percent per year) because of the way in which it has been achieved (mainly by increasing the area of land under cultivation).

Nevertheless it shows that more intensive farming methods are needed in Africa and that this approach has the potential to significantly increase agricultural output. When the FAO states that the "load capacity of the land" in many countries has now been exceeded, it is basing its conclusions on the use of traditional production methods, such as use of the hoe more often than not, lack of fertilizers, lack of irrigation, and use of a diverse range of traditional varieties of grain to compensate for climatic and pedological constraints, with but modest results. The average yield for Africa remains 1,000 kilograms per hectare of millet and corn, which shows how much room for improvement there would be if African governments were to decide to treat their farmers a little better and to invest in their agricultural potential.

What are some of the ways in which agricultural production can be increased? It is interesting to note that food problems do not occur in countries that are at peace, that enjoy democracy, and in which farmers operate under conditions of relative legal and administrative security, even when these countries are densely populated and located in unfavorable climatic zones. It is better to live in Burkina Faso than in the Democratic Republic of Congo, even though the Congo is infinitely better endowed than Burkina Faso in terms of rainfall and available land. Similarly, the "white revolution" in Mali is revitalizing those regions that produce rice, millet, and cotton and enriching their farmers, even as hunger still plagues Madagascar, the former breadbasket of southern Africa, which has been making error after economic error over the last quarter of a century.

In order to bring about peace and security in Africa, a resumption of cooperation is necessary. However, the level of official development assistance has never been lower. Will the renegotiation of the Lomé Convention relaunch the partnership between Europe and Africa for the concerted development of agriculture?

Janet Raloff

Global Food Trends

Last year, for the third year in four, world per-capita grain production fell. Even more disturbing in a world where people still go hungry, at 294 kilograms, last year's per capita grain yield was the lowest in more than 30 years. Indeed, the global grain harvest has not met demand for 4 years, causing governments and food companies to mine stocks of these commodities that they were holding in reserve.

This is just one of the sobering observations about world food trends offered by researchers with the Worldwatch Institute, an Earth-resources think tank in Washington, D.C.

Each year, Worldwatch reads several key indicators of our planet's environmental health. The organization's latest 153-page almanac, *Vital Signs 2003*, issued May 22 in cooperation with the United Nations Environment Programme, notes that production of the world's three major cereals fell in absolute terms in 2002: wheat by 3 percent, to 562 million metric tons (Mt); corn by almost 2 percent, to 598 Mt; and rice by 2 percent, to 391 Mt. Together, these three crops make up 85 percent of the world's grain harvest, notes Worldwatch's Brian Halweil.

Throughout the early 1960s, world grain reserves were equal to at least 1 year's global demand for these commodities. By last year, that excess had fallen to just 20 percent of what's now consumed annually.

What makes these trends so dangerous, Halweil reports, is that despite increasing dietary diversity, most people around the world "still primarily eat foods made from grain." Globally, people derive 48 percent of their calories from grain-based foods. Moreover, Halweil points out, grains—especially corn—serve as "the primary feedstock for industrial livestock production."

Indeed, he told *Science News Online*, "livestock consume 35 percent of the world's grain, over 90 percent of the soybeans, and millions of tons of other oilseeds, roots, and tubers each year." In the United States, the share of these plant-based foods going to livestock is even higher: 50 percent of all grains (including 60 percent of corn) and virtually all soy.

Parched Bread Baskets

Drought in Australia and the United States last year explains much of the drop in world cereal harvests, Halweil says.

New data issued by the U.S. Department of Agriculture flesh out the picture. They show that record or near-record droughts last year throughout much of the western and midwestern United States—the nation's breadbasket and corn belt, respectively—accounted for shortfalls in wheat and corn. It was so bad throughout much of the West that earlier this month the USDA announced a new $53 million program to help farmers and ranchers mitigate the effect of drought. The initiative will provide money for implementing new water-conserving technologies and farming practices.

Agricultural economists see no sign that U.S. grain production will recover soon. A map in the USDA's May 20 *Weekly Weather and Crop Bulletin* depicts much of the intermountain West in the throes of "extreme drought" with large surrounding areas—spanning from Mexico to Canada and from Nevada through middle-Nebraska—in only a slightly better situation: suffering merely a "severe drought."

Data reported in a May 1 water forecast by the National Drought Mitigation Center in Lincoln, Neb., show "spring and summer stream flows [at] less than 50 percent of average in parts of the Intermountain West." Water reservoirs mirror the problem. Despite a cool, wet spring throughout much of the West, fall and winter drought conditions have left water supplies well below average—in some cases at half of average amounts—in many Western basins.

The bottom line: No one expects bumper grain crops even in the United States. Attaining just average yields may prove difficult.

Feed's Growing Demand on Grains

In a second *Vital Signs 2003* report, Danielle Nierenberg highlights a related food trend, the world's growing appetite for meat. Last year, livestock growers raised some 242 million metric tons of meat. That's five times what was produced in 1950 and double the yield in 1977.

Because meat production is relatively expensive—it requires 11 to 17 calories of feed to produce each calorie of beef, pork, or chicken—wealthy industrialized countries have led in demand for these foods. But Nierenberg reports that "two-thirds of the gains in meat consumption in 2002 occurred in developing countries, where urbanization, rising incomes, and the globalization of trade are changing diets." In fact, she finds, developing countries have recently surpassed industrialized ones as producers of meat by total weight.

Still, there's a huge disparity in the amounts of meat consumed per capita in rich and poor countries. In industrial nations, the average person eats some 80 kilograms per year—or 2.8 times that in the developing world. Most people eat pork, which accounts for 38 percent of world meat production, followed by poultry at 30 percent and beef at 25 percent.

To help raise some 5 billion hoofed and 16 billion winged animals for meat, farmers have increasingly turned to raising animals in factorylike conditions. Today, industrial feedlots account for 43 percent of the world's beef and more than 50 percent of all pork and poultry. These confined setups also concentrate the noise, stink, and wastes associated with livestock into industrial operations. As unpopular as these are, they will probably dominate world meat production if it continues to grow. And the United Nations projects that it will grow, to 300 Mt by 2020.

Nevertheless, many people of the world won't have the luxury of choosing between grains or meats, even by 2020. Today, Halweil notes, more than 800 million people regularly go to bed hungry—a number that's greater than 2.5 times the combined population of the United States, Canada, and Mexico. It's tempting to speculate that if humanity ate less meat, there would be more grain available to feed these people, Halweil says. However, he says, "you can't necessarily make that leap," since most hunger today stems not from a shortage of food as much as a shortage of funds to pay for it.

On the other hand, Halweil notes that there is one food for which the industrialized world's consumption directly robs the developing world: fish. Recent reports have chronicled how overfishing by commercial fleets are decimating fish stocks around the world. "Because fishing is really a global industry—that is, you have American, Japanese, and Norwegian ships crisscrossing the globe and plucking fish from all over the planet," Halweil says, "meeting the demands of diners in New York or Tokyo can mean there's less available for someone in Bangkok or Bombay."

"I think the take-home message" on worldwide production and consumption patterns for grain and meat, Nierenberg says, is that people in the industrial world "are overconsuming and setting a bad example."

References

2003. USDA provides $53 million to farmers and ranchers in 17 states to help with drought recovery. U.S. Department of Agriculture press release. May 9. Available at http://www.usda.gov/news/releases/2003/05/0148.htm.

Halweil, B. 2003. Grain production drops. In *Vital Signs 2003*, L. Starke, ed. New York: W.W. Norton. See http://www.worldwatch.org/pubs/vs/2003.

Nierenberg, D. 2003. Meat production and consumption grow. In *Vital Signs 2003*, L. Starke, ed. New York: W.W. Norton. See http://www.worldwatch.org/pubs/vs/2003.

U.S. Department of Agriculture. 2003. U.S. drought monitor. *Weekly Weather and Crop Bulletin* 90(May 20):5. Available at http://www.usda.gov/oce/waob/jawf/wwcb/p_5.pdf.

_____. 2003. Water supply forecast for the western United States. *Weekly Weather and Crop Bulletin* 90(May 20):2-3. Available at http://www.usda.gov/oce/waob/jawf/wwcb/p_2.pdf.

POSTSCRIPT

Will the World Be Able to Feed Itself in the Foreseeable Future?

Presumably, economist Thomas Robert Malthus was not the first to address the question of the planet's ability to feed its population. But his 1789 *Essay on the Principle of Population* is the most quoted of early writings on the subject. Malthus's basic proposition was that population, if left unchecked, would grow geometrically, while subsistence resources could grow only arithmetically. Malthus, who wrote his essay in response to an argument with his father about the ability of the human race to produce sufficient resources vital for life, created a stir back in the late eighteenth century. The same debate holds the public's attention today as population grows at a rate unparalleled at any other time in human history.

Syvie Brunel argues that an examination of world agricultural production over a long period of time reveals that, in fact, it outpaced population growth. In the last 30 years of the twentieth century, for example, each individual's amount of food, in theory, grew by 20 percent. This equated to 2,700 calories per person per day, more than a person's estimated need. There is a catch, however. Half of the current grain production does not benefit humans. Approximately 20 percent is used to feed cattle, 5 percent is keep for seeds, and the remaining 25 percent is lost. In short, ecological factors are not at fault; humans, particularly from the developed world, are the culprits. Therein lies the solution for Brunel.

Janet Raloff, a writer for *Science News*, looks at a number of factors—declining per capita grain harvests, the world's growing appetite for meat, the declining availability of fish for the developing world, and continuing individual poverty. Lester R. Brown adds another dimension to the problem in *State of the World 2001* (W.W. Norton & Company, 2001). His pessimism about future food supplies arises from the belief that world leaders have not come forward with a comprehensive master plan to address the problem and are extremely unlikely to do so in the foreseeable future.

A balanced look at the planet's capacity to feed the UN's projected 2050 population is L.T. Evans' *Feeding the Ten Billion* (Cambridge University Press, 1998). In it, the author takes the reader through the ages, showing how the human race has addressed the agricultural needs of each succeeding billion people. The biggest challenge during the next half century, according to Evans, is to solve two problems: producing enough food for a 67 percent increase in the population, and eliminating the chronic undernutrition afflicting so many people. Solving the first problem requires a focus on the main components of increased food supply. For Evans, these include: "(1) increase in land under cultivation; (2) increase in yield per hectare per crop; (3) increase in the number of crops per hectare per year; (4) displacement of lower yielding crops by higher yielding ones; (5) reduction of post-harvest losses; (and) (6) reduced use as feed for animals."

The second problem brings into play many socioeconomic factors beyond those that are typically associated with agricultural production. Growing enough food worldwide is a necessary but not sufficient condition for eliminating problems associated with hunger. Many studies have observed that the root cause of hunger is poverty. Addressing poverty, therefore, is a prerequisite for ensuring that the world's food supply is distributed such that the challenge of global hunger is met.

Three other sources are reports by the UN's Food and Agricultural Organization: *World Agriculture: Towards 2000; The State of Food and Agriculture 2002*; and *World Agriculture: Towards 2015/2030*. The central message of these studies is that the planet will be able to feed a growing population in the foreseeable future, if certain conditions are met. Another FAO report, *The State of Food Insecurity in the World 2005 (2005)*, analyzes the latest data on hunger in the context of the Millennium Development Goals.

Another optimistic viewpoint about future food prospects is *The World Food Outlook* by Donald Mitchell, Merlinda D. Ingco, and Ronald C. Duncan (Cambridge University Press, 1997). Their basic conclusion is that the world food situation has improved dramatically for most of the regions of the globe and will continue to do so. The only exception is sub-Saharan Africa, but there the problems go far beyond agriculture.

For Julian L. Simon in "What Are the Limits on Food Production," *The Ultimate Resource 2* (Princeton University Press,1996) the answer is simple. The world can produce vastly more food than it currently does, even in those places that rely on conventional methods. The essence of his argument is this: More people with higher incomes cause scarcity problems in the short run, which, in turn, results in raised prices. Into the picture then come inventors and entrepreneurs, some of whom will be successful in finding appropriate solutions. To Simon, we are better off in the long run because of the potential for profit on the part of the problem solvers.

David Pimentel et al., "Impact of Population Growth on Food Supplies and Environment," *Population and Environment* (1997) allude to the warnings of a number of impressive groups concerning the world's future food situation. They cite the Royal Society, the U.S. National Academy of Sciences, the UN Food and Agricultural Organization, and numerous other international organizations, as well as scientific research to support their view of the pending danger. For them, two issues are significant. The first is the existence of enough agricultural inputs: water, land, energy, and the like. The second is the global economic system that treats food as a commodity rather than a nutrient. For these authors, the bottom line is that population must be curtailed.

Two other sources are worth mentioning. *Halving Global Hunger* (Sara Scherr, Background Paper of the Millennium Project's Task Force 2 on Hunger, April 18, 2003) provides an overview of existing knowledge relating to the reduction of hunger. Joachim von Braun and associates analyze two future policy scenarios for ensuring food security by 2050 (*New Risks and Opportunities for Food Security: Scenario Analyses for 2015 and 2050*, International Food Policy Research Institute, February 2005).

ISSUE 8

Is the Threat of Global Warming Real?

YES: Intergovernmental Panel on Climate Change, from "Climate Change 2001: The Scientific Basis," A Report of Working Group I of the Intergovernmental Panel on Climate Change (2001)

NO: Christopher Essex and Ross McKitrick, from *Taken By Storm: The Troubled Science, Policy and Politics of Global Warming* (Key Porter Books, 2002)

ISSUE SUMMARY

YES: The summary of the most recent assessment of climatic change by a UN-sponsored group of scientists concludes that an increasing set of observations reveals that the world is warming and much of it is due to human intervention.

NO: Christopher Essex and Ross McKitrick, Canadian university professors of applied mathematics and economics, respectively, attempt to prove wrong the popularly held assumption that scientists know what is happening with respect to climate and weather, and thus understand the phenomenon of global warming.

At the UN-sponsored Earth Summit in Rio de Janeiro in 1992, a Global Climate Treaty was signed. According to S. Fred Singer, in *Hot Talks, Cold Science: Global Warming's Unfinished Debate* (Independent Institute, 1998), the treaty rested on three basic assumptions. First, global warming has been detected in the records of climate of the last 100 years. Second, a substantial warming in the future will produce catastrophic consequences—droughts, floods, storms, a rapid and significant rise in sea level, agricultural collapse, and the spread of tropical disease. And third, the scientific and policy-making communities know: (1) which atmospheric concentrations of greenhouse gases are dangerous and which ones are not, (2) that drastic reductions of carbon dioxide (CO_2) emissions as well as energy use in general by industrialized countries will stabilize CO_2 concentrations at close to current levels, and (3) that such economically damaging measures can be justified politically despite no significant scientific support for global warming as a threat.

Since the Earth Summit, it appears that scientists have opted for placement into one of three camps. The first camp buys into the three assumptions outlined above. In late 1995, 2,500 leading climate scientists announced in the first Intergovernmental Panel on Climatic Change (IPCC) report that the planet was warming due to coal and gas emissions. Scientists in a second camp suggest that while global warming has occurred and continues at the present, the source of such temperature rise cannot be ascertained yet. The conclusions of the Earth Summit were misunderstood by many in the scientific community, the second camp would suggest. For these scientists, computer models, the basis of much evidence for the first group, have not yet linked global warming to human activities.

A third group of scientists, representing a minority, argues that we cannot be certain that global warming is taking place, yet alone determine its cause. They present a number of arguments in support of their position. Among them is the contention that pre-satellite data (pre-1979) showing a century-long pattern of warming is an illusion because satellite data (post-1979) reveal no such warming. Furthermore, when warming was present, it did not occur at the same time as a rise in greenhouse gases. Scientists in the third camp are also skeptical of studying global warming in the laboratory. They suggest, moreover, that most of the scientists who have opted for one of the first two camps have done so as a consequence of laboratory experiments, rather than of evidence from the real world.

Despite what appear to be wide differences in scientific thinking about the existence of global warming and its origins, the global community has moved forward with attempts to achieve consensus among the nations of the world for taking appropriate action to curtail human activities thought to affect warming. A 1997 international meeting in Kyoto, Japan, concluded with an agreement for reaching goals established at the earlier Earth Summit. Thirty-eight industrialized countries, including the United States, agreed to reduction levels outlined in the treaty. However, the U.S. Senate never ratified the treaty, and the Bush administration decided not to support it. Nonetheless, the two basic criteria for going into effect—the required number of countries (55) with the required levels of carbon dioxide's emissions (55 percent of carbon dioxide emissions from developed countries) must sign the treaty—were met when Russia ratified the treaty on November 18, 2004. The treaty went into effect on February 19, 2005.

The first selection is the recent report by the IPCC, an international body of scientists created by the United Nations to address the issue of global warming. It provides the most recent analysis by a broad scientific community on the twin issues of global warming's existence and causes. It predicts that global temperature will likely rise by 1.4 to 5.8°C by the year 2100. The second selection by Christopher Essex and Ross McKitrick suggests a different conclusion to the question posed in this issue. To them, important components of what they term the "doctrine" of global warming—the conventional view described above—are either unproven or simply wrong. In their judgment, the science used by advocates of global warming may be good enough for popular journals read by interested citizens but it has not yet made a strong enough case for global policymakers.

YES

Summary for Policymakers

This Summary for Policymakers (SPM), which was approved by IPCC [Intergovernmental Panel on Climate Change] member governments in Shanghai in January 2001,[1] describes the current state of understanding of the climate system and provides estimates of its projected future evolution and their uncertainties. Further details can be found in the underlying report, and the appended Source Information provides cross references to the report's chapters.

> An increasing body of observations gives a collective picture of a warming world and other changes in the climate system.

Since the release of the Second Assessment Report (SAR[2]), additional data from new studies of current and palaeoclimates, improved analysis of data sets, more rigorous evaluation of their quality, and comparisons among data from different sources have led to greater understanding of climate change.

> The global average surface temperature has increased over the 20th century by about 0.6°C.

- The global average surface temperature (the average of near surface air temperature over land, and sea surface temperature) has increased since 1861. Over the 20th century the increase has been $0.6 \pm 0.2°C$[3,4]. This value is about 0.15°C larger than that estimated by the SAR for the period up to 1994, owing to the relatively high temperatures of the additional years (1995 to 2000) and improved methods of processing the data. These numbers take into account various adjustments, including urban heat island effects. The record shows a great deal of variability; for example, most of the warming occurred during the 20th century, during two periods, 1910 to 1945 and 1976 to 2000.
- Globally, it is very likely[5] that the 1990s was the warmest decade and 1998 the warmest year in the instrumental record, since 1861.
- New analyses of proxy data for the Northern Hemisphere indicate that the increase in temperature in the 20th century is likely[5] to have been the largest of any century during the past 1,000 years. It is also likely[5] that, in the Northern Hemisphere, the 1990s was the warmest decade and 1998 the warmest year. Because less data are available, less is known about annual averages prior to 1,000 years before present and for conditions prevailing in most of the Southern Hemisphere prior to 1861.

From Intergovernmental Panel on Climate Change "Climate Change 2001: The Scientific Basis," A Report of Working Group I of the Intergovernmental Panel on Climate Change (2001).

- On average, between 1950 and 1993, night-time daily minimum air temperatures over land increased by about 0.2°C per decade. This is about twice the rate of increase in daytime daily maximum air temperatures (0.1°C per decade). This has lengthened the freeze-free season in many mid- and high latitude regions. The increase in sea surface temperature over this period is about half that of the mean land surface air temperature.

Temperatures have risen during the past four decades in the lowest 8 kilometres of the atmosphere.

- Since the late 1950s (the period of adequate observations from weather balloons), the overall global temperature increases in the lowest 8 kilometres of the atmosphere and in surface temperature have been similar at 0.1°C per decade.
- Since the start of the satellite record in 1979, both satellite and weather balloon measurements show that the global average temperature of the lowest 8 kilometres of the atmosphere has changed by +0.05 ± 0.10°C per decade, but the global average surface temperature has increased significantly by +0.15 ± 0.05°C per decade. The difference in the warming rates is statistically significant. This difference occurs primarily over the tropical and sub-tropical regions.
- The lowest 8 kilometres of the atmosphere and the surface are influenced differently by factors such as stratospheric ozone depletion, atmospheric aerosols, and the El Niño phenomenon. Hence, it is physically plausible to expect that over a short time period (e.g., 20 years) there may be differences in temperature trends. In addition, spatial sampling techniques can also explain some of the differences in trends, but these differences are not fully resolved.

Snow cover and ice extent have decreased.

- Satellite data show that there are very likely[5] to have been decreases of about 10% in the extent of snow cover since the late 1960s, and ground-based observations show that there is very likely[5] to have been a reduction of about two weeks in the annual duration of lake and river ice cover in the mid- and high latitudes of the Northern Hemisphere, over the 20th century.
- There has been a widespread retreat of mountain glaciers in non-polar regions during the 20th century.
- Northern Hemisphere spring and summer sea-ice extent has decreased by about 10 to 15% since the 1950s. It is likely[5] that there has been about a 40% decline in Arctic sea-ice thickness during late summer to early autumn in recent decades and a considerably slower decline in winter sea-ice thickness.

Global average sea level has risen and ocean heat content has increased.

- Tide gauge data show that global average sea level rose between 0.1 and 0.2 metres during the 20th century.

- Global ocean heat content has increased since the late 1950s, the period for which adequate observations of sub-surface ocean temperatures have been available.

Changes have also occurred in other important aspects of climate.

- It is very likely[5] that precipitation has increased by 0.5 to 1% per decade in the 20th century over most mid- and high latitudes of the Northern Hemisphere continents, and it is likely[5] that rainfall has increased by 0.2 to 0.3% per decade over the tropical (10°N to 10°S) land areas. Increases in the tropics are not evident over the past few decades. It is also likely[5] that rainfall has decreased over much of the Northern Hemisphere sub-tropical (10°N to 30°N) land areas during the 20th century by about 0.3% per decade. In contrast to the Northern Hemisphere, no comparable systematic changes have been detected in broad latitudinal averages over the Southern Hemisphere. There are insufficient data to establish trends in precipitation over the oceans.
- In the mid- and high latitudes of the Northern Hemisphere over the latter half of the 20th century, it is likely[5] that there has been a 2 to 4% increase in the frequency of heavy precipitation events. Increases in heavy precipitation events can arise from a number of causes, e.g., changes in atmospheric moisture, thunderstorm activity and largescale storm activity.
- It is likely[5] that there has been a 2% increase in cloud cover over mid- to high latitude land areas during the 20th century. In most areas the trends relate well to the observed decrease in daily temperature range.
- Since 1950 it is very likely[5] that there has been a reduction in the frequency of extreme low temperatures, with a smaller increase in the frequency of extreme high temperatures.
- Warm episodes of the El Niño-Southern Oscillation (ENSO) phenomenon (which consistently affects regional variations of precipitation and temperature over much of the tropics, sub-tropics and some mid-latitude areas) have been more frequent, persistent and intense since the mid-1970s, compared with the previous 100 years.
- Over the 20th century (1900 to 1995), there were relatively small increases in global land areas experiencing severe drought or severe wetness. In many regions, these changes are dominated by interdecadal and multi-decadal climate variability, such as the shift in ENSO towards more warm events.
- In some regions, such as parts of Asia and Africa, the frequency and intensity of droughts have been observed to increase in recent decades.

Some important aspects of climate appear not to have changed.

- A few areas of the globe have not warmed in recent decades, mainly over some parts of the Southern Hemisphere oceans and parts of Antarctica.
- No significant trends of Antarctic sea-ice extent are apparent since 1978, the period of reliable satellite measurements.

- Changes globally in tropical and extra-tropical storm intensity and frequency are dominated by inter-decadal to multi-decadal variations, with no significant trends evident over the 20th century. Conflicting analyses make it difficult to draw definitive conclusions about changes in storm activity, especially in the extra-tropics.
- No systematic changes in the frequency of tornadoes, thunder days, or hail events are evident in the limited areas analysed.

Emissions of greenhouse gases and aerosols due to human activities continue to alter the atmosphere in ways that are expected to affect the climate.

Changes in climate occur as a result of both internal variability within the climate system and external factors (both natural and anthropogenic). The influence of external factors on climate can be broadly compared using the concept of radiative forcing.[6] A positive radiative forcing, such as that produced by increasing concentrations of greenhouse gases, tends to warm the surface. A negative radiative forcing, which can arise from an increase in some types of aerosols (microscopic airborne particles) tends to cool the surface. Natural factors, such as changes in solar output or explosive volcanic activity, can also cause radiative forcing. Characterisation of these climate forcing agents and their changes over time is required to understand past climate changes in the context of natural variations and to project what climate changes could lie ahead. . . .

Concentrations of atmospheric greenhouse gases and their radiative forcing have continued to increase as a result of human activities.

- The atmospheric concentration of carbon dioxide (CO_2) has increased by 31% since 1750. The present CO_2 concentration has not been exceeded during the past 420,000 years and likely[5] not during the past 20 million years. The current rate of increase is unprecedented during at least the past 20,000 years.
- About three-quarters of the anthropogenic emissions of CO_2 to the atmosphere during the past 20 years is due to fossil fuel burning. The rest is predominantly due to land-use change, especially deforestation.
- Currently the ocean and the land together are taking up about half of the anthropogenic CO_2 emissions. On land, the uptake of anthropogenic CO_2 very likely[5] exceeded the release of CO_2 by deforestation during the 1990s.
- The rate of increase of atmospheric CO_2 concentration has been about 1.5 ppm[7] (0.4%) per year over the past two decades. During the 1990s the year to year increase varied from 0.9 ppm (0.2%) to 2.8 ppm (0.8%). A large part of this variability is due to the effect of climate variability (e.g., El Niño events) on CO_2 uptake and release by land and oceans.
- The atmospheric concentration of methane (CH_4) has increased by 1060 ppb[7] (151%) since 1750 and continues to increase. The present CH_4 concentration has not been exceeded during the past 420,000 years. The annual growth in CH_4 concentration slowed and became more variable in the 1990s, compared with the 1980s. Slightly more

than half of current CH_4 emissions are anthropogenic (e.g., use of fossil fuels, cattle, rice agriculture and landfills). In addition, carbon monoxide (CO) emissions have recently been identified as a cause of increasing CH_4 concentration.

- The atmospheric concentration of nitrous oxide (N_2O) has increased by 46 ppb (17%) since 1750 and continues to increase. The present N_2O concentration has not been exceeded during at least the past thousand years. About a third of current N_2O emissions are anthropogenic (e.g., agricultural soils, cattle feed lots and chemical industry).

- Since 1995, the atmospheric concentrations of many of those halocarbon gases that are both ozone-depleting and greenhouse gases (e.g., $CFCl_3$ and CF_2Cl_2), are either increasing more slowly or decreasing, both in response to reduced emissions under the regulations of the Montreal Protocol and its Amendments. Their substitute compounds (e.g., CHF_2Cl and CF_3CH_2F) and some other synthetic compounds (e.g., perfluorocarbons (PFCs) and sulphur hexafluoride (SF_6)) are also greenhouse gases, and their concentrations are currently increasing.

- The radiative forcing due to increases of the well-mixed greenhouse gases from 1750 to 2000 is estimated to be 2.43 Wm^{-2}: 1.46 Wm^{-2} from CO_2; 0.48 Wm^{-2} from CH_4; 0.34 Wm^{-2} from the halocarbons; and 0.15 Wm^{-2} from N_2O. . . .

- The observed depletion of the stratospheric ozone (O_3) layer from 1979 to 2000 is estimated to have caused a negative radiative forcing (–0.15 Wm^{-2}). Assuming full compliance with current halocarbon regulations, the positive forcing of the halocarbons will be reduced as will the magnitude of the negative forcing from stratospheric ozone depletion as the ozone layer recovers over the 21st century.

- The total amount of O_3 in the troposphere is estimated to have increased by 36% since 1750, due primarily to anthropogenic emissions of several O_3-forming gases. This corresponds to a positive radiative forcing of 0.35 Wm^{-2}. O_3 forcing varies considerably by region and responds much more quickly to changes in emissions than the long-lived greenhouse gases, such as CO_2.

Anthropogenic aerosols are short-lived and mostly produce negative radiative forcing.

- The major sources of anthropogenic aerosols are fossil fuel and biomass burning. These sources are also linked to degradation of air quality and acid deposition.

- Since the SAR, significant progress has been achieved in better characterising the direct radiative roles of different types of aerosols. Direct radiative forcing is estimated to be -0.4 Wm^{-2} for sulphate, -0.2 Wm^{-2} for biomass burning aerosols, -0.1 Wm^{-2} for fossil fuel organic carbon and +0.2 Wm^{-2} for fossil fuel black carbon aerosols. There is much less confidence in the ability to quantify the total aerosol direct effect, and its evolution over time, than that for the gases listed above. Aerosols also vary considerably by region and respond quickly to changes in emissions.

- In addition to their direct radiative forcing, aerosols have an indirect radiative forcing through their effects on clouds. There is now more evidence for this indirect effect, which is negative, although of very uncertain magnitude.

Natural factors have made small contributions to radiative forcing over the past century.

- The radiative forcing due to changes in solar irradiance for the period since 1750 is estimated to be about +0.3 Wm^{-2}, most of which occurred during the first half of the 20th century. Since the late 1970s, satellite instruments have observed small oscillations due to the 11-year solar cycle. Mechanisms for the amplification of solar effects on climate have been proposed, but currently lack a rigorous theoretical or observational basis.
- Stratospheric aerosols from explosive volcanic eruptions lead to negative forcing, which lasts a few years. Several major eruptions occurred in the periods 1880 to 1920 and 1960 to 1991.
- The combined change in radiative forcing of the two major natural factors (solar variation and volcanic aerosols) is estimated to be negative for the past two, and possibly the past four, decades. . . .

There is new and stronger evidence that most of the warming observed over the last 50 years is attributable to human activities.

The SAR concluded: "The balance of evidence suggests a discernible human influence on global climate." That report also noted that the anthropogenic signal was still emerging from the background of natural climate variability. Since the SAR, progress has been made in reducing uncertainty, particularly with respect to distinguishing and quantifying the magnitude of responses to different external influences. Although many of the sources of uncertainty identified in the SAR still remain to some degree, new evidence and improved understanding support an updated conclusion.

- There is a longer and more closely scrutinised temperature record and new model estimates of variability. The warming over the past 100 years is very unlikely[5] to be due to internal variability alone, as estimated by current models. Reconstructions of climate data for the past 1,000 years also indicate that this warming was unusual and is unlikely[5] to be entirely natural in origin.
- There are new estimates of the climate response to natural and anthropogenic forcing, and new detection techniques have been applied. Detection and attribution studies consistently find evidence for an anthropogenic signal in the climate record of the last 35 to 50 years.
- Simulations of the response to natural forcings alone (i.e., the response to variability in solar irradiance and volcanic eruptions) do not explain the warming in the second half of the 20th century. However, they indicate that natural forcings may have contributed to the observed warming in the first half of the 20th century.
- The warming over the last 50 years due to anthropogenic greenhouse gases can be identified despite uncertainties in forcing due to anthropogenic sulphate aerosol and natural factors (volcanoes and

solar irradiance). The anthropogenic sulphate aerosol forcing, while uncertain, is negative over this period and therefore cannot explain the warming. Changes in natural forcing during most of this period are also estimated to be negative and are unlikely[5] to explain the warming.

- Detection and attribution studies comparing model simulated changes with the observed record can now take into account uncertainty in the magnitude of modelled response to external forcing, in particular that due to uncertainty in climate sensitivity.
- Most of these studies find that, over the last 50 years, the estimated rate and magnitude of warming due to increasing concentrations of greenhouse gases alone are comparable with, or larger than, the observed warming. Furthermore, most model estimates that take into account both greenhouse gases and sulphate aerosols are consistent with observations over this period.
- The best agreement between model simulations and observations over the last 140 years has been found when all the above anthropogenic and natural forcing factors are combined. These results show that the forcings included are sufficient to explain the observed changes, but do not exclude the possibility that other forcings may also have contributed.

In the light of new evidence and taking into account the remaining uncertainties, most of the observed warming over the last 50 years is likely[5] to have been due to the increase in greenhouse gas concentrations.

Furthermore, it is very likely[5] that the 20th century warming has contributed significantly to the observed sea level rise, through thermal expansion of sea water and widespread loss of land ice. Within present uncertainties, observations and models are both consistent with a lack of significant acceleration of sea level rise during the 20th century.

Human influences will continue to change atmospheric composition throughout the 21st century.

Models have been used to make projections of atmospheric concentrations of greenhouse gases and aerosols, and hence of future climate, based upon emissions scenarios from the IPCC Special Report on Emission Scenarios (SRES). These scenarios were developed to update the IS92 series, which were used in the SAR and are shown for comparison here in some cases.

Greenhouse gases

- Emissions of CO_2 due to fossil fuel burning are virtually certain[5] to be the dominant influence on the trends in atmospheric CO_2 concentration during the 21st century.
- As the CO_2 concentration of the atmosphere increases, ocean and land will take up a decreasing fraction of anthropogenic CO_2 emissions. The net effect of land and ocean climate feedbacks as indicated by models is to further increase projected atmospheric CO_2 concentrations, by reducing both the ocean and land uptake of CO_2.

- By 2100, carbon cycle models project atmospheric CO_2 concentrations of 540 to 970 ppm for the illustrative SRES scenarios (90 to 250% above the concentration of 280 ppm in the year 1750). These projections include the land and ocean climate feedbacks. Uncertainties, especially about the magnitude of the climate feedback from the terrestrial biosphere, cause a variation of about –10 to +30% around each scenario. The total range is 490 to 1260 ppm (75 to 350% above the 1750 concentration).
- Changing land use could influence atmospheric CO_2 concentration. Hypothetically, if all of the carbon released by historical land-use changes could be restored to the terrestrial biosphere over the course of the century (e.g., by reforestation), CO_2 concentration would be reduced by 40 to 70 ppm.
- Model calculations of the concentrations of the non-CO_2 greenhouse gases by 2100 vary considerably across the SRES illustrative scenarios, with CH_4 changing by –190 to +1,970 ppb (present concentration 1,760 ppb), N_2O changing by +38 to +144 ppb (present concentration 316 ppb), total tropospheric O_3 changing by –12 to +62%, and a wide range of changes in concentrations of HFCs, PFCs and SF_6, all relative to the year 2000. In some scenarios, total tropospheric O_3 would become as important a radiative forcing agent as CH_4 and, over much of the Northern Hemisphere, would threaten the attainment of current air quality targets.
- Reductions in greenhouse gas emissions and the gases that control their concentration would be necessary to stabilise radiative forcing. For example, for the most important anthropogenic greenhouse gas, carbon cycle models indicate that stabilisation of atmospheric CO_2 concentrations at 450, 650 or 1,000 ppm would require global anthropogenic CO_2 emissions to drop below 1990 levels, within a few decades, about a century, or about two centuries, respectively, and continue to decrease steadily thereafter. Eventually CO_2 emissions would need to decline to a very small fraction of current emissions.

Aerosols

- The SRES scenarios include the possibility of either increases or decreases in anthropogenic aerosols (e.g., sulphate aerosols, biomass aerosols, black and organic carbon aerosols) depending on the extent of fossil fuel use and policies to abate polluting emissions. In addition, natural aerosols (e.g., sea salt, dust and emissions leading to the production of sulphate and carbon aerosols) are projected to increase as a result of changes in climate.

Radiative forcing over the 21st century

- For the SRES illustrative scenarios, relative to the year 2000, the global mean radiative forcing due to greenhouse gases continues to increase through the 21st century, with the fraction due to CO_2 projected to increase from slightly more than half to about three quarters. The change in the direct plus indirect aerosol radiative forcing is projected to be smaller in magnitude than that of CO_2.

Global average temperature and sea level are projected to rise under all IPCC SRES scenarios.

In order to make projections of future climate, models incorporate past, as well as future emissions of greenhouse gases and aerosols. Hence, they include estimates of warming to date and the commitment to future warming from past emissions.

Temperature

- The globally averaged surface temperature is projected to increase by 1.4 to 5.8°C over the period 1990 to 2100. These results are for the full range of 35 SRES scenarios, based on a number of climate models.[8,9]
- Temperature increases are projected to be greater than those in the SAR, which were about 1.0 to 3.5°C based on the six IS92 scenarios. The higher projected temperatures and the wider range are due primarily to the lower projected sulphur dioxide emissions in the SRES scenarios relative to the IS92 scenarios.
- The projected rate of warming is much larger than the observed changes during the 20th century and is very likely[5] to be without precedent during at least the last 10,000 years, based on palaeoclimate data.
- By 2100, the range in the surface temperature response across the group of climate models run with a given scenario is comparable to the range obtained from a single model run with the different SRES scenarios.
- On timescales of a few decades, the current observed rate of warming can be used to constrain the projected response to a given emissions scenario despite uncertainty in climate sensitivity. This approach suggests that anthropogenic warming is likely[5] to lie in the range of 0.1 to 0.2°C per decade over the next few decades under the IS92a scenario. . . .
- Based on recent global model simulations, it is very likely[5] that nearly all land areas will warm more rapidly than the global average, particularly those at northern high latitudes in the cold season. Most notable of these is the warming in the northern regions of North America, and northern and central Asia, which exceeds global mean warming in each model by more than 40%. In contrast, the warming is less than the global mean change in south and southeast Asia in summer and in southern South America in winter.
- Recent trends for surface temperature to become more El Niño-like in the tropical Pacific, with the eastern tropical Pacific warming more than the western tropical Pacific, with a corresponding eastward shift of precipitation, are projected to continue in many models.

Precipitation

- Based on global model simulations and for a wide range of scenarios, global average water vapour concentration and precipitation are projected to increase during the 21st century. By the second half of the 21st century, it is likely[5] that precipitation will have increased over northern mid- to high latitudes and Antarctica in winter. At low latitudes there are both regional increases and decreases over land areas. Larger year to year variations in precipitation are very likely[5] over most areas where an increase in mean precipitation is projected.

Notes

1. Delegations of 99 IPCC member countries participated in the Eighth Session of Working Group I in Shanghai on 17 to 20 January 2001.

2. The IPCC Second Assessment Report is referred to in this Summary for Policymakers as the SAR.

3. Generally temperature trends are rounded to the nearest 0.05°C per unit time, the periods often being limited by data availability.

4. In general, a 5% statistical significance level is used, and a 95% confidence level.

5. In this Summary for Policymakers and in the Technical Summary, the following words have been used where appropriate to indicate judgmental estimates of confidence: *virtually certain* (greater than 99% chance that a result is true); *very likely* (90–99% chance); *likely* (66–90% chance); *medium likelihood* (33–66% chance); *unlikely* (10–33% chance); *very unlikely* (1–10% chance); *exceptionally unlikely* (less than 1% chance). . . .

6. *Radiative forcing* is a measure of the influence a factor has in altering the balance of incoming and outgoing energy in the Earth-atmosphere system, and is an index of the importance of the factor as a potential climate change mechanism. It is expressed in Watts per square metre (Wm^{-2}).

7. ppm (parts per million) or ppb (parts per billion, 1 billion = 1,000 million) is the ratio of the number of greenhouse gas molecules to the total number of molecules of dry air. For example: 300 ppm means 300 molecules of a greenhouse gas per million molecules of dry air.

8. Complex physically based climate models are the main tool for projecting future climate change. In order to explore the full range of scenarios, these are complemented by simple climate models calibrated to yield an equivalent response in temperature and sea level to complex climate models. These projections are obtained using a simple climate model whose climate sensitivity and ocean heat uptake are calibrated to each of seven complex climate models. The climate sensitivity used in the simple model ranges from 1.7 to 4.2°C, which is comparable to the commonly accepted range of 1.5 to 4.5°C.

9. This range does not include uncertainties in the modelling of radiative forcing, e.g., aerosol forcing uncertainties. A small carbon-cycle climate feedback is included.

**Christopher Essex and
Ross McKitrick**

 **NO**

Taken by Storm: The Troubled Science, Policy and Politics of Global Warming

After Doctrine: Making Policy Amid Uncertainty and Nescience

After Doctrine

. . . [The Doctrine] it consists of the following familiar ideas.

1. The earth is warming.
2. Warming has already been observed.
3. Humans are causing it.
4. All but a handful of scientists on the fringe believe it.
5. Warming is bad.
6. Action is required immediately.
7. Any action is better than none.
8. Uncertainty only covers the ulterior motives of individuals aiming to stop needed action.
9. Those who defend uncertainty are bad people.

Let's go over these points one by one:

Is the Earth Warming?

. . . This cannot seem like a very sensible question. It sets up a simplistic context in which to view climate change. No direct answer to the question is true; yes and no are both wrong. Someone who skipped straight to this chapter might be puzzled, but it is really a simple concept in the end. There is no single physical variable that describes warming and cooling, for the whole Earth; there is single global temperature. This is just a basic truth of thermodynamics.

If there were some climate changes in the category of our sun going nova, or even something more moderate like a major ice age, then all of the infinity of local temperatures would be saying the same thing; namely, that it is heating or cooling everywhere and everywhen. But even the most strident and extreme scenarios put forward by the Panel of the United Nations (PUN) are Lilliputian in comparison. Contrary movements in the local temperature field everywhere would swamp any potential "signal."

The only way to force a simple warming or cooling picture onto a field of local temperatures is to say that, in some way, the places where and when it is cooling outweigh the places where and when it is warming. There is a problem doing this with temperature because there is no way to weight temperature at one point against temperature at another. Mathematicians would say it has no integration measure in space. It is an intensive thermodynamic variable, one of many quantities in thermodynamics that represents a condition and not an amount of something.

So if you want to take some average of the temperature field, it has very little physical meaning, unlike an average over energy, or the height of people in a classroom. Lots of—infinitely many—different averages over temperature are mathematically possible, but none of them has direct physical significance for the Earth. Any resulting average does not represent the temperature of anything at all. This is complicated by the huge variations in local temperatures up and down in comparison to what PUN claims to be looking for. You might get away with calling some average an index, but this huge variation virtually guarantees that different averages will behave quite differently from each other.

We have already seen that different published averages conducted over different pieces of the Earth's temperature field in different ways behave very differently from each other. This is to be expected. It would be remarkable if it were otherwise. Indeed, . . . a simple example of a temperature field in which two equally plausible mathematical averages over the same data from a temperature field showed opposite trends. It would not be hard to construct families of averages over the Earth's data that decline instead of the more popular trend in the opposite direction. This is not physics; it is mathematics.

Plain-vanilla warming or cooling is not how climate change actually works. There is a lot more going on than temperature in climate change. And that is why the big climate models are not just thermometers, but must also treat a huge spectrum of fiendishly complex dynamics as best they can.

Has Warming Been Observed Already?

Sure. You can find lots of places in the world where temperatures have gone up recently. You can find lots of places where temperatures have gone down, too. But what people have in mind here is that an "unnatural" warming has been observed here or there. To conclude that would require some idea of what the natural temperature in a location is, but there is no such single thing. Nor do we have any theory of the climate that would define natural levels or rates of change in temperatures in any location.

Are Humans Causing It?

Causing what? Causing someone's favourite global average temperature statistic (i.e., T-Rex) to go up, or causing climate change? The former doesn't make much sense, while the latter is more complicated. Commonly used methods that purport to show humans causing climate change rely on climate models, which, . . . are not suitable for this kind of purpose. And the absence of an underlying theory means that any statistical method, . . . must fall short because they compare changes in some arbitrarily defined statistic to an arbitrarily defined alternative.

Moreover, as the signal is generated from a turbulent and chaotic deterministic dynamical process, we ultimately must be looking for a signal on an unknown chaotic carrier. In chaos cryptography, classic statistical methods cannot crack an encrypted signal carried on a chaotic carrier. The only way to do so is to have the equations of the chaotic carrier to work with. No one knows how to get hold of them for climate.

Do All But a Handful of Scientists on the Fringe Believe It?

This is the sort of claim that governments have found very useful to spread around. It certainly makes their job easier, since they can avoid the tough business of having to think through a complicated issue. . . . The interaction of government and Official Science has led to great declarations of certainty, as well as political marginalization of those who try to voice their doubts.

So it is understandable that people think only a handful of "marginal" critics have doubts. The critics are often referred to as "skeptics." A skeptic is someone that true believers do not want to invite to a seance. Neither are they generally invited to Big Panel meetings, though the two events are not quite the same.

One of the ways people have tried to argue that there is more than a handful of critics is to circulate petitions and get scientists to publicly own up to their doubts. We have not argued along this line. Instead we have simply shown, page after page, that certainty on the subject of the future direction of climate is impossible. There is no theoretical basis for predicting climate; models do not provide a substitute for theory, and the profound complexity of such things as turbulence and chaotic systems means that anyone who thinks we can predict the climate only courts the laughter of the gods. So if this part of the Doctrine is correct, then the scientific community is in a sorry state.

The truth of a scientific fact does not depend upon a democratic vote among "qualified" people. But that is what governments, Official Science and advocacy groups have pushed. Moreover, there is no consensus in the scientific community anyway. The existence of the PUN report does not establish that at all. It is like a collection of short stories on many topics, with a small summary attached in which a small group of people offers its interpretation. What each scientist wrote was neither conclusive nor comprehensive in terms of the overall position set out in the SPAMs. There was no simple proposition that the authors of the scientific report endorsed, and many would not agree with many aspects of the SPAMs.

There was no consensus among scientists, who are always debating debatable issues. And climate change is debatable, to say the least! What genuine scientists all agree on is that science is a personal journey. It is not an exercise in authoritarianism in which edicts on the truth sever the world into patriots and dissidents. Such a picture invokes aspects of politics offensive to free and thinking people. It is not science.

Is It Bad? Should We Act Immediately?

Items 5 and 6 founder on the basic problem that if we cannot forecast something, we do not know whether it will be bad. Moreover, we cannot say it is bad or not if we aren't even sure what "it" is. Clearly we cannot say that "it" is all bad, and getting even more bad, as some seem to want to do. We cannot function rationally this way. It degenerates into nothing more than modern-day soothsaying without proper skepticism about the connections of bad events to climate change.

We may never even know if "it" is happening! As we discussed in Chapter 8, the genre of "impacts" studies is marked by an inordinate focus on temperature (often naively understood), a false belief that we can predict climate on a local scale far into the future, and a persistent inability to recognize the ingenuity with which people adapt to, and prosper under, changes of all kinds.

Is Any Action Better Than None?

Item 7 gets to the heart of thinking about policy. [We earlier] talked about factors like costs and benefits and the question of whether Kyoto is worth doing. We think not. And there are many other related policy ideas that have been floated over the past few years that make equally little sense. Consequently item 7 cannot be true. Not every action is worth the expense.

Are Critics of the Doctrine Bad People With Ulterior Motives?

As for items 8 and 9, you are on your own with those. If you think the present book has been offered to the public as part of a plot by bad people who are just covering ulterior and malicious motives, well, so be it. Nothing we say now will convince you otherwise. And as we talked about in Chapter 7, uncertainty is not really the problem. Nescience is the problem. The difference between uncertainty and nescience is like the difference between looking for your lost puppy in your neighbourhood and looking for a purple dinosaur in an undefined parallel dimension.

Once that distinction between nescience and uncertainty is understood, we are in a position to have a realistic discussion about policy.

What Should We Do About Global Warming?

If you have skipped directly to this page, go back to the beginning of the book. This section will make no sense to you unless you have read what came before. It is certainly not a "Summary for Policymakers." Start at page 1 and take the time to read what follows. After all, a big part of the answer to the question above is that we ought to stop relying on quick summaries and actually take the time to think through this issue carefully.

The question as posed assumes that we are not already "doing" something when we are studying the issue. But at this point, study is precisely what we ought to be doing. There is a lot of science on which we can draw when trying to understand the climate change issue. But bear in mind a distinction: There is science that is good enough for, say, *Science* magazine, and there is science that is good enough for making policy. They are not the same things, even though they have been muddled together in this problem and many others involving big issues in science.

Articles that are good enough to publish do not have to be right or rock-solid or magisterial. They just have to be interesting to other scientists and free of any obvious errors, in the opinion of the editor and reviewers. Their contribution toward knowledge can be no more than that. An article like the one in which the hockey stick diagram (Chapter 5) was drawn fits these criteria. But that does not make it a sensible basis on which to change society.

The fact that an article was published does not relieve readers of the obligation to look critically upon its assertions and ask if they are really true or not, or to understand what the limitations are on the conclusions. Very few conclusions in such an article are free of a great many limitations and most authors go to great trouble to lay them out. However, there are also limitations implicit to a whole field that are not laid out. A reader is just supposed to know. It can happen that even some within a field lose track of these things.

There is no shortcut for decisionmakers that avoids the need to think carefully about the decisions they make. People who propose costly policy changes like the Kyoto Protocol cannot shield themselves from the responsibility attached to making such proposals simply by citing a published article or a scholarly authority. The peer review process provides a service to the scientific community for the purposes internal to science. This is not the same as providing a service to society for the purpose of ensuring policy decisions are sound. While we hope the two processes are not fundamentally contradictory, they are not the same. Peer review does not absolve policymakers of the responsibility for making bad decisions.

Nor should we suppose that we have climate change basically figured out, and all that remains is to shrink some bounds of uncertainty. The problem is much deeper. On the core issues we are confronted by an absence of knowledge. Under the circumstances, it is impossible to recommend a detailed plan of action when we do not know if action is even needed, nor what its effects would be. Therefore the only sensible recommendation is not to take action except for further study.

But study alone will not amount to progress if it continues along the same lines as those organized by the Big Panel. The fact that the Doctrine emerged on their watch, and that it has generated such deep public confusion and professional malaise over the global warming issue, is surely sufficient cause for us to consider a different way of studying the topic.

We will shortly propose a new method for public study of global warming, based on a different way of thinking about how science and public policy should be related. It requires that we shed some illusions about the easy presumption of certainty on complex topics, and the danger of applying political models of authority to scientific endeavours.

But first we must finish with the question in the heading. While familiar, it is badly posed. An improvement would be: *What, if anything, should we do about the fact that some scientists think adding carbon dioxide to the atmosphere will cause a deleterious change in the climate system?* The answer is that we should take the concern seriously enough to try and clarify the issue by posing some additional questions. How would we find out if this is true? Would it matter? Can we do anything about it? And would we want to, even if we could?

As to the first, the problem is that in the absence of a theory of climate, we would have no way of knowing if human-caused climate change were taking place, even if it were going on right before our eyes. Nor could we look back in time and determine that it had happened and that it was responsible for particular changes at any particular location, as opposed to natural long-term changes. We cannot expect to identify human-caused climate change even after the fact!

What we do know is that prosperous and free societies are best able to deal with changes, including adverse ones, regardless of cause. Therefore we ought to encourage freedom and prosperity around the world. All but a handful of scientists on the fringe believe this.

On the second question, we do not know how this or that particular average of temperatures ought to behave in the absence of fossil fuel use, so whether T-Rex is rising or falling means nothing in terms of interpreting our influence on the world. And T-Rex only exists on paper anyway. We experience an infinite number of temperatures at every location on Earth, not an average cobbled together from white boxes and good intentions. If things are changing in the climate, we have to look at them directly in all their particularity and complexity, rather than trying to reduce the problem to the artificial simplicity of a one-dimensional character in a B-movie.

As to the third question, the climate is not a clock we can take apart and put back together or manipulate at will. Even if the world were as sensitive to CO_2 as the Big Panel has tried to argue, all the king's horses and all the king's men have not succeeded in coming up with anything better than the Kyoto Protocol, which, with its compromises at Bonn and prostrations at Marrakech, make it pointless with regard to any effect on atmospheric carbon dioxide concentrations. T-Rex will do what it was going to do despite it.

The world is full of surprises. "Surprises" was precisely the word used to describe it in Section 14.2.2 in PUN's scientific report. But too few actually read the actual scientific report. There the real scientists were able to send us a message in a bottle through the sea of politics to tell us that they too do not know. For all we know, we may yet be on the verge of an ice age, in which case we won't think global warming such a bad thing.

No one knows why some versions of T-Rex are rising and others are not. Temperatures rise and fall in response to changes in solar output, fluctuations in the earth's orbit and the lunar pull on tides, volcanic activity, long-term ocean dynamics and many other factors we do not understand, including every flap of a seagull's wings. And sometimes it rises and falls for no external reason at all, just because nonlinear, turbulent and chaotic dynamics can generate large sudden uncaused changes in the state of a system. That is what they said in 14.2.2 and that is what we have shown in this book. The presumption that every marked change has a cause is simply wrong. On Earth, these things are beyond our understanding, let alone our control.

As to the question of whether we would *want* to change the climate even if we could, we must look at the costs and benefits of small steps. The steps that involve carbon dioxide emission reductions (such as Kyoto) cost more than any benefits they are likely to generate. So they are not worth taking.

That said, some kinds of economic changes, driven by the internal comparison of costs and benefits, lead to emission reductions. For instance, there is a continuing drive toward energy efficiency in industrial economies. Every year we get a bit more efficient in how we use fuels, because it is economical to do so. But we cannot point to those actions and say that *therefore* all carbon dioxide emission reduction proposals are warranted. The question is not whether beneficial emission reductions are warranted: They will be undertaken with or without a push from government. The question is whether costly emission reductions are warranted, and the answer is no.

Even so, some people will rebel against this counsel. Over the past couple of years, as we have talked to people in and out of government, we keep hearing the message, *Yes but we have to do something.* Whether the rationale is the "precautionary principle" (which doesn't apply in this case) or some other dressing up of a simple gut reaction, there is a prevailing assumption out there that it is unacceptable to simply "do nothing."

At some point, we hit the wall in terms of what reason can accomplish. For those who really feel that way, despite all we have argued, we have run out of arguments. We could keep rephrasing what we have already written, but we won't. What is needed is for the reader to climb over the psychological hurdle of admitting that we should not burden society with carbon dioxide emission reduction requirements.

Regardless of how much effort and ingenuity has been expended developing the Kyoto Accord, it is the wrong thing to do. The right thing to do is to muddle along, focusing on basic priorities like economic development, wealth creation, education and the spread of freedom. We know these are good things to do. Policies like Kyoto do nobody any good and only take resources away from those real priorities.

So the best policy on global warming is to make sure science is free to investigate it, without having to prove constantly that this or that is relevant to policy issues. This has created a climate in the research community that has systematically pushed out the absolutely essential basic research that is all but disappearing because of it. Adequate funds must be made available to those who are studying it just because the research is good, not because it is producing fodder for political action.

Otherwise, the best policy is to do nothing unless future information indicates otherwise. This has been regarded as a political position. "More study" does not appeal to impatient political ideologues. But it is the only course that reflects a proper humility in the face of the enormity of the scientific problem we are facing. . . .

POSTSCRIPT

Is the Threat of Global Warming Real?

The issue of global warming is to this decade what acid rain was to the 1970s. Just as the blighted trees and polluted lakes of Scandinavia captured the hearts of the then newly emerging group of environmentalists, the issue of global warming has been front page news for over a decade and fodder for environmentalists and policymakers everywhere. Library citations abound, making it the most often written about global issue today. Web sites pop up, public interest groups emerge, and scientists and nonscientists pick up the rallying cry for one side or another. "Googling" the words "global warming" on the Internet yields 105 million responses.

In a sense, the issue of global warming is a prototype of the contemporary issue making its way onto the global agenda. Recall that a global issue is characterized by disagreement over the extent of the condition, disagreement over the causes of the condition, disagreement over desirable future alternatives to the present condition, and disagreement over appropriate policies to reach desired end states.

All of these characteristics are present in global warming. Both sides of the issue can find a substantial number of scientists, measured in the thousands, to support their case that the Earth is or is not warming. Both sides can find hundreds of experts who will attest that the warming is either a cyclical phenomenon or the consequence of human behavior. Both sides can find a substantial number of policymakers and policy observers who will say that the Kyoto Treaty is humankind's best hope to reverse the global warming trend or that the treaty is seriously flawed with substantial negative consequences for the United States. It is an issue whose debate heats up on occasion as the international community grapples with answers to the various disagreements summarized above. Finally, it is an issue whose potential solutions will impact different sectors of the economy and different countries differently.

The IPCC 2001 report concludes the following. First, the global average surface temperature has risen over the twentieth century by about 0.6 °C. Two, temperatures have risen during the past four decades in the lowest 8 kilometers of the atmosphere. Third, snow cover and ice extent have decreased. Fourth, global average sea level has risen, and ocean heat has increased. Fifth, changes have occurred in other important aspects of climate. Sixth, concentrations of atmospheric greenhouse gases have continued to increase as a result of human activity. Seventh, there is new and stronger evidence that most of the warming observed over the last 50 years is attributable to human activities. Eighth, human influences will continue to change atmospheric composition throughout the twenty-first century. Ninth, global average temperature and sea level are projected to rise under all IPCC scenarios. And tenth, climate change will persist for many centuries. The list is impressive.

Yet others argue the opposite position. In their comprehensive book on the subject, Essex and McKitrick lay out a series of nine statements that comprise the "Doctrine" of global warming. In their analysis of these components, they conclude that the evidence simply does not warrant the global community's undertaking policy making at this time to address a problem that may not exist or may have been caused by some other phenomena.

This view is a shared one. Brian Tucker's 1997 article in *The National Interest* ("Science Fiction: The Politics of Global Warming") suggests that science "does not support the conclusion that calamitous effects from global warming are nigh upon us." He continues: "There is no scientific justification for such a view." Tucker raises the stakes by asserting that the global warming controversy is much more than a debate about the causes and extent of the phenomenon. It is also about global development, power, and morality in the struggle between the rich and poor countries, with population "control" a central issue.

This view is echoed by John R. Christy in "The Global Warming Fiasco" (*Global Warming and Other Eco-Myths*, Ronald Bailey, Forum, 2002). He accuses those who suggest global warming is a major problem of adhering to the science of "calamitology" rather than the science of "climatology." His bottom line assessment is that "No global climate disaster is looming."

The Heartland Institute (www.heartland.org/studies/ieguide.htm) suggests "seven facts" to counteract observations such as those of the recent and earlier IPCC studies. First, "most scientists do not believe human activities threaten to disrupt the earth's climate." Second, "the most reliable temperature data show no global warming trend." Third, "global computer models are too crude to predict future climate changes." Fourth, the IPCC did not prove that human activities are causing global warming" (a reference to the 1995 study). Fifth, "a modest amount of global warming, should it occur, would be beneficial to the natural world and to human civilization." Sixth, "reducing our greenhouse gas emissions would be costly and would not stop global warming." And seventh, "the best strategy to pursue is one of 'no regrets'." The latter refers to the idea that it is it is not better to be safe than sorry (suggested by the other side), as immediate action will not make us safer, just poorer. Another sharp critique of the 2001 IPCC report is a Web-based book by Vincent Gray, *The Greenhouse Delusion: Critique of "Climate Change 2001": The Scientific Basis* (http://www.john-daly.com/tar-2000/summary.htm).

The above references capture the extreme distance between the two sides in the debate. Other sources are equally certain of their position. Ross Gelbson in *The Heat Is On: The High Stakes Battle Over Earth's Threatened Climate* (Addison-Wesley, 1997) examines "the campaign of deception by big coal and big oil" that is keeping the global warming issue off the public policy agenda.

Three other studies make valuable reading, each one of which takes a different position (one each at the extreme ends of the debate and a third one that suggests moderate climate change). S. Fred Singer's *Hot Talk; Cold Science: Global Warming's Unfinished Debate* (Independent Institute, 1998) enhances the author's reputation as one of the leading opponents of global warming's adverse consequences. At the other extreme, John Houghton in *Global Warming: The Complete Briefing,* 2d ed. (Cambridge University Press, 1997) accepts global warming as a significant concern and describes how it can be reversed in the future. S. George Philander, in *Is the Temperature Rising?* (1998), concludes that the global temperatures will rise 2°C over several decades, creating the prospect of some regional climate changes, with major consequences. Finally, Roy W. Spencer's "How Do We Know the Temperature of the Earth?" (Ronald Bailey, ed., *Earth Report 2000,* 2000), presents a basic argument with evidence that the popular perception of global warming as an environmental catastrophe cannot be supported with evidence. Finally, an objective analysis of the issue can be found in Chapter 5 ("Is the Earth Warming?") of Jack M. Hollander's *The Real Environmental Crisis* (University of California Press, 2003).

What are we to make of all of this? Simply put, whether or not global warming exists and is caused by human behavior, the issue will remain on the front page until agreement can be reached on these two fundamental questions.

ISSUE 9

Is the Threat of a Global Water Shortage Real?

YES: Mark W. Rosegrant, Ximing Cai, and Sarah A. Cline, from "Global Water Outlook to 2025: Averting an Impending Crisis," A Report of the International Food Policy Research Institute and the International Water Management Institute (September 2002)

NO: Bjørn Lomborg, from *The Skeptical Environmentalist: Measuring the Real State of the World* (Cambridge University Press, 2001)

ISSUE SUMMARY

YES: Rosegrant and colleagues conclude that if current water policies continue, farmers will find it difficult to grow sufficient food to meet the world's needs.

NO: Water is not only plentiful but is a renewable resource that, if properly treated as valuable, should not pose a future problem.

Water shortages and other water problems are occurring with greater frequency, particularly in large cities. Some observers have speculated that the situation is reminiscent of the fate that befell ancient glorious cities like Rome. Recognition that the supply of water is a growing problem is not new. As early as 1964, the United Nations Environmental Programme (UNEP) indicated that close to a billion people were at risk from desertification. At the Earth Summit in Rio in 1992, world leaders reaffirmed that desertification was of serious concern.

Moreover, in conference after conference and in study after study, increasing population growth and declining water supplies and quality are being linked together, as is the relationship between the planet's ability to meet its growing food needs and available water. Lester R. Brown, in "Water Deficits Growing in Many Countries: Water Shortages May Cause Food Shortages," *Eco-Economy Update 2002-11* (August 6, 2002), sums up the problem this way: "The world is incurring a vast water deficit. It is largely invisible, historically recent, and growing fast." The World Water Council's study, "World Water Actions Report, Third Draft" (October 31, 2002), describes the problem in much the same way: "Water is no longer taken for granted as a plentiful resource that will always be available when we need it." The report continues with the observation that increasing numbers of people in more and more countries are discovering that water is a "limited resource that must be managed for the benefit of people and the environment, in the present and in the future." In short, water is fast becoming both a food-related issue and a health-related problem. Some scholars are now arguing that water shortage is likely to become the twenty-first century's analog to the oil crisis of the last half of the previous century. The one major difference, as scholars are quick to point out, is that water is not like oil; there is no substitute.

Proclamations of impending water problems abound. Peter Gleick, in *The World's Water 1998-99: The Biennial Report on Freshwater Resources* (Island Press, 1998), reports that the demand for freshwater increased six-fold between 1900 and 1995, twice the rate of population growth. The UN study "United Nations Comprehensive Assessment of Freshwater Resources of the World" (1997) suggests that one-third of the world's population live in countries that have medium to high water stress. One 2001 headline reporting the release of a new study proclaimed that "Global thirst 'will turn millions into water refugees'" (The Millennium Environment Debate). News reports released by the UN Food and Agricultural Organization in conjunction with World Food Day 2002 asserted that water scarcity could result in millions of people having inadequate access to clean water or sufficient food. And the World Meteorological Organization predicts that two out of every three people will live in water-stressed conditions by 2050 if consumption patterns remain the same.

Sandra Postel, in *Pillar of Sand: Can the Irrigation Miracle Last?* (W.W. Norton, 1999), suggests another variant of the water problem. For her, the time-tested method of maximizing water usage in the past, irrigation, may not be feasible as world population marches toward seven billion. She points to the inadequacy of surface water supplies, increasing depletion of groundwater supplies, the salinization of the land, and the conversion of traditional agricultural land to other uses as reasons for the likely inability of irrigation to be a continuing panacea. Yet the 1997 UN study concluded that annual irrigation use would need to increase 30 percent for annual food production to double, necessary for meeting food demands of 2025.

The issue of water quality is also in the news. The World Health Organization reports that in some parts of the world, up to 80 percent of all transmittable diseases are attributable to the consumption of contaminated water. Also, a UNEP-sponsored study, *Global Environment Outlook 2000*, reported that 200 scientists from 50 countries pointed to the shortage of clean water as one of the most pressing global issues.

In the following selection, Mark W. Rosegrant, Ximing Cai, and Sarah A. Cline project that by 2025, water scarcity will result in annual global losses of 350 million metric tons, equivalent to approximately the entire U.S. crop. In the second selection, Bjørn Lomborg takes issue with the prevailing wind in the global water debate. His argument can be summed up in his simple quote: "Basically we have sufficient water." Lomborg maintains that water supplies rose during the twentieth century and that we have gained access to more water through technology.

YES

Mark W. Rosegrant, Ximing Cai,
and Sarah A. Cline

Global Water Outlook to 2025:
Averting an Impending Crisis

Introduction

Demand for the world's increasingly scarce water supply is rising rapidly, challenging its availability for food production and putting global food security at risk. Agriculture, upon which a burgeoning population depends for food, is competing with industrial, household, and environmental uses for this scarce water supply. Even as demand for water by all users grows, groundwater is being depleted, other water ecosystems are becoming polluted and degraded, and developing new sources of water is getting more costly.

A Thirsty World

Water development underpins food security, people's livelihoods, industrial growth, and environmental sustainability throughout the world. In 1995 the world withdrew 3,906 cubic kilometers (km^3) of water for these purposes (Figure 1). By 2025 water withdrawal for most uses (domestic, industrial, and livestock) is projected to increase by at least 50 percent. This will severely limit irrigation water withdrawal, which will increase by only 4 percent, constraining food production in turn.

About 250 million hectares are irrigated worldwide today, nearly five times more than at the beginning of the 20th century. Irrigation has helped boost agricultural yields and outputs and stabilize food production and prices. But growth in population and income will only increase the demand for irrigation water to meet food production requirements (Figure 2). Although the achievements of irrigation have been impressive, in many regions poor irrigation management has markedly lowered groundwater tables, damaged soils, and reduced water quality.

Water is also essential for drinking and household uses and for industrial production. Access to safe drinking water and sanitation is critical to maintain health, particularly for children. But more than 1 billion people across the globe lack enough safe water to meet minimum levels of health and income. Although the domestic and industrial sectors use far less water than agriculture, the growth in water consumption in these sectors has been rapid. Globally, withdrawals for domestic and industrial uses

From Mark W. Rosegrant, Ximing Cai, and Sarah A. Cline, "Global Water Outlook to 2025: Averting an Impending Crisis," A Report of the International Food Policy Research Institute and the International Water Management Institute (September 2002). Copyright © 2002 by The International Food Policy Research Institute.

Figure 1

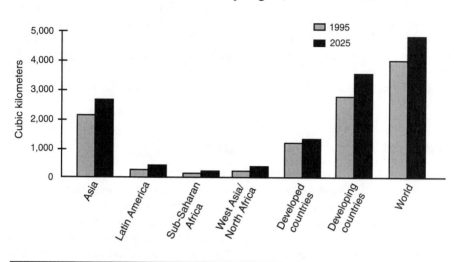

Total Water Withdrawal by Region, 1995 and 2025

Source: Authors' estimates and IMPACT-WATER projections, June 2002.

Note: Projections for 2025 are for the business as usual scenario.

Figure 2

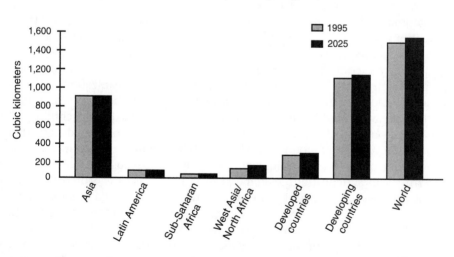

Total Irrigation Water Consumption by Region, 1995 and 2025

Source: Authors' estimates and IMPACT-WATER projections, June 2002.

Note: Projections for 2025 are for the business as usual scenario.

quadrupled between 1950 and 1995, compared with agricultural uses, for which withdrawals slightly more than doubled.[1]

Water is integrally linked to the health of the environment. Water is vital to the survival of ecosystems and the plants and animals that live in them, and in turn ecosystems help to regulate the quantity and quality of water. Wetlands retain water during high rainfall, release it during dry periods, and purify it of many contaminants. Forests reduce erosion and sedimentation of rivers and recharge groundwater. The importance of reserving water for environmental purposes has only recently been recognized: during the 20th century, more than half of the world's wetlands were lost.[2]

Alternative Futures for Water

The future of water and food is highly uncertain. Some of this uncertainty is due to relatively uncontrollable factors such as weather. But other critical factors can be influenced by the choices made collectively by the world's people. These factors include income and population growth, investment in water infrastructure, allocation of water to various uses, reform in water management, and technological changes in agriculture. Policy decisions—and the actions of billions of individuals—determine these fundamental, long-term drivers of water and food supply and demand.

To show the very different outcomes that policy choices produce, we present three alternative futures for global water.[3] . . .

Business As Usual Scenario

In the business as usual scenario current trends in water and food policy, management, and investment remain as they are. International donors and national governments, complacent about agriculture and irrigation, cut their investments in these sectors. Governments and water users implement institutional and management reforms in a limited and piecemeal fashion. These conditions leave the world ill prepared to meet major challenges to the water and food sectors.

Over the coming decades the area of land devoted to cultivating food crops will grow slowly in most of the world because of urbanization, soil degradation, and slow growth in irrigation investment, and because a high proportion of arable land is already cultivated. Moreover, steady or declining real prices for cereals will make it unprofitable for farmers to expand harvested area. As a result, greater food production will depend primarily on increases in yield. Yet growth in crop yields will also diminish because of falling public investment in agricultural research and rural infrastructure. Moreover, many of the actions that produced yield gains in recent decades, such as increasing the density of crop planting, introducing strains that are more responsive to fertilizer, and improving management practices, cannot easily be repeated.

In the water sector, the management of river basin and irrigation water will become more efficient, but slowly. Governments will continue to transfer management of irrigation systems to farmer organizations and water-user associations. Such transfers will increase water efficiency when they are built upon existing patterns of cooperation and backed by a supportive policy and legal environment. But these conditions are often lacking.

In some regions farmers will adopt more efficient irrigation practices. Economic incentives to induce more efficient water management, however, will still face political opposition from those concerned about the impact of higher water prices on farmers' income and from entrenched interests that benefit from existing systems of allocating water. Water management will also improve slowly in rainfed agriculture as a result of small advances in water harvesting, better on-farm management techniques, and the development of crop varieties with shorter growing seasons.

Public investment in expanding irrigation systems and reservoir storage will decline as the financial, environmental, and social costs of building new irrigation systems escalate and the prices of cereals and other irrigated crops drop. Nevertheless, where benefits outweigh costs, many governments will construct dams, and reservoir water for irrigation will increase moderately.

With slow growth in irrigation from surface water, farmers will expand pumping from groundwater, which is subject to low prices and little regulation. Regions that currently pump groundwater faster than aquifers can recharge, such as the western United States, northern China, northern and western India, Egypt, and West Asia and North Africa, will continue to do so.

The cost of supplying water to domestic and industrial users will rise dramatically. Better delivery and more efficient home water use will lead to some increase in the proportion of households connected to piped water. Many households, however, will remain unconnected. Small price increases for industrial water, improvements in pollution control regulation and enforcement, and new industrial technologies will cut industrial water use intensity (water demand per $1,000 of gross domestic product). Yet industrial water prices will remain relatively low and pollution regulations will often be poorly enforced. Thus, significant potential gains will be lost.

Environmental and other interest groups will press to increase the amount of water allocated to preserving wetlands, diluting pollutants, maintaining riparian flora and other aquatic species, and supporting tourism and recreation. Yet because of competition for water for other uses, the share of water devoted to environmental uses will not increase.

The Water Situation Almost all users will place heavy demands on the world's water supply under the business as usual scenario. Total global water withdrawals in 2025 are projected to increase by 22 percent above 1995 withdrawals, to 4,772 km^3 (see Figure 1).[4] Projected withdrawals in developing countries will increase 27 percent over the 30-year period, while developed-country withdrawals will increase by 11 percent.[5]

Together, consumption of water for domestic, industrial, and livestock uses—that is, all nonirrigation uses—will increase dramatically, rising by 62 percent from 1995 to 2025. Because of rapid population growth and rising per capita water use, total domestic consumption will increase by 71 percent, of which more than 90 percent will be in developing countries. Conservation and technological improvements will lower per capita domestic water use in developed countries with the highest per capita water consumption.

Industrial water use will grow much faster in developing countries than in developed countries. In 1995 industries in developed countries consumed much more water than industries in the developing world. By 2025, however, developing-world industrial water demand is projected to increase to 121 km^3, 7 km^3 greater than in the developed world. The intensity of industrial water use will decrease worldwide, especially in developing countries (where initial intensity levels are very high), thanks to improvements in water-saving technology and demand policy. Nonetheless, the sheer size of the increase in the world's industrial production will still lead to an increase in total industrial water demand.

Direct water consumption by livestock is very small compared with other sectors. But the rapid increase of livestock production, particularly in developing countries, means that livestock water demand is projected to increase 71 percent between 1995 and 2025. Whereas livestock water demand will increase only 19 percent in the developed world between 1995 and 2025, it is projected to more than double in the developing world, from 22 to 45 km^3.

Although irrigation is by far the largest user of the world's water, use of irrigation water is projected to rise much more slowly than other sectors. For irrigation water, we have computed both potential demand and actual consumption. Potential demand is the demand for irrigation water in the absence of any water supply constraints, whereas actual consumption of irrigation water is the realized water demand, given the limitations of water supply for irrigation. The proportion of potential demand that is realized in actual consumption is the irrigation water supply reliability index (IWSR).[6] An IWSR of 1.0 would mean that all potential demand is being met.

Potential irrigation demand will grow by 12 percent in developing countries, while it will actually decline in developed countries by 1.5 percent. The fastest growth in potential demand for irrigation water will occur in Sub-Saharan Africa, with an increase of 27 percent, and in Latin America, with an increase of 21 percent. Each of these regions has a high percentage increase in irrigated area from a relatively low 1995 level. India is projected to have the highest absolute growth in potential irrigation water demand, 66 km^3 (17 percent), owing to relatively rapid growth in irrigated area from an already high level in 1995. West Asia and North Africa will increase by 18 percent (28 km^3, mainly in Turkey), while China will experience a much smaller increase of 4 percent (12 km^3). In Asia as a region, potential irrigation water demand will increase by 8 percent (100 km^3).

Water scarcity for irrigation will intensify, with actual consumption of irrigation water worldwide projected to grow more slowly than potential consumption, increasing only 4 percent between 1995 and 2025. In developing countries a declining fraction of potential demand will be met over time. The IWSR for developing countries will decline from 0.81 in 1995 to 0.75 in 2025, and in dry river basins the decline will be steeper. For example, in the Haihe River Basin in China, which is an important wheat and maize producer and serves major metropolitan areas, the IWSR is projected to decline from 0.78 to 0.62, and in the Ganges of India, the IWSR will decline from 0.83 to 0.67.

In the developed world, the situation is the reverse: the supply of irrigation water is projected to grow faster than potential demand (although certain basins will face increasing water scarcity). Increases in river basin efficiency will more than offset the very small increase in irrigated area. As a result, after initially declining from 0.87 to 0.85 in 2010, the IWSR will improve to 0.90 in 2025 thanks to slowing growth of domestic and industrial demand (and actual declines in total domestic and industrial water use in the United States and Europe) and more efficient use of irrigation water. . . .

Water Crisis Scenario

A moderate worsening of many of the current trends in water and food policy and in investment could build to a genuine water crisis. In the water crisis scenario, government budget problems worsen. Governments further cut their spending on irrigation systems and accelerate the turnover of irrigation systems to farmers and farmer groups but without the necessary reforms in water rights. Attempts to fund operations and maintenance in the main water system, still operated by public agencies, cause water prices to irrigators to rise. Water users fight price increases, and conflict spills over to local management and cost-sharing arrangements. Spending on the operation and maintenance of secondary and tertiary systems falls dramatically, and deteriorating infrastructure and poor management lead to falling water use efficiency. Likewise, attempts to organize river basin organizations to coordinate water management fail because of inadequate funding and high levels of conflict among water stakeholders within the basin.

National governments and international donors will reduce their investments in crop breeding for rainfed agriculture in developing countries, especially for staple crops such as rice, wheat, maize, other coarse grains, potatoes, cassava, yams, and sweet potatoes. Private agricultural research will fail to fill the investment gap for these commodities. This loss of research funding will lead to further declines in productivity growth in rainfed crop areas, particularly in more marginal areas. In search of improved incomes, people will turn to slash-and-burn agriculture, thereby deforesting the upper watersheds of many basins. Erosion and sediment loads in rivers will rise, in turn causing faster sedimentation of reservoir storage. People will increasingly encroach on wetlands for both land and water, and the integrity and health of aquatic ecosystems will be compromised. The amount of water reserved for environmental purposes will decline as unregulated and illegal withdrawals increase.

The cost of building new dams will soar, discouraging new investment in many proposed dam sites. At other sites indigenous groups and nongovernmental organizations (NGOs) will mount opposition, often violent, over the environmental and human impacts of new dams. These protests and high costs will virtually halt new investment in medium and large dams and storage reservoirs. Net reservoir storage will decline in developing countries and remain constant in developed countries.

In the attempt to get enough water to grow their crops, farmers will extract increasing amounts of groundwater for several years, driving down water tables. But because of the accelerated pumping, after 2010 key aquifers in northern China, northern and northwestern India, and West Asia and North Africa will begin to fail. With declining water tables, farmers will find the cost of extracting water too high, and a big drop in groundwater extraction from these regions will further reduce water availability for all uses.

As in the business as usual scenario, the rapid increase in urban populations will quickly raise demand for domestic water. But governments will lack the funds to extend piped water and sewage disposal to newcomers. Governments will respond by privatizing urban water and sanitation services in a rushed and poorly planned fashion. The new private water and sanitation firms will be undercapitalized and able to do little to connect additional populations to piped water. An increasing number and percentage of the urban population must rely on high-priced water from vendors or spend many hours fetching often-dirty water from standpipes and wells.

The Water Situation   The developing world will pay the highest price for the water crisis scenario. Total worldwide water consumption in 2025 will be 261 km^3 higher than under the business as usual scenario—a 13 percent increase—but much of this water will be wasted, of no benefit to anyone. Virtually all of the increase will go to irrigation, mainly because farmers will use water less efficiently and withdraw more water to compensate for water losses. The supply of irrigation water will be less reliable, except in regions where so much water is diverted from environmental uses to irrigation that it compensates for the lower water use efficiency.

For most regions, per capita demand for domestic water will be significantly lower than under the business as usual scenario, in both rural and urban areas. The result is that people will not have access to the water they need for drinking and sanitation. The total domestic demand under the water crisis scenario will be 162 km^3 in developing countries, 28 percent less than under business as usual; 64 km^3 in developed countries, 7 percent less than under business as usual; and 226 km^3 in the world, 23 percent less than under business as usual.

Demand for industrial water, on the other hand, will increase, owing to failed technological improvements and economic measures. In 2025 the total industrial water demand worldwide will be 80 km^3 higher than under the business as usual scenario—a 33 percent rise—without generating additional industrial production.

With water diverted to make up for less efficient water use, the water crisis scenario will hit environmental uses particularly hard. Compared with business as usual, environmental flows will drop significantly by 2025, with 380 km^3 less environmental flow in the developing world, 80 km^3 less in the developed world, and 460 km^3 less globally. . . .

Sustainable Water Scenario

A sustainable water scenario would dramatically increase the amount of water allocated to environmental uses, connect all urban households to piped water, and achieve higher per capita domestic water consumption, while maintaining food production at the levels described in the business as usual scenario. It would achieve greater social equity and environmental protection through both careful reform in the water sector and sound government action.

Governments and international donors will increase their investments in crop research, technological change, and reform of water management to boost water productivity and the growth of crop yields in rainfed agriculture. Accumulating evidence shows that even drought-prone and high-temperature rainfed environments have the potential for dramatic increases in yield. Breeding strategies will directly target these rainfed areas. Improved policies and increased investment in rural infrastructure will help link remote farmers to markets and reduce the risks of rainfed farming.

To stimulate water conservation and free up agricultural water for environmental, domestic, and industrial uses, the effective price of water to the agricultural sector will be gradually increased. Agricultural water price increases will be implemented through incentive programs that provide farmers income for the water that they save, such as charge-subsidy schemes that pay farmers for reducing water use, and through the establishment, purchase, and trading of water use rights. By 2025 agricultural water prices will be twice as high in developed countries and three times as high in developing countries as in the business as usual scenario. The government will simultaneously transfer water rights and the responsibility for operation and management of irrigation systems to communities and water user associations in many countries and regions. The transfer of rights and systems will be facilitated with an improved legal and institutional environment for preventing and eliminating conflict and with technical and organizational training and support. As a result, farmers will increase their on-farm investments in irrigation and water management technology, and the efficiency of irrigation systems and basin water use will improve significantly.

River basin organizations will be established in many water-scarce basins to allocate mainstream water among stakeholder interests. Higher funding and reduced conflict over water, thanks to better water management, will facilitate effective stakeholder participation in these organizations.

Farmers will be able to make more effective use of rainfall in crop production, thanks to breakthroughs in water harvesting systems and the adoption of advanced farming techniques, like precision agriculture, contour plowing, precision land leveling, and minimum-till and no-till technologies. These technologies will increase the share of rainfall that goes to infiltration and evapotranspiration.

Spurred by the rapidly escalating costs of building new dams and the increasingly apparent environmental and human resettlement costs, developing and developed countries will reassess their reservoir construction plans, with comprehensive analysis of the costs and benefits, including environmental and social effects, of proposed projects. As a result, many planned storage projects will be canceled, but others will proceed with support from civil society groups. Yet new storage capacity will be less necessary because rapid growth in rainfed crop yields will help reduce rates of reservoir sedimentation from erosion due to slash-and-burn cultivation.

Policy toward groundwater extraction will change significantly. Market-based approaches will assign rights to groundwater based on both annual withdrawals and the renewable stock of groundwater. This step will be combined with stricter regulations and better enforcement of these regulations. Groundwater overdrafts will be phased out in countries and regions that previously pumped groundwater unsustainably.

Domestic and industrial water use will also be subject to reforms in pricing and regulation. Water prices for connected households will double, with targeted subsidies for low-income households. Revenues from price increases will be invested to reduce water losses in existing systems and to extend piped water to previously unconnected households. By 2025 all households will be connected. Industries will respond to higher prices, particularly in developing countries, by increasing in-plant recycling of water, which reduces consumption of water.

With strong societal pressure for improved environmental quality, allocations for environmental uses of water will increase. Moreover, the reforms in agricultural and nonagricultural water sectors will reduce pressure on wetlands and other environmental uses of water. Greater investments and better water management will improve the efficiency of water use, leaving more water instream for environmental purposes. All reductions in domestic and urban water use, due to higher water prices, will be allocated to instream environmental uses.

The Water Situation   In the sustainable water scenario the world consumes less water but reaps greater benefits than under business as usual, especially in developing countries. In 2025 total worldwide water consumption is 408 km^3, or 20 percent, lower under the sustainable scenario than under business as usual. This reduction in consumption frees up water for environmental uses. Higher water prices and higher water use efficiency reduces consumption of irrigation water by 296 km^3 compared with business as usual. The reliability of irrigation water supply is reduced slightly in the sustainable scenario compared with business as usual, because this scenario places a high priority on environmental flows. Over time, however, more efficient water use in this scenario counterbalances the transfer of water to the environment and results in an improvement in the reliability of supply of irrigation water by 2025.

This scenario will improve the domestic water supply through universal access to piped water for rural and urban households. Globally, potential domestic water demand under the sustainable water scenario will decrease 9 percent compared with business as usual, owing to higher water prices. However, potential per capita domestic demand for connected households in rural areas will be 12 percent higher than that under business as usual in the developing world, and 5 percent higher in the developed world. This increase is accomplished by expanding universal access to piped water in rural areas even with higher prices for water. And in urban areas, potential per capita water consumption for poor households sharply improves through connection to piped water, while the initially connected households reduce consumption in response to higher prices and improved water-saving technology.

Through technological improvements and effective economic incentives, the sustainable water scenario will reduce industrial water demand. In 2025 total industrial water demand worldwide under the sustainable scenario will be 85 km^3, or 35 percent, lower than under business as usual.

The environment is a major beneficiary of the sustainable water scenario, with large increases in the amount of water reserved for wetlands, instream flows, and other environmental purposes. Compared with the business as usual scenario, the sustainable scenario will also result in an increase in the environmental flow of 850 km^3 in the developing world, 180 km^3 in the developed world, and 1,030 km^3 globally. This is the equivalent of transferring 22 percent of global water withdrawals under business as usual to environmental purposes.

Notes

1. W. J. Cosgrove and F. Rijsberman, *World Water Vision: Making Water Everybody's Business* (London: World Water Council and World Water Vision and Earthscan, 2000); I. A. Shiklomanov, "Electronic Data Provided to the Scenario Development Panel, World Commission on Water for the 21st Century" (State Hydrological Institute, St. Petersburg, Russia, 1999), mimeo.

2. E. Bos and G. Bergkamp, "Water and the Environment," in *Overcoming Water Scarcity and Quality Constraints,* 2020 Focus 9, ed. R. S. Meinzen-Dick and M. W. Rosegrant (Washington, D.C.: International Food Policy Research Institute, 2001).

3. The business as usual, crisis, and sustainable scenarios are compared using average 2025 results generated from 30 hydrologic scenarios. The other scenarios are compared with business as usual based on a single 30-year hydrologic sequence drawn from 1961–90, and results are shown as the average of the years 2021–25.

4. Water demand can be defined and measured in terms of withdrawals and actual consumption. While water withdrawal is the most commonly estimated figure, consumption best captures actual water use, and most of our analysis will utilize this concept.

5. The global projection is broadly consistent with other recent projections to 2025, including the 4,580 km^3 in the medium scenario of J. Alcamo, P. Döll, F. Kaspar, and S. Sieberg, *Global Change and Global Scenarios of Water Use and Availability: An Application of Water GAP 1.0* (Kassel, Germany: Center for Environmental System Research, University of Kassel, 1998), the 4,569 km^3 in the "business-as-usual" scenario of D. Seckler, U. Amarasinghe, D. Molden, S. Rhadika, and R. Barker, *World Water Demand and Supply, 1990 to 2025: Scenarios and Issues,* Research Report Number 19 (Colombo, Sri Lanka: International Water Management Institute, 1998), and the forecast of 4,966 km^3 (not including reservoir evaporation) of Shiklomanov, "Electronic Data."

6. Compared with other sectors, the growth of irrigation water potential demand is much lower, with 12 percent growth in potential demand between 1995 and 2025 in developing countries and a slight decline in potential demand in developed countries.

Bjørn Lomborg

Water

There is a resource which we often take for granted but which increasingly has been touted as a harbinger of future trouble. Water.

Ever more people live on Earth and they use ever more water. Our water consumption has almost quadrupled since 1940. The obvious argument runs that "this cannot go on." This has caused government agencies to worry that "a threatening water crisis awaits just around the corner." The UN environmental report *GEO 2000* claims that the water shortage constitutes a "full-scale emergency," where "the world water cycle seems unlikely to be able to cope with the demands that will be made of it in the coming decades. Severe water shortages already hamper development in many parts of the world, and the situation is deteriorating."

The same basic argument is invoked when WWF [World Wildlife Fund] states that "freshwater is essential to human health, agriculture, industry, and natural ecosystems, but is now running scarce in many regions of the world." *Population Reports* states unequivocally that "freshwater is emerging as one of the most critical natural resource issues facing humanity." Environmental discussions are replete with buzz words like "water crisis" and "time bomb: water shortages," and *Time* magazine summarizes the global water outlook with the title "Wells running dry." The UN organizations for meteorology and education simply refer to the problem as "a world running out of water."

The water shortages are also supposed to increase the likelihood of conflicts over the last drops—and scores of articles are written about the coming "water wars." Worldwatch Institute sums up the worries nicely, claiming that "water scarcity may be to the nineties what the oil price shocks were to the seventies—a source of international conflicts and major shifts in national economies."

But these headlines are misleading. True, there may be *regional* and *logistic* problems with water. We will need to get better at using it. But basically we have sufficient water.

How Much Water in the World?

Water is absolutely decisive for human survival, and the Earth is called the Blue Planet precisely because most of it is covered by water: 71 percent of the Earth's surface is covered by water, and the total amount is estimated at the unfathomably large 13.6 billion cubic kilometers. Of all this water, oceans make up 97.2 percent and the polar ice contains 2.15 percent. Unfortunately sea water is too saline for direct human consumption, and while polar ice contains potable water it is hardly within easy reach. Consequently, humans are primarily dependent on the last 0.65 percent water, of which 0.62 percent is groundwater.

Fresh water in the groundwater often takes centuries or millennia to build up—it has been estimated that it would require 150 years to recharge all of the groundwater in the United States totally to a depth of 750 meters if it were all removed. Thus, thoughtlessly exploiting the groundwater could be compared to mining any other non-renewable natural resource. But groundwater is continuously replenished by the constant movement of water through oceans, air, soil, rivers, and lakes, in the so-called hydrological cycle. The sun makes water from the oceans evaporate, the wind moves parts of the vapor as clouds over land, where the water is released as rain and snow. The precipitated water then either evaporates again, flows back into the sea through rivers and lakes, or finds its way into the groundwater.

The total amount of precipitation on land is about 113,000 km^3, and taking into account an evaporation of 72,000 km^3 we are left with a net fresh water influx of 41,000 km^3 each year or the equivalent of 30 cm (1 foot) of water across the entire land mass. Since part of this water falls in rather remote areas, such as the basins of the Amazon, the Congo, and the remote North American and Eurasian rivers, a more reasonable, geographically accessible estimate of water is 32,900 km^3. Moreover, a large part of this water comes within short periods of time. In Asia, typically 80 percent of the runoff occurs from May to October, and globally the flood runoff is estimated at about three-quarters of the total runoff. This leaves about 9,000 km^3 to be captured. Dams capture an additional 3,500 km^3 from floods, bringing the total accessible runoff to 12,500 km^3. This is equivalent to about 5,700 liters of water for every single person on Earth *every single day.* For comparison, the average citizen in the EU uses about 566 liters of water per day. This is about 10 percent of the global level of available water and some 5 percent of the available EU water. An American, however, uses about three times as much water, or 1,442 liters every day.

Looking at global water consumption, as seen in Figure 1, it is important to distinguish between water withdrawal and water use. Water withdrawal is the amount of water physically removed, but this concept is less useful in a discussion of limits on the total amount of water, since much of the withdrawn water is later returned to the water cycle. In the EU and the US, about 46 percent of the withdrawn water is used merely as cooling water for power generation and is immediately released for further use downstream. Likewise, most industrial uses return 80–90 percent of the water, and even in irrigation 30–70 percent of the water runs back into lakes and rivers or percolates into aquifers, whence it can be reused. Thus, a more useful measure of water consumption is the amount of water this consumption causes to be irretrievably lost through evaporation or transpiration from plants. This is called water use.

Over the twentieth century, Earth's water use has grown from about 330 km^3 to about 2,100 km^3. As can be seen from Figure 1 there is some uncertainty about the future use and withdrawal (mainly depending on the development of irrigation),

Figure 1

Global, Annual Water Withdrawal and Use, in Thousand km³ and Percentage of Accessible Runoff, 1900–95, and Predictions for 2025

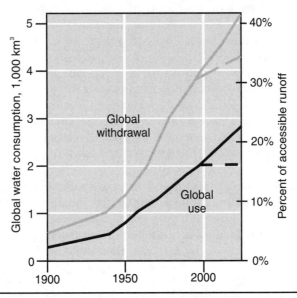

Source: Shiklomanov 2000:22 (high prediction), World Water Council 2000:26 (low prediction).

but until now most predictions have tended to overestimate the actual water consumption by up to 100 percent. Nevertheless, total use is still less than 17 percent of the accessible water and even with the high prediction it will require just 22 percent of the readily accessible, annually renewed water in 2025.

At the same time, we have gained access to more and more water, as indicated in Figure 2. Per person we have gone from using about 1,000 liters per day to almost 2,000 liters over the past 100 years. Particularly, this is due to an approximately 50 percent increase in water use in agriculture, allowing irrigated farms to feed us better and to decrease the number of starving people. Agricultural water usage seems, however, to have stabilized below 2,000 liters per capita, mainly owing to higher efficiency and less water consumption in agriculture since 1980. This pattern is also found in the EU and the US, where consumption has increased dramatically over the twentieth century, but is now leveling off. At the same time, personal consumption (approximated by the municipal withdrawal) has more than quadrupled over the century, reflecting an increase in welfare with more easily accessible water. In developing countries, this is in large part a question of health—avoiding sickness through better access to clean drinking water and sanitation, whereas in developed countries higher water use is an indication of an increased number of domestic amenities such as dishwashers and better-looking lawns.

So, if the global use is less than 17 percent of the readily accessible and renewable water and the increased use has brought us more food, less starvation, more health and increased wealth, why do we worry?

Figure 2

Global Withdrawal of Water for Agriculture, Industry and Municipal Use, and Total Use, in Liters and Gallons Per Capita Per Day, 1900–95

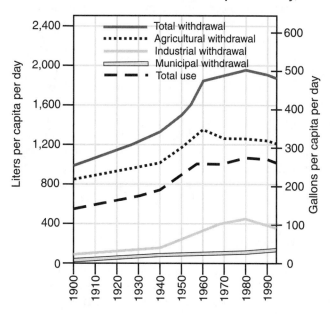

Source: Shiklomanov 2000:24.

The Three Central Problems

There are three decisive problems. First, precipitation is by no means equally distributed all over the globe. This means that not all have equal access to water resources and that some countries have much less accessible water than the global average would seem to indicate. The question is whether water shortages are already severe in some places today. Second, there will be more and more people on Earth. Since precipitation levels will remain more or less constant this will mean fewer water resources for each person. The question is whether we will see more severe shortages in the future. Third, many countries receive a large part of their water resources from rivers; 261 river systems, draining just less than half of the planet's land area, are shared by two or more countries, and at least ten rivers flow through half a dozen or more countries. Most Middle Eastern countries share aquifers. This means that the water question also has an international perspective and—if cooperation breaks down—an international conflict potential.

Beyond these three problems there are two other issues, which are often articulated in connection with the water shortage problem, but which are really conceptually quite separate. One is the worry about water pollution, particularly of potable water. While it is of course important to avoid water pollution in part because pollution restricts the presently available amount of freshwater, it is not related to the problem of water shortage *per se*. . . .

The second issue is about the shortage of *access* to water in the Third World. . . . This problem, while getting smaller, is still a major obstacle for global welfare. In discussing water shortage, reference to the lack of universal access to drinking water and sanitation is often thrown in for good measure, but of course this issue is entirely separate from the question of shortages. First, the cause is *not* lack of water (since human requirements constitute just 50–100 liters a day which any country but Kuwait can deliver, cf. Table 1) but rather a lack of investment in infrastructure. Second, the solution lies not in cutting back on existing consumption but actually in increasing future consumption.

Finally, we should just mention global warming . . . and its connection to water use. Intuitively, we might be tempted to think that a warmer world would mean more evaporation, less water, more problems. But more evaporation also means more precipitation. Essentially, global climate models seem to change *where* water shortages appear (pushing some countries above or below the threshold) but the total changes are small (1–5 percent) and go both ways.

Not Enough Water?

Precipitation is not distributed equally. Some countries such as Iceland have almost 2 million liters of water for each inhabitant every day, whereas Kuwait must make do with just 30 liters. The question, of course, is when does a country not have *enough* water.

It is estimated that a human being needs about 2 liters of water a day, so clearly this is not the restrictive requirement. The most common approach is to use the so-called *water stress index* proposed by the hydrologist Malin Falkenmark. This index tries to establish an approximate minimum level of water per capita to maintain an adequate quality of life in a moderately developed country in an arid zone. This approach has been used by many organizations including the World Bank, in the standard literature on environmental science, and in the water scarcity discussion in *World Resources*. With this index, human beings are assessed to need about 100 liters per day for drinking, household needs and personal hygiene, and an additional 500–2,000 liters for agriculture, industry and energy production. Since water is often most needed in the dry season, the water stress level is then set even higher—if a country has less than 4,660 liters per person available it is expected to experience periodic or regular water stress. Should the accessible runoff drop to less than 2,740 liters the country is said to experience chronic water scarcity. Below 1,370 liters, the country experiences absolute water scarcity, outright shortages and acute scarcity.

Table 1 shows the 15 countries comprising 3.7 percent of humanity in 2000 suffering chronic water scarcity according to the above definition. Many of these countries probably come as no surprise. But the question is whether we are facing a serious problem.

How does Kuwait actually get by with just 30 liters per day? The point is, it doesn't. Kuwait, Libya and Saudi Arabia all cover a large part of their water demand by exploiting the largest water resource of all—through desalination of sea water.

Table 1

Countries With Chronic Water Scarcity (Below 2,740 Liters Per Capita Per Day) in 2000, 2025, and 2050, Compared to a Number of Other Countries

Available water, liters per capita per day	2000	2025	2050
Kuwait	30	20	17
United Arab Emirates	174	129	116
Libya	275	136	92
Saudi Arabia	325	166	118
Jordan	381	203	145
Singapore	471	401	403
Yemen	665	304	197
Israel	969	738	644
Oman	1,077	448	268
Tunisia	1,147	834	709
Algeria	1,239	827	664
Burundi	1,496	845	616
Egypt	2,343	1,667	1,382
Rwanda	2,642	1,562	1,197
Kenya	2,725	1,647	1,252
Morocco	2,932	2,129	1,798
South Africa	2,959	1,911	1,497
Somalia	3,206	1,562	1,015
Lebanon	3,996	2,971	2,533
Haiti	3,997	2,497	1,783
Burkina Faso	4,202	2,160	1,430
Zimbabwe	4,408	2,830	2,199
Peru	4,416	3,191	2,680
Malawi	4,656	2,508	1,715
Ethiopia	4,849	2,354	1,508
Iran, Islamic Rep.	4,926	2,935	2,211
Nigeria	5,952	3,216	2,265
Eritrea	6,325	3,704	2,735
Lesotho	6,556	3,731	2,665
Togo	7,026	3,750	2,596
Uganda	8,046	4,017	2,725
Niger	8,235	3,975	2,573
Percent people with chronic scarcity	**3.7%**	**8.6%**	**17.8%**
United Kingdom	3,337	3,270	3,315
India	5,670	4,291	3,724
China	6,108	5,266	5,140
Italy	7,994	8,836	10,862
United States	24,420	20,405	19,521
Botswana	24,859	15,624	12,122
Indonesia	33,540	25,902	22,401
Bangladesh	50,293	35,855	29,576
Australia	50,913	40,077	37,930
Russian Federation	84,235	93,724	107,725
Iceland	1,660,502	1,393,635	1,289,976

Source: WRI 1998a.

Kuwait in fact covers more than half its total use through desalination. Desalting requires a large amount of energy (through either freezing or evaporating water), but all of these countries also have great energy resources. The price today to desalt sea water is down to 50-80 ¢/m^3 and just 20-35 ¢/m^3 for brackish water, which makes desalted water a more expensive resource than fresh water, but definitely not out of reach.

This shows two things. First, we can have sufficient water, if we can pay for it. Once again, this underscores that *poverty* and not the environment is the primary limitation for solutions to our problems. Second, desalination puts an upper boundary on the degree of water problems in the world. In principle, we could produce the Earth's entire present water consumption with a single desalination facility in the Sahara, powered by solar cells. The total area needed for the solar cells would take up less than 0.3 percent of the Sahara.

Today, desalted water makes up just 0.2 percent of all water or 2.4 percent of municipal water. Making desalination cover the total municipal water withdrawal would cost about 0.5 percent of the global GDP. This would definitely be a waste of resources, since most areas have abundant water supplies and all areas have some access to water, but it underscores the upper boundary of the water problem.

Also, there's a fundamental problem when you only look at the total water resources and yet try to answer whether there are sufficient supplies of water. The trouble is that we do not necessarily know *how* and *how wisely* the water is used. Many countries get by just fine with very limited water resources because these resources are exploited very effectively. Israel is a prime example of efficient water use. It achieves a high degree of efficiency in its agriculture, partly because it uses the very efficient drip irrigation system to green the desert, and partly because it recycles household wastewater for irrigation. Nevertheless, with just 969 liters per person per day, Israel should according to the classification be experiencing absolute water scarcity. Consequently, one of the authors in a background report for the 1997 UN document on water points out that the 2,740 liters water bench-mark is "misguidedly considered by some authorities as a critical minimum amount of water for the survival of a modern society."

Of course, the problem of faulty classification increases, the higher the limit is set. The European Environmental Agency (EEA) in its 1998 assessment somewhat incredibly suggested that countries below 13,690 liters per person per day should be classified as "low availability," making not only more than half the EU low on water but indeed more than 70 percent of the globe. Denmark receives 6,750 liters of fresh water per day and is one of the many countries well below this suggested limit and actually close to EEA's "very low" limit. Nevertheless, national withdrawal is just 11 percent of the available water, and it is estimated that the consumption could be almost doubled without negative environmental consequences. The director of the Danish EPA has stated that, "from the hand of nature, Denmark has access to good and clean groundwater far in excess of what we actually use."

By far the largest part of all water is used for agriculture—globally, agriculture uses 69 percent, compared to 23 percent for industry and 8 percent for households. Consequently, the greatest gains in water use come from cutting down on agricultural use. Many of the countries with low water availability therefore compensate by importing a large amount of their grain. Since a ton of grain uses about 1,000 tons of water, this is in effect a very efficient way of importing water. Israel imports about 87 percent of its grain consumption, Jordan 91 percent, Saudi Arabia 50 percent.

Summing up, more than 96 percent of all nations have at present sufficient water resources. On all continents, water accessibility has *increased* per person, and at the same time an ever higher proportion of people have gained access to clean drinking water and sanitation. While water accessibility has been getting *better* this is not to deny that there are still widespread shortages and limitations of basic services, such as access to clean drinking water, and that local and regional scarcities occur. But these problems are primarily related not to physical water scarcity but to a lack of proper water management and in the end often to lack of money—money to desalt sea water or to increase cereal imports, thereby freeing up domestic water resources.

Will It Get Worse in the Future?

The concerns for the water supply are very much concerns that the current problems will become worse over time. As world population grows, and as precipitation remains constant, there will be less water per person, and using Falkenmark's water stress criterion, there will be more nations experiencing water scarcity. In Figure 3 it is clear that the proportion of people in water stressed nations will increase from 3.7 percent in 2000 to 8.6 percent in 2025 and 17.8 percent in 2050.

It is typically pointed out that although more people by definition means more water stress, such "projections are neither forecasts nor predictions."

Figure 3

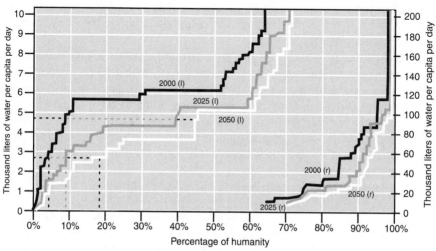

Share of Humanity With Maximum Water Availability in the Year 2000, 2025, and 2050, Using UN Medium Variant Population Data

The left side uses the left axis, the right side the right axis.

Source: WRI 1998a.

Indeed, the projections merely mean that if we do not improve our handling of water resources, water will become more scarce. But it is unlikely that we will not become better at utilizing and distributing water. Since agriculture takes up the largest part of water consumption, it is also here that the largest opportunities for improving efficiency are to be found. It is estimated that many irrigation systems waste 60–80 percent of all water. Following the example of Israel, drip irrigation in countries as diverse as India, Jordan, Spain and the US has consistently been shown to cut water use by 30–70 percent while increasing yields by 20–90 percent. Several studies have also indicated that industry almost without additional costs could save anywhere from 30 to 90 percent of its water consumption. Even in domestic distribution there is great potential for water savings. EEA estimates that the leakage rates in Europe vary from 10 percent in Austria and Denmark up to 28 percent in the UK and 33 percent in the Czech Republic.

The problem of water waste occurs because water in many places is not well priced. The great majority of the world's irrigation systems are based on an annual flat rate, and not on charges according to the amount of water consumed. The obvious effect is that participants are not forced to consider whether all in all it pays to use the last liter of water—when you have first paid to be in, water is free. So even if there is only very little private utility from the last liter of water, it is still used because it is free. . . .

This is particularly a problem for the poor countries. The poorest countries use 90 percent of their water for irrigation compared to just 37 percent in the rich countries. Consequently, it will be necessary to redistribute water from agriculture to industry and households, and this will probably involve a minor decline in the potential agricultural production (i.e. a diminished increase in the actual production). The World Bank estimates that this reduction will be very limited and that water redistribution definitely will be profitable for the countries involved. Of course, this will mean increased imports of grain by the most water stressed countries, but a study from the International Water Management Institute indicates that it should be possible to cover these extra imports by extra production in the water abundant countries, particularly the US.

At the same time there are also large advantages to be reaped by focusing on more efficient household water consumption. In Manila 58 percent of all water disappears (lost in distribution or stolen), and in Latin America the figure is about 40 percent. And on average households in the Third World pay only 35 percent of the actual price of water. Naturally, this encourages overconsumption. We know that pricing and metering reduces demand, and that consumers use less water if they have to pay for each unit instead of just paying a flat rate.

Actually, it is likely that more sensible pricing will not only secure future water supplies but also increase the total social efficiency. When agriculture is given cheap or even free water, this often implies a hidden and very large subsidy—in the United States the water subsidy to farmers is estimated to be above 90 percent or $3.5 billion. For the developing countries this figure is even larger: it is estimated that the hidden water subsidy to cities is about $22 billion, and the hidden subsidy to agriculture around $20–25 billion.

Thus, although an increasing population will increase water demands and put extra water stress on almost 20 percent of humanity, it is likely that this scarcity can be solved. Part of the solution will come from higher water prices, which will cut down on inefficient water use. Increased cereal imports will form another part of the solution, freeing up agricultural water to be used in more valuable areas of industry or domestic consumption. Finally, desalting will again constitute a backstop process which can produce virtually unlimited amounts of drinking water given sufficient financial backing.

POSTSCRIPT

Is the Threat of a Global Water Shortage Real?

The authors of the two selections agree that something must be done—that is, major public policy making must occur—if the water future is going to be acceptable. They disagree, however, on the urgency of the task and the level of optimism (or pessimism) that they bring to their analysis of likely success. Rosegrant and his colleagues approach the global water problem from the perspective of its role in food production. They argue that water is among one of the main factors that will limit food production in the future. Competition for water comes from many quarters, and farmers find themselves in an increasingly competitive situation as they attempt to keep pace with agricultural needs.

Rosegrant and his colleagues' view of past water trends—whether for irrigation or for drinking and household uses—leads them to conclude that the future of water is "highly uncertain," at best. The authors' pessimism stems from the failure of governments to address the issue adequately to date. Their model used to analyze future trends posits three basic scenarios: business as usual, water crisis, and sustainable water scenarios.

In the business-as-usual scenario, farmers will be unable to grow food as quickly as in the past. The cost of supplying water to users will increase, as heavier demands are placed on the global water supply. Industrial water use in the developing world is expected to grow faster than in the developed world as the developing countries push toward industrialization. Water availability for irrigation in the poorer sectors of the globe will decline significantly. Sub-Saharan Africa will, as it has for three decades, be the region of the globe hardest hit by the business-as-usual scenario. The water crisis scenario posits that "increasing water scarcity, combined with poor water policies and inadequate investment in water, has the potential to generate sharp increases in cereal food prices over the coming decades." Finally, the sustainable water scenario would result in a dramatic increase in water availability, improve environmental uses, lead to higher per capita domestic water consumption, and maintain food production at the business-as-usual scenario levels.

Lomborg argues that our ability to find more water has resulted in a global usage rate of 17 percent of the readily accessible and renewable water. The consequence, according to Lomborg, is that the world has "more food, less starvation, more health and increased wealth [so] why do we worry?" He does suggest that there are significant problems and identifies three: (1) the unequal distribution of precipitation throughout the globe, (2) increasing global population, and (3) the fact that many countries receive their water through shared river systems. Additionally, water pollution and the shortage of access to water in the developing world are issues, but of a different sort.

Both readings point out the need for aggressive policy action on the part of governments and other actors in the global water regime. This is not to suggest, however, that the world's leaders have been idle. At least 10 major international conferences since 1977 have addressed water issues, resulting in significant action-oriented proposals. For example, the 1992 International Conference on Water and Environment in Dublin established four basic ideas (known as the Dublin Principles). In addition, the 1992 Earth Summit highlighted water as an integral part of the ecosystem.

The current cry is for aggressive global water management. Every major study related to water concludes with the observation that governments need to do much more if future water crises are to be avoided. The Web is replete with public (and private) interest groups urging more global action. The Third World Traveler site titles one such plea from the International Forum on Globalization "The Failure of Governments" in a discussion of the crisis. The World Water Council's recent draft of its *World Water Actions Report* presents an overview of global actions to improve water management and to spell out priorities for future efforts. The report centers on a new paradigm for looking at water. No longer to be viewed as a physical product, the paradigm calls for water to be seen as an ecological process "that connects the glass of water on the table with the upper reaches of the watershed." The key ideas are "scarcity, conservation, and awareness of water's life cycle from rain to capture, consumption, and disposal."

The good news for the World Water Council is that the international community has broad consensus, forged at the 10 conferences mentioned above, about the need for policy action. The creation of the World Water Council itself in 1996 and the Global Water Partnership the same year is an example of such institutional action. But it was the Second World Water Forum, held at The Hague in The Netherlands in March 2000, that was a landmark event in raising global water consciousness. Among the many research reports and documents that emerged from the conference, two stand out: *Vision for Water, Life, and the Environment in the 21st Century* and *Towards Water Security: Framework for Action.*

Vision proposed five key actions: "Involve all stakeholders in integrated management, move towards full-cost pricing of all water services, increase public funding for research and innovation in the public interest, increase cooperation in international water basins, (and) massively increase investments in water." The *Framework for Action* document addresses the question of where do we go from here, suggesting four basic steps: "generating water wisdom . . ., expanding and deepening dialogue among diverse stakeholders, strengthening the capabilities of the organizations involved in water management, and ensuring adequate financial resources to pay for the many actions required."

A final important source for understanding the global water problem, particularly its relationship to food and the environment, is *Dialogue on Water, Food and Environment* (2002), published by 10 important actors in the field (FAO, GWP, ICID, IFAP, IWMI, IUCN, UNEP, WHO, WWC, and WWF). Perhaps it is fitting to end this discussion with the quote from UN Secretary General Kofi Annan on the cover of this report: "We need a blue revolution in agriculture that focuses on increasing productivity per unit of water—more crop per drop."

On the Internet . . .

United Nations Office on Drugs and Crime

Established in 1997, this UN organization assists members in their struggle against illicit drugs and human trafficking. It focuses on research, assistance with treaties, and field-based technical assistance. It is headquartered in Vienna with 21 field offices.

http://www.unodc.org

Beckley Foundation Drug Policy Programme (BFDPP)

This British foundation aims to promote objective debate about national and international drug policies. It lists many other Web sites as well.

http://www.internationaldrugpolicy.net

World Health Organization

This international organization's Web site provides substantial information about current and potential pandemics as well as other Web site links. See also http://www.globalhealthreporting.org and http://www.globalhealthfacts.org for additional information.

http://www.who.int/en/

Council of Europe

The Council of Europe established a campaign in 2006 to combat trafficking of human beings. It focuses on creating awareness among governments, NGOs, and civil society about the problem, as well as promoting global public policy to combat the problem.

http://www.coe.int/T/E/Human_Rights/Trafficking

Globalization Guide.Org

This Web site lists around 40 pro- and anti-globalization Web sites as well as other sources on both globalization and cultural imperialism.

http://globalizationguide.org

Globalization: Threat or Opportunity?

This Web site contains the article "Globalization: Threat or Opportunity?" by the staff of the International Monetary Fund (IMF). This article discusses such aspects of globalization as current trends, positive and negative outcomes, and the role of institutions and organizations. "The Challenge of Globalization in Africa," by IMF acting managing director Stanley Fischer, and "Factors Driving Global Economic Integration," by Michael Mussa, IMF's director of research, are also included on this site.

http://www.imf.org/external/np/exr/ib/2000/041200.htm

PART 3

Expanding Global Forces and Movements

*O*ur *ability to travel from one part of the globe to another in a short amount of time has expanded dramatically since the Wright brothers first lifted an airplane off the sand dunes of North Carolina's Outer Banks. The decline of national borders has also been made possible by the explosion of global technology.*

Many consequences flow from this realization, including the global spread of health pandemics and illegal drugs, along with the trafficking of human beings against their will. In addition, flows of money, information, and ideas that connect people around the world also create fissures of conflict that heighten anxieties and cause increased tensions between rich and poor, connected and disconnected, cultures, and regimes. Much of this flow originates in the United States, which has created a fear of American cultural imperialism. The impact of these new and emerging patterns of access has yet to be fully calculated or realized, but we do know that billions are feeling their impact and the result is both exhilarating and frightening.

- Can the Global Community "Win" the Drug War?

- Is the International Community Adequately Prepared to Address Global Health Pandemics?

- Has the International Community Designed an Adequate Strategy to Address Human Trafficking?

- Is Globalization a Positive Development for the World Community?

- Is the World a Victim of American Cultural Imperialism?

- Do Global Financial Institutions and Multinational Corporations Exploit the Developing World?

ISSUE 10

Can the Global Community "Win" the Drug War?

YES: Federico Mayor in collaboration with Jérôme Bindé, from *The World Ahead: Our Future in the Making* (UNESCO, 2001)

NO: Harry G. Levine, from "The Secret of Worldwide Drug Prohibition," *The Independent Review* (Fall 2002)

ISSUE SUMMARY

YES: Mr. Mayor suggests that drug trafficking and consumption "constitute one of the most serious threats to our planet" and the world must dry up the demand and attack the financial power of organized crime.

NO: Harry G. Levine, professor of sociology at Queens College, City University of New York, argues that the emphasis on drug prohibition should be replaced by a focus on "harm reduction," creating mechanisms to address tolerance, regulation, and public health.

In 1999, the United Nations pegged the world illicit drug trade at $400 billion, about the size of the Spanish economy. Such activity takes place as part of a global supply chain that "uses everything from passenger jets that can carry shipments of cocaine worth $500 million in a single trip to custom-built submarines that ply the waters between Colombia and Puerto Rico." In June 2004, the United Nations Office on Drugs and Crime (UNODC) reported that 3 percent of the world population (and 4.7 percent of those aged 15 to 64), approximately 185 million people, had "abused drugs during the previous 12 months." This includes people from virtually every country on the planet and from every walk of life. The *2004 World Drug Report* suggested that opiates remained the most serious problem in terms of treatment, with 67 percent of the drug treatment in Asia, 61 percent in Europe and 47 percent in Oceania. Cocaine is still the drug of choice in the Americas, and the number of admissions for treatment is now higher for cocaine in the United States. In Africa, cannabis tops the treatment list (65 percent). One-fifth of the world's population (and 29 percent of those above age 15) also use tobacco.

While the UN report suggests that the spread of drug abuse may be losing some momentum, the use of cannabis is growing at an accelerated rate. After cannabis, the highest increases in the past decade have been ATS (mainly ecstasy), then cocaine and the opiates. The amount of drugs seized has also increased over the last decade, rising from 14 billion doses in 1990 to 26 billion in 2000. The bottom line statement from the United Nations report is that "the second half of the twentieth century . . . witnessed an epidemic of illicit drug use."

The report also included some good news. Especially important has been the emergence of a consensus among governments and global public opinion that the current level of illegal drug use is unacceptable. In two drug-producing regions, declines in production have actually occurred. In Southeast Asia, opium poppy cultivation continues to drop in Myanmar and Laos. In the Andean region, coca cultivation has declined for four straight years in the three leading producing countries (Colombia, Peru, and Bolivia).

The illegal movement of drugs across national borders is accompanied by the same kind of movement for illegal weapons. They go hand in hand with one another. It is estimated by the UN that only 3 percent of such weapons (18 million of a total of 550 million in circulation) are used by government, the military, or police.

This increase in drug use has occurred despite a rather long history of government attempts to control the illegal international drug trade. Beginning in 1961, such efforts have been part of the social policies of governments' worldwide. Precisely because drug policy crosses over into social policy, policymakers and scholars have been at odds over how best to deal with this ever growing problem, whether talking about national policy or international policy. Simply stated, the debate has centered on legalization vs. prohibition, and treatment vs. prevention.

Policies of the United States have always had the goal of drug use reduction and punishment for abusers, resulting in less attention to treatment. This includes a number of important elements, as outlined in a Congressional Research Service Brief for Congress (2003): "(1) eradication of narcotic crops, (2) interdiction and law enforcement activities in drug-producing and drug-transmitting countries, (3) international cooperation, (4) sanctions/economic assistance, and (5) institution development." Many have charged the United States and those other countries that share its fundamental philosophy of drug wars of using the issue to expand its national power in other domains. On the other hand, other countries, particularly those in Western Europe, have been shifting attention for some time away from "repressive policies" and toward those associated with harm reduction and treatment.

The two selections in this section contribute to the debate over the proper approach to "winning" the drug war. Federico Mayor understands the need for creating a declining market for drugs. He argues that although the control of narcotics is preferable, it is not likely to happen. The major reason for such pessimism is due to the overwhelming power of criminal organizations. Harry G. Levine suggests that drug prohibition has had wide acceptability throughout the world because of pressure from the United States. These countries have embraced this position because it is useful for other purposes associated with national power.

**Federico Mayor in collaboration
with Jérôme Bindé**

Winning the Fight Against Drugs: Education, Development and Purpose

A Rapidly Expanding Market

While many economists are rejoicing at the sustained growth of the world economy, there is one market in particular that is undergoing uninterrupted expansion throughout the world: the drug market, the cause of the most radical marginalization of human beings, since drugs abolish all notions of self or of other human beings.[1] According to the United Nations, profits derived from drug trafficking amount to $400 billion annually, that is to say 8% of world trade or, at a rough estimate, the equivalent of international trade in textiles in 1994,[2] or 1% of world GNP, or the GNP of the whole of Africa.[3] The production and consumption of drugs are rising constantly. According to the 1996 report of the International Narcotics Control Board (INCB), 'in spite of increased repression, production of and trafficking in drugs, together with drug addiction, have now reached hitherto preserved regions of the world.' As emphasized by an expert, 'Narco-states and narco-democracies, narco-terrorism and narco-guerrilla activities, narco-tourism and narco-dollars, are all signs that drugs have penetrated every sphere of political, economic and social life. The expansion of drug trafficking now goes hand in hand with the globalization of the economy and free market democracy.'[4] The highly efficient organization of drug trafficking, through a worldwide network based on extremely flexible and constantly changing units, has rendered any control of narcotics particularly difficult.[5] Production as such is still highly concentrated, as 90% of the illicit production of narcotics derived from opium worldwide originates from two major areas: the 'Golden Crescent' (Afghanistan, Iran and Pakistan) and the 'Golden Triangle' (Laos, Burma and Thailand) and 98% of world supplies of cocaine come from the Andean countries (Peru, Colombia and Bolivia).[6] But new connections have developed, production and trafficking areas have expanded and new synthetic drugs have appeared on the market. Drug trafficking is the hidden face of globalization. It also happens to be one of its main beneficiaries, owing to increasingly porous national frontiers, the volatile nature of financial operations and the contagion of lifestyles, and even what might be called 'death styles.' INTERPOL estimates that only 5% to 15% of banned drugs are actually seized, which means that at least 85% of narcotics escape repression and circulate freely, in a clandestine market controlled by criminals.[7]

According to OECD estimations, $85 billion derived from profits from such trafficking are laundered every year on the financial markets—that sum is greater than the GNP of three-quarters of the 207 economies in the world, according to a group of G7 experts.[8] The wealth accumulated by drug traffickers over the last 10 to 15 years could amount to as much as 'several trillion dollars.'[9] The drugs trade first and foremost 'benefits' the industrialized countries, if we may venture to say so: 90% of these sums is thought to be reinvested in the Western countries.[10] Many experts are increasingly concerned at the growing expansion of 'grey areas' within the world economy, which enable major organized crime networks to penetrate the very heart of some strategic spheres of the international economy such as the major world financial exchanges.[11] As pointed out by an expert, 'in every country, the banking system is actively involved in recycling drugs revenue, particularly through subsidiaries and correspondents established in the worldwide constellation of tax havens,' which means, sometimes, that 'laundered money from drugs enables debt instalments to be paid or funds structural adjustment plans.'[12] The growing sophistication of financial operations, the globalization of the banking system, which is no longer hampered by frontiers and is able to operate round the clock, and the rapid emergence of unrecorded 'cyber-payments' require increased vigilance on the part of regulatory bodies and the extension of their partnership to the whole range of world financial institutions. The USA, for its part, recently advocated regulation of the non-banking financial sector, ranging from currency exchange and brokerage houses to casinos, as well as express delivery services, insurance firms and the precious-metal trade.[13] Multinational corporations and transnational finance companies should abide by codes of conduct in order to prevent the laundering of money derived from crime, whether it comes from drug trafficking, arms dealing or any form of criminal trade or mafia activity (embezzlement of public funds, racketeering, prostitution, illicit gambling, etc.).

The influence, whether overt or covert, of major criminal organizations, seemingly on the increase in many countries in both North and South, means that serious dangers are threatening economic ethics and the rule of law. According to the UNDP, expenditure on consumption of narcotics in the USA alone exceeds the accumulated GDP of more than eighty developing countries. Furthermore, organized crime has considerably extended its geographical areas of influence thanks to globalization and the development of drug trafficking, which often occurs in symbiosis with other criminal activities (arms trafficking, prostitution and the slave trade, the embezzlement of public funds, illegal gambling and penetration of the casino network by the mafia, etc.). The gigantic scale of illicit profits from drugs, together with the 'penetration' of entire sectors of the legal economy now controlled through money-laundering, could ultimately lead, through the dynamics of accumulation and concentration observed during the last two decades, to an irreversible situation whereby no state or organized force would be in a position to react as, through the laundering process, a substantial part of the economy and of pressure groups, in both North and South, would fall under the influence of drug trafficking. Keeping quiet in this matter amounts to observing the very principle that underlies the power of criminal organizations, namely the law of silence.

In our opinion, drug trafficking and consumption constitute one of the most serious threats to our planet, with disastrous consequences for health, development and society. We are all the more sensitive to this problem as we know all too well what effects drug addiction has on the brain's receptors and how irreversible lesions are caused above a certain level of consumption.[14] To these evils are added the effects of the spread of AIDS among drug addicts who absorb drugs intravenously. Young people, education and human values are affected first but drug addiction makes life unbearable for the whole family of the addict and, sooner or later, it is democracy itself that is threatened and, with it, peace. As pointed out by a specialist in this field, 'there is virtually no local conflict today that is not linked to a greater or lesser extent to drug trafficking.'[15] Drugs have become a form of violence not only towards the individual but also towards the whole of society. . . .

Eliminate Supply or Dry up Demand?

If effective action is to be taken against drugs, we must first of all open our eyes wider and open those of others. We need to discuss the problem with greater scientific rigour and critical awareness within the institutions that disseminate knowledge, namely schools, universities and all the channels of mass communication. Our task must be to show both clearly and unceasingly the real damage caused by the various drugs, the moral and physical servitude and the destruction of mind and body, attitudes and values, of which they are the cause. We must shed light on the harm caused to both society—starting with the immense suffering inflicted on the addict's family—and the individual. We must cease to portray drugs as malevolent yet attractive, and thus avoid demonizing them, as is too often the case, unaware that, in so doing, we turn them into a symbol of the urge to transgress social rules. If we cry 'Wolf!' too often, we are more likely to push young people towards drugs than to put them off. A survey recently conducted by a French opinion poll agency showed that for 52% and 44% respectively of drug-takers, pleasure and curiosity were the prime motive.[16] What we need to do is to demystify drugs by explaining to children that they are first and foremost a denial of existence, as stressed by Rita Levi-Montalcini, Nobel prizewinner in medicine, in her fine essay dedicated to young people, aptly entitled 'Your future.' Drugs will stop attracting adolescents once they have understood that they constitute, above all, a 'lessening of the power to act,' as Spinoza might have put it, when he defined sadness in such terms in his work *Ethics*.

Some people believe that drugs should no longer be banned. According to these liberal-minded opponents of prohibition, it is through re-establishing control of narcotics by the legal economy that we can best fight the plagues which are the result of the illegal nature of the market, namely the enrichment of dealers and middlemen (which stimulates expansion of this trade), crime, violence, the marginalization of addicts, arms trafficking, terrorism and the suspicion of corruption that in many countries weighs on wide areas of public and political life. Nevertheless, we are among those who believe that we must firmly oppose the legalization of the non-therapeutic use of drugs. In that regard, the report of the International Narcotics Control Board (INCB) for 1997 deplored the existence of 'an overall climate of acceptance that is favourable to or at least tolerant of drug abuse.'[17] We can no more play around with narcotics than we can with weapons or medicinal drugs. The risk of a real surge in consumption is far too great. That is why drugs cannot be left to market forces, whether legal or illegal. In this domain, we cannot afford to be sorcerers' apprentices.

Conversely, the idea that it is possible to eliminate the production and consumption of drugs, however commendable it may be, strikes us as being scarcely credible. Many commentators are now advocating the replacement of prohibition policies by measures to reduce the harmful effects of drugs, in the belief that 'drugs are here to stay and we have no choice but to learn to live with them so that they cause the least possible harm.'[18] The policy of all-round repression can fail because, by making drugs scarcer, it merely makes drug trafficking more lucrative for new networks that have replaced earlier ones. While we thought that we were fighting organized crime, we were actually strengthening its financial power and capacity to corrupt, and dragging down large sections of society into delinquency.

A recent report states that 'Despite regional successes supply suppression is not a prescription for solving the world's illicit-drug problem. It is a prescription for funding drug mafias, peasant growers, petty traffickers and smugglers.'[19] There is naturally a price to pay for such a policy: several tens of billions of dollars are spent every year on repression, with results that, to put it mildly, are hardly convincing: delinquency in a growing part of society (the inner cities, ghettos, minorities, the younger generation, the interaction between consumption and 'petty trafficking,' etc.), and the corresponding excessive growth of the repressive and penitentiary system, which ultimately penalizes addicts rather than traffickers. Everything, therefore, needs to be changed. On the one hand, an efficient judicial and penitentiary system must be set up to deal with drug dealers. On the other hand, preventive and curative measures should be introduced on a large scale. We must stress that addicts need to be given treatment—and adequate funds should be allocated for that purpose—by bringing into play all the means required, whether they be medical, scientific or of another nature.

We must therefore learn how to deal with the problems of drug addiction in our societies while reducing this phenomenon to a minimum and avoiding the criminalization of addicts. In terms of public health alone, realistic public policies 'for reducing the harmful effects' are required to respond to the fact that the total absence of control over the drugs market is a powerful vector for the development of AIDS and other epidemics: the distribution of free syringes helped in several countries to lower the contamination of drug addicts by HIV. Moreover, we believe the time has come to contemplate an international agreement whereby it would be possible, under medical supervision, to distribute a limited quantity of drugs to addicts who are not able to break out of addiction.[20] Drug addicts should be treated as patients rather than delinquents. As such, patients have a right to benefit from medical supervision and social assistance in much the same way as any human being suffering from a curable pathology. What is more, such a measure could reduce violence and delinquency and contribute to dismantling the illegal drugs market and, therefore, the major source of profits for organized crime. The difficulty underlying such a policy can be summarized as follows: any agreement would have to be international, as policies for fighting against drugs can no longer be conceived in purely national terms. Harmonizing the policies of various states would be the key to effectiveness in this field where interdependence is particularly marked.[21]

Clearly, such a policy will have to be accompanied by an in-depth survey on the specific harmfulness of drugs by the scientific and medical community, in close cooperation with the relevant national and international authorities. Such concertation might, as suggested by a French consultative body, make it possible to establish regulations for each substance, 'taking account of its toxicity, the risks of dependence relating to its consumption, the danger of desocialization it might entail and the risks to which its consumption might expose other people.'[22]

Prevention by educating and informing the public would also be indispensable for this project. We need the help of the media, as well as municipal and local authorities, to foster appropriate awareness, commitment and participation and to ensure that drug addiction does not become commonplace, the lame excuse of a society that tolerates the degeneration and distress of those who symbolize its future, namely, its younger generations. The Youth Charter for a Twenty-first Century Free from Drugs (1997), which received the support of UNESCO and the United Nations Programme for International Narcotics Control (UNPINC), rightly states that 'the first experiences with drugs are often motivated by curiosity, idleness, lack of self-confidence, indifference and violence in our immediate surroundings, but also by the difficulties and trials of everyday life.'[23]

Many experts and institutions nevertheless continue to give priority to reducing the supply of drugs. One of the solutions for reducing the production and therefore the supply of drugs would be to develop sufficiently lucrative alternative crops and new markets for the peasant farmers whose livelihood depends on poppy and coca cultivation. To do that, the cooperation of the peasant farmers concerned is all the more important in the choice of new crops as the cultivation of toxic plants is often related to cultural traditions. They need to be made aware of the dangers of drugs for their health and for the life of their community, as well as for the well-being of the whole of humanity. Unfortunately, policies for the eradication of plantations and help for the substitution of illegal crops have often failed and have had adverse effects through ignorance of the cultural factors of development, the local social environment, the requirements of sustainable development, as well as through anthropocentric naivety.

In such circumstances, the establishment of a scheme for subsidies and guaranteed prices for new crops bringing into play national and international resources would seem indispensable. In this type of situation, the international community, by lending its financial support, could invest in the future with success, benefit and a sense of long-term vision.

In addition to these difficulties, replacing drugs with alternative crops, in the absence of accompanying structural measures and international support, would appear virtually condemned to failure for four main reasons. No government would seem ready to pay the very high price of replacing crops worldwide, if the operation had to be subsidized. No agricultural production, on the basis of market prices, would be competitive with the price of base plants. Furthermore, the economy of many producing countries, which are often very poor, is now increasingly based on drugs. It is no secret, as illustrated by various reports produced by the United Nations, that the drug industry accounts for as much as 20% of GNP for some countries.[24] Worse still, once crops have actually been eradicated in a particular country, the outcome is disappointing and deceptive as production moves elsewhere, often to a neighbouring country.[25]

A global threat requires a global solution. A large-scale threat calls for large-scale solutions. Many countries are members of major international alliances that guarantee their borders and security. If we wish to mark the advent of the new century by making a fresh start, we must then have comparable alliances at our disposal for combating drugs, as would be required for combating global catastrophes of all kinds. The drugs problem should not be confined to protracted, trivial discussions on the respective responsibility of the producer and consumer countries: such a dispute is all the more futile as the frontier between the former and the latter has become increasingly blurred since the explosion of consumption in the South. Let us endeavour instead to fight the causes of supply and demand by offering acceptable living conditions and a better future to the peasant farmers attracted by illicit but more profitable crops, to the middle-men who, in many underprivileged countries or communities, often have no alternative income and to the consumers who, through lack of education and information, are ignorant of the dangers of drugs or who feel excluded from a society in which they are not able to fashion their own lives. Let us give hope and a future to them all. The terrible effects of drug addiction on human dignity constitute a powerful illustration of the importance of preventive action, which involves educating young people as early as possible.

Let us recall that international action against drug trafficking began some 80 years ago when the opium trade came under international jurisdiction. Since then, the multilateral system has devised many conventions and plans of action for combating this traffic which, at the highest levels of responsibility, may be considered to be a crime against humanity. In fact, the United Nations General Assembly proclaimed the 1990s as the 'United Nations Decade against Drugs.'

The most efficient means of fighting against drug trafficking is, as stated by the Italian judge, Giovanni Falcone, shortly before he was assassinated by the Mafia, 'the destruction of the financial power of organized crime, which would presuppose powerful international collaboration.'[26] This alone can help to prevent the emergence of the 'chain of connivance' composed of obscure acts of corruption and unavowed links described by the great Sicilian writer, Leonardo Sciascia, in his novel *The Context,* published in 1971. It is that chain of connivance that undermines democratic institutions and threatens their legitimate representatives.[27]

Judge Falcone added that it was necessary, with that aim in mind, to encourage and coordinate 'efforts aimed at identifying and confiscating wealth of illegal origin,' which requires 'adapting international laws and achieving constant international collaboration.' Giovanni Falcone advocated 'first and foremost, the elimination of tax havens which, up to now, have countered the most serious attempts of various countries to identify financial flows originating from illegal trafficking.' According to Judge Falcone, 'this is a fight that concerns all members of the international community because its outcome will determine whether organized crime is destroyed or at least limited in such a way as to be no longer a serious peril for society.' The advice and sacrifice of Judge Falcone would not seem to have been totally pointless: since then, the Italian magistracy has intensified the seizure of assets of illegal origin, while the profits of the four major Italian criminal organizations (estimated by the Italian Anti-Mafia Investigatory Department at 10 trillion lire, or 30% of a turnover estimated at 30 trillion in 1994), seem to have shrunk massively in the same year.[28]

Furthermore, the production of narcotics is by no means limited to substances of natural origin. In its 1996 and 1997 reports, the INCB highlighted the preoccupying expansion throughout the world of synthetic drugs, particularly amphetamines or by-products, such as ecstasy, produced in clandestine laboratories. They supply a very lucrative illicit market for dealers and meet with alarming success among young people. The US Department of State believes that amphetamines, on account of their simple manufacturing process and the sudden growth in demand, are about to become 'the drug-control nightmare of the next century.'[29] In the face of this growing threat, which, paradoxically, has benefited from progress achieved in the field of pharmaceutical research, new measures for control, information, research and education are required, particularly for the benefit of young people.

There is therefore no miracle solution to the drugs problem. As long as there is demand, there will be supply. As noted by the UNDP in a lucid report, the real solution requires tackling the causes of drug addiction and eliminating the poverty that leads farmers to become involved in producing narcotics.[30]

We are particularly concerned with the consequences of drug consumption on the fate of street children who, today, number more than 100 million and who are fighting every day to survive in conditions of total deprivation. These children are those who are most threatened by violence, sexual and economic exploitation, AIDS, hunger, solitude and the scourges of exclusion, illiteracy and drugs. They are the 'golden fish' referred to by the French novelist Le Clézio which the ill-intentioned fishermen in search of innocent prey attempt to catch in their nets. Everything must be done to ensure that these children are fully integrated into society, that they learn to live in it, that they have access to education and that they are no longer manipulated by criminals who make them serve their evil purposes. The latter deserve to be punished all the more severely as by destroying innocence, they attempt to eradicate faith and confidence in the future.

The fight against drug addiction—like that against AIDS or against the collective shame represented by street children and children who are sexually or professionally exploited—will not be truly effective unless it is based on a major alliance between all countries, translated practically into a political will not to abandon the cause, just as we defend our country when national sovereignty is in jeopardy. In fact, in all the cases we have just referred to, it is national dignity that is threatened, which cannot be defended simply through charity or by organizing tombolas and galas. The best way of celebrating human rights, the fiftieth anniversary of the declaration of which we celebrated in 1998, would be an internationally reached decision aimed at ensuring its effective exercise by all human beings. The rest is no more than ceremony and empty rhetoric. The United Nations International Drug Control Programme (UNDCP) should be one of the most powerful in the entire United Nations system, in terms both of its authority and of its resources. This should also be the case, in a different field, for the United Nations of Environment Programme (UNEP). The limited means available to these programmes reflect a lack of political will and of public awareness as to what is required for combating drugs or preserving the global environment. We are, to varying degrees, all responsible for this twofold deficiency.

To fight against drug-related problems, the causes of marginalization and exclusion have to be tackled by investing in the welfare of young people, particularly through sports and training activities. It is UNESCO's responsibility to fight against the demand for drugs through education, and more especially preventive education. While education may be the main victim of drugs, it is also its best antidote. In fact, it is thanks to education that young people can become aware of the real dangers of narcotics, that they can escape 'the blues' and find their true path in society, and that they can acquire the knowledge and ethical attitudes that will enable them to assert their own personality and take their destiny in hand. Instead of paying the price for war and over-investing in armed defence, let us invest in the peaceful defence of individuals and young people, in cultural security and in genuine spiritual freedom which access to the world of knowledge and freedom from any servitude, can provide. Once education is widely perceived as having the objective of 'ensuring that people have control over their lives,' then it is through education that any form of dependency can be combated, such as dependency on alcohol, tobacco, drugs and sects, etc. Through education we can learn to be free and responsible.

More than ever before, the vital issue is the political will of governments to agree on effective solutions and on implementing them. More than ever, UNESCO has a major role to play in the context of its fields of competence: education and information against drug abuse, communication activities among the populations, and the contribution of the social sciences and scientific research in order to fine tune action plans and national and international strategies, and to assess the specific harmfulness of drugs, which are a subject of debate.[31]

To be perfectly frank, however, education and information alone cannot, even in the long term, be the only solutions in this field, nor can development if it is reduced to mere economic prosperity. Without referring to the great minds who succumbed to 'artificial paradises,' it is striking to observe the number of people with a high level of education and a comfortable income, who consume drugs in a number of countries. Psychological malaise as reflected by drug addiction cannot be cured merely by knowledge as, to paraphrase Henri Michaux, the poet, knowledge itself may lead to an abyss. If the twenty-first century is to win the battle against drug addiction (which some experts in science fiction doubt, imagining, on the contrary, the expansion of a form of 'addiction to soft drugs' controlled by the neurosciences and pharmacology), it will have to win the battle against nihilism, consumerism, and the fruitless pursuit of intoxication and ecstasy. We shall have to bring about a 'global mobilization' of governments, parliaments, the media, industry and society as a whole against drugs and addiction. The next century will have to give a new meaning to life.

Education, economic development and material well-being will probably not suffice to eliminate drugs even if they are the major instruments of prevention. To think so would be to imagine that human beings can be prevented from walking along the edge of precipices, from seeking to experience ecstasy or trance or, quite simply, from wanting to poison themselves. What therefore has to be done is to construct humanity's defences in people's minds, all the more so as drugs tempt the minds as much as they do the body. Thus new forms of wisdom and ethics will have to develop. What must also be done is to build humanity's defences by investing in human dignity, by reducing dire poverty, racism and exclusion. Fighting against drugs, the source of destruction, suffering and war, also means responding to the aims of the founders of UNESCO. It simply means building peace and development on the basis of the intellectual and moral solidarity of humanity. Fighting against drugs, in a united effort, with human and financial means that correspond to the scale of the plague, means protecting young people, our children and our future. It also means speeding up the transition from a culture of violence, war and indignity to a culture of peace, non-violence and dignity for all.

Pointers and Recommendations

- Reduce the demand for drugs in consumer countries, particularly through education, prevention and treatment.
- Educate and inform children and young people about the risks of drug consumption.
- Mobilize the international community against the main causes of drug consumption which are marginalization and poverty, in both urban and rural environments.
- Develop specific machinery, on international, regional and national scales, for fighting against corruption, the laundering of money from

drugs and organized crime. Encourage ratification and implementation of international treaties related to narcotics control and the conclusion of international agreements aimed at destroying the financial power of organized crime.

- Help drug addicts to overcome their dependency and to adopt a sustainable lifestyle without drug consumption, through appropriate educational, rehabilitation and vocational training programmes.
- Reduce the perverse effects of drug trafficking and consumption (financial development of organized crime, criminalization, delinquency and social pathologies) by studying the feasibility, on an international scale, of an agreement that, under medical supervision, would allow addicts who are in need and incapable of giving up their habit a limited supply of drugs.
- Give serious consideration to the adoption of 'reduction of the harmful effects' policies implemented in various countries under the terms of a policy and a world programme for narcotics control.
- Encourage scientific and medical research on the specific harmfulness of drugs and scientific research on the environmental impact of drug cultivation.
- Plan for the convening of a world summit on drugs, organized by the United Nations (United Nations International Drug Control Programme and the World Health Organization), which would take account of all aspects, whether old or new, of the problem of drug consumption and trafficking. Strengthen the means and authority of the UNDCP.
- Bring about a 'global mobilization' of governments, parliaments, the media, industry and society against drugs and addiction.

Notes

1. The term 'drugs' refers to narcotics and psychotropic substances as defined by the relevant international conventions.

2. UNDCP, *1997 World Drug Report,* Oxford University Press, New York, 1997; *Le Monde,* 27 June 1997.

3. Eric de la Maisonneuve, *La Violence qui vient,* Éditions Arléa, Paris, 1997.

4. Christian de Brie, 'La drogue dopée par le marché,' *Le Monde Diplomatique,* April 1996.

5. Hugues de Jouvenel, 'L'inextricable marché des drogues illicites,' *Futuribles,* No. 185, March 1994 (special issue 'Géopolitique et économie politique de la drogue').

6. UNDCP, *1997 World Drug Report.*

7. Address by the director-general of UNESCO at the 58th session of the United Nations International Narcotics Control Board (INCB), Vienna, 9 May 1995.

8. Quoted in UNDP, *1994 World Report on Human Development,* UNDP, Economica, Paris, 1994, p. 37; *UN Chronicle,* No. 3, 1996.

9. Alain Labrousse, *Les Idées en mouvement,* No. 35, January 1996.

10. Alain Labrousse, interview in *Le Nouvel Observateur,* 19–25 September 1996.

11. See 'The Mob on Wall Street,' *Business Week,* 16 December 1996; Laurent Zecchini, 'La "pieuvre" mafieuse prolifère à Wall Street,' *Le Monde,* 3 January 1997; Brie, 'La drogue dopée par le marché'; LaMond Tullis, *Unintended Consequences: Illegal Drugs and Drug Policies in Nine Countries,* Studies on the Impact

of the Illegal Drug Trade, Vol. 4, series editor LaMond Tullis, United Nations University and United Nations Research Institute for Social Development, Lynne Rienner, Boulder, CO and London, 1995.

12. De Brie, *Monde Diplomatique* feature, *Le Monde Diplomatique,* February 1996.

13. US Department of State, *International Narcotics Control Strategy Report, 1996,* March 1997.

14. Address by the director-general of UNESCO at the 58th session of the United Nations International Narcotics Control Board (INCB), Vienna, 9 May 1995.

15. Labrousse, *Les Idées en mouvement.*

16. Patrick Piro, *Les Idées en mouvement,* No. 35, January 1996.

17. Report of the International Narcotics Control Board, 1997, para. 20.

18. Ethan A. Nadelmann, 'Commonsense drug policy,' *Foreign Affairs,* January/ February 1998, p. 112; Anthony Lewis, 'The war on drugs is being lost,' *International Herald Tribune,* 6 January 1998.

19. LaMond Tullis, *Unintended Consequences,* p. 183.

20. See address by the director-general of UNESCO at the 58th session of the United Nations International Narcotics Control Board (INCB), Vienna, 9 May 1995.

21. Kopp and Schiray, 'Les sciences sociales.'

22. Conclusions of the Comité consultatif national d'éthique (CCNE), November 1994, quoted in Observatoire géopolitique des drogues, *Géopolitique des Drogues 1995,* Paris, 1995.

23. Charter co-ordinated by the NGO Environnement sans frontière, with the support of UNESCO and the United Nations Programme for International Narcotics Control (UNPINC). The signatories also emphasized that 'drug trafficking and use are a threat to the development and progress of our societies, they invariably cause greater violence, crime, exploitation and other infringements of our rights' and that the fight against drugs hinges on guaranteeing 'peace, freedom, democracy, solidarity, justice, protection of the environment and access to employment'.

24. UNDP, *World Report on Human Development,* 1994.

25. Ibid.

26. Giovanni Falcone, 'What is the mafia?' lecture to the Bundeskriminalamt (Wiesbaden), 1990, *Frankfurter Allgemeine Zeitung,* 27 May 1992, reproduced in *Esprit,* No. 185, October 1992, pp. 111–18.

27. See *Le Monde,* 10 February 1998, 'Cadavres exquis.' 'Cadavres exquis' is the French title of Francesco Rosi's film *Cadaveri Eccellenti,* based on Leonardo Sciascia's novel *The Context.*

28. *Libération,* 13 September 1995.

29. US Department of State, *International Narcotics Control Strategy Report, 1996,* March 1997.

30. UNDP, *World Report on Human Development,* 1994.

31. See for example, 'Marijuana: special report,' *New Scientist,* 21 February 1998.

Harry G. Levine

The Secret of Worldwide Drug Prohibition: The Varieties and Uses of Drug Prohibition

What percentage of countries in the world have drug prohibition? Is it 100 percent, 75 percent, 50 percent, or 25 percent? I recently asked many people I know to guess the answer to this question. Most people in the United States, especially avid readers and the politically aware, guess 25 or 50 percent. More suspicious individuals guess 75 percent. The correct answer is 100 percent, but *almost no one* guesses that figure. Most readers of this paragraph will not have heard that every country in the world has drug prohibition. Surprising as it seems, almost nobody knows about the existence of worldwide drug prohibition.

In the last decade of the twentieth century, men and women in many countries became aware of *national* drug prohibition. They came to understand that the narcotic or drug policies of the United States and some other countries are properly termed *drug prohibition.* Even as this understanding spread, the fact that drug prohibition covers the entire world remained a kind of "hidden-in-plain-view" secret. Now, in the twenty-first century, that situation, too, is changing. As "global drug prohibition" becomes more visible, it loses some of its ideological and political powers.

In this article, I briefly describe the varieties and uses of drug prohibition and the growing crisis of the worldwide drug prohibition regime. . . .

Drug Prohibition Is Useful to All Types of Governments

There is no doubt that governments throughout the world have accepted drug prohibition because of enormous pressure from the U.S. government and a few powerful allies, but U.S. power alone cannot explain the global acceptance of drug prohibition. Governments of all types, all over the world, have found drug prohibition *useful for their own purposes,* for several reasons.

The Police and Military Powers of Drug Prohibition

Drug prohibition has given all types of governments additional police and military powers. Police and military narcotics units can go undercover almost

The Article is reprinted with permission of the publisher from *The Independent Review: A Journal of Political Economy* (Fall 2002, vol. V11, no. 2). @ Copyright 2002, The Independent Institute, 100 Swan Way, Oakland, California 94621-1428 USA; info@independent.org; www.independent.org

anywhere to investigate—after all, almost anybody might be in the drug business. More undercover police in the United States are in narcotics squads than are in any other branch of police work. Antidrug units within city, county, and state police departments are comparatively large and often receive federal subsidies. Police antidrug units have regular contact with informers; they can make secret recordings and photographs; they have cash for buying drugs and information. In the United States, police antidrug units sometimes are allowed to keep money, cars, houses, and other property that they seize. Top politicians and government officials in many countries may have believed deeply in the cause of drug prohibition, but other health-oriented causes could not have produced for them so much police, coast guard, and military power.

Government officials throughout the world have used antidrug squads to conduct surveillance operations and military raids that they would not otherwise have been able to justify. Many times these antidrug forces have been deployed against targets other than drug dealers and users—as was the case with President Richard Nixon's own special White House antidrug team, led by former CIA agents, which later became famous as the Watergate burglars. Nixon was brought down by his squad's mistakes, but over the years government antidrug forces all over the world have carried out countless successful nondrug operations.

The Usefulness of Antidrug Messages and of Drug Demonization

Drug prohibition also has been useful for governments and politicians of all types because it has required at least some antidrug crusades and what is properly called *drug demonization.* Antidrug crusades articulate a moral ideology that depicts "drugs" as extremely dangerous and destructive substances. Under drug prohibition, police departments, the media, and religious and health authorities tend to describe the risks and problems of drug use in extreme and exaggerated terms. "Drugs" are dangerous enemies. "Drugs" are evil, vile, threatening, and powerfully addicting. Politicians and governments crusade against "drugs," declare war on them, and blame them for many unhappy conditions and events. Antidrug crusades and drug scares popularize images of "drugs" as highly contagious, invading evils. Words such as *plague, epidemic, scourge,* and *pestilence* are used to describe psychoactive substances, drug use, and moderate, recreational drug users.

Government officials, the media, and other authorities have found that almost anyone at any time can blame drug addiction, abuse, and even use for long-standing problems, recent problems, and the worsening of almost anything. Theft, robbery, rape, malingering, fraud, corruption, physical violence, shoplifting, juvenile delinquency, sloth, sloppiness, sexual promiscuity, low productivity, and all around irresponsibility—anything at all—can be and has been blamed on "drugs." Almost any social problem is said to be made worse—often much worse—by "drugs."

In a war on "drugs," as in other wars, defining the enemy necessarily involves defining and teaching about morality, ethics, and the good things to be defended. Since the temperance or antialcohol campaigns of the nineteenth century, antidrug messages, especially those aimed at children and their parents, have had recognizable themes. Currently in the United States, these antidrug messages stress individual responsibility for health and economic success, respect for police, resistance to peer-group pressure, the value of God or a higher power in recovering from drug abuse, parents' knowledge of where their children are, sports and exercise as alternatives to drug use, drug testing of sports heroes, low grades as evidence of drug use, abstinence as the cause of good grades, and the need for parents to set good examples for their children. Almost anyone—police, politicians, schools, medical authorities, religious leaders—can find some value that can be defended or taught while attacking "drugs." (See the U.S. government-sponsored antidrug Web site at www.theantidrug.com).

In the United States, newspapers, magazines, and other media have long found that supporting antidrug campaigns is good for public relations. The media regularly endorse government antidrug campaigns and favorably cover antidrug efforts as a "public service." For doing so, they receive praise from government officials and prominent organizations. No doubt many publishers and editors deeply believe in the "war on drugs" and in defending the criminalized, prison-centered tradition of U.S. drug policies. But few of the other causes that people in the media support can be turned so easily into stories that are good for public relations and, simultaneously, that are very good for attracting customers and business.

Since at least the 1920s, top editors in the news media have recognized as an economic fact of their business that an alarming front-page antidrug story will likely increase sales of magazines and newspapers, especially when it is about a potential drug epidemic threatening to destroy middle-class teenagers, families, and neighborhoods. Editors know that a frightening story about a new, tempting, addicting drug attracts more TV viewers and radio listeners than most other kinds of news stories, including nonscary drug stories. In short, whatever their personal values, publishers, editors, and journalists give prominent space to scary antidrug articles because they know the stories attract customers.

Consider the case of crack cocaine and the still active U.S. war on drugs. In the 1980s, the media popularized the image of crack cocaine as "the most addicting drug known to man." Politicians from both parties then used that image to explain the deteriorating conditions in America's impoverished city neighborhoods and schools. Front-page stories in the *New York Times* and other publications warned that crack addiction was rapidly spreading to the suburbs and the middle class. In the election years of 1986 and 1988, politicians from both parties enthusiastically voted major increases in funding for police, prisons, and the military to save America's children from crack cocaine.

Even if crack was as bad as Republicans, Democrats, and the media said, it still probably could not have caused all the enduring problems they blamed on it, but the truth about crack cocaine is as startling as the myths. Crack cocaine, "the most addicting drug known to man," turned out to be a drug that very few people used continuously for long. Many Americans tried crack, but not many kept on using it heavily and steadily for a long time, mainly because most people cannot physically tolerate, much less enjoy, frequent encounters with crack's brutally brief and extreme ups and downs. Nor has crack become popular anywhere else in the world. Heavy, long-term crack smoking appeals only to a small number of deeply troubled people, most of whom are also impoverished. Because frequent bingeing on the drug is so thoroughly unappealing, it was extremely unlikely that an epidemic or plague of crack addiction would spread across America to the middle class and the suburbs.

Nonetheless, the contradictions between the drug war's myths about crack and the reality of crack cocaine's very limited appeal have not undermined the credibility or usefulness of antidrug messages, news stories, or political statements. In this respect, drug war propaganda is like the propaganda from other wars: the claims often remain useful even though they are patently false or do not make logical sense. In the 1990s, when crack cocaine finally ceased to be a useful enemy, American politicians, media, and police did not acknowledge their exaggerations and falsehoods about crack cocaine. They simply claimed victory, stopped discussing crack, and focused on other scary drugs, most recently MDMA (ecstasy) and prescription narcotics.

Additional Political and Ideological Support for Drug Prohibition

In many countries, popular and political support for drug prohibition also has been rooted in the widespread faith in the capacity of the state to penetrate and police many aspects of daily life for the "common good." This romantic or utopian view of the coercive state became especially strong and pervasive in the twentieth century. Unlike, say, the dissenters who insisted on the U.S. Bill of Rights in the eighteenth century, and unlike the members of many nineteenth-century political movements, in the twentieth century liberals, conservatives, fascists, communists, socialists, populists, left-wingers, and right-wingers shared this vision of the benevolent national state—if *they* controlled it. Drug prohibition was one of the few things on which they could all agree. Drug prohibition has been part of what I think it is appropriate to call the twentieth century's "romance with the state."

Because politicians in many countries, from one end of the political spectrum to the other, have shared this positive view of the powerful, coercive state, they could all agree on drug prohibition as sound nonpartisan government policy. In the United States during the 1980s and 1990s, Democrats feared and detested Presidents Reagan and Bush, and Republicans feared and detested President Clinton, but the parties united to fight the war on drugs. They even competed to enact more punitive antidrug laws, build more prisons, hire more drug police, expand antidrug military forces, and fund many more government-sponsored antidrug messages and crusades for a "drug-free" America. Opposing political parties around the world have fought about many things, but until recently they have often united in endorsing efforts to fight "drugs."

Finally, drug prohibition has enjoyed widespread support and legitimacy because the United States has used the UN as the international agency to create, spread, and supervise worldwide prohibition. Other than the U.S. government, the UN has done more to defend and extend drug prohibition than any other organization in the world. The UN currently identifies a "drug-free world" as the goal of its antidrug efforts (http://www.odccp.org/adhoc/gass/ga9411.htm). . . .

The Place of Harm Reduction Within Drug Prohibition

Since the early 1980s, harm-reduction workers and activists in Europe and increasingly throughout the world have sought to provide drug users and addicts with a range of services aimed at reducing the harmful effects of drug use. In the United States, conservative pundits and liberal journalists have accused harm-reduction advocates of being "drug legalizers" in disguise, but in most other countries many prominent politicians, public-health professionals, and police officials who are strong defenders of drug prohibition also have supported harm-reduction programs as practical public-health policies. Even the UN agencies that supervise worldwide drug prohibition have come to recognize the public-health benefits of harm-reduction services *within* current drug prohibition regimes.

A better understanding of the varieties and scope of worldwide drug prohibition helps us to see better the place of the "harm-reduction movement" within the history of drug prohibition. I suggest that harm reduction is a movement within drug prohibition that shifts drug polices from the criminalized and punitive end to the more decriminalized and openly regulated end of the drug policy continuum. Harm reduction is the name of the movement within drug prohibition that in effect (though not always in intent) moves drug policies away from punishment, coercion, and repression and toward tolerance, regulation, and public health. Harm reduction is not inherently an enemy of drug prohibition. However, in the course of pursuing public-health goals, harm reduction necessarily seeks to reduce the criminalized and punitive character of U.S.-style drug prohibition.

Consider the many programs identified as part of harm reduction: needle exchange and distribution, methadone maintenance, injection rooms, heroin clinics, medical use of marijuana by cancer and AIDS patients, truthful drug education aimed at users, drug-testing services at raves, and so on. Harm-reduction programs have pursued all these ways to increase public health and to help users reduce the harms of drug use. In order to carry out their stated objectives, these programs have often required laws, policies, or funding that reduce the harshness of drug prohibition. The reforms seek to reduce the punitive character of drug prohibition without necessarily challenging drug prohibition itself.

Harm-reduction advocates' stance toward drug prohibition is exactly the same as their stance toward drug use. Harm reduction seeks to reduce the harmful effects of drug use without requiring users to be drug free. It also seeks to reduce the harmful effects of drug prohibition without requiring governments to be prohibition free. Harm-reduction organizations say to drug users: "We are not asking you to give up drug use; we just ask you to do some things (such as using clean syringes) to reduce the harmfulness of drug use (including the spread of AIDS) to you and the people close to you." In precisely the same way, these organizations say to governments: "We are not asking you to give up drug prohibition; we just ask you to do some things (such as making clean syringes and methadone available) to reduce the harmfulness of drug prohibition."

Harm reduction offers a radically tolerant and pragmatic approach to both drug use and drug prohibition. It assumes that neither is going away soon and suggests therefore that reasonable and responsible people try to persuade both those who use drugs and those who use drug prohibition to minimize the harms that their activities produce. . . .

The Future of Global Drug Prohibition

Global drug prohibition is in crisis. The fact that it is at long last becoming visible is one symptom of that crisis. In the long run, the more criminalized and punitive forms of drug prohibition almost certainly are doomed. In the short run, the ever-growing drug-law and drug-policy reform movements make it likely that criminalized drug prohibition will find itself confronted with new opponents. (This prediction is already becoming a reality, in Switzerland, Australia, Germany, Portugal, Canada, the Netherlands, Spain, the United Kingdom, the United States, and other counties.)

In the twentieth century, for specific practical and ideological reasons, the nations of the world constructed a global system of drug prohibition. In the twenty-first century, because of the spread of democracy, trade, and information and for other practical and ideological reasons, the peoples of the world will likely dismantle and end worldwide drug prohibition.

It is important to understand that *this process of dismantling global drug prohibition will not end local drug prohibition.* The end of *global* drug prohibition will not (and cannot) be the end of *all* national drug prohibition. Advocating the end of worldwide drug prohibition is not the same as advocating worldwide drug legalization. Long after the demise of the UN's Single Convention, communities, regions, and some democratic nations will choose to retain forms of drug prohibition. Many places in the world will also continue to support vigorous antidrug crusades.

However, as accurate information about drug effects and alternative drug policies becomes more widespread, increasing numbers of countries, especially democratic ones, will likely choose not to retain full-scale criminalized drug prohibition. Most places eventually will develop their own varied local forms of regulated personal cultivation and use of the once-prohibited plants and substances. Many places also eventually will allow some forms of commercial production and sale—of cannabis, first of all and above all, because it is by far the most widely grown, traded, sold, and used illegal drug in the world.

These changes will take time. Prohibitionists and drug warriors in every country will fight tenaciously to maintain their local regimes, and enormous power will be employed to prevent the Single Convention of 1961 and its related treaties from being repealed or even modified. As a result, in coming years, all around the world there will be even greater public discussion and debate about drug prohibition, about criminalized drug policies, and about the worldwide movement within drug prohibition to decriminalize the possession and use of cannabis, cocaine, heroin, and other substances.

As part of that process of conversation and debate, many more people will discover—often with considerable astonishment—that they have lived for decades within a regime of worldwide drug prohibition. That growing understanding will itself push worldwide drug prohibition closer to its end. Here in the twenty-first century, it may turn out that the most powerful three holding global drug prohibition in place is the secret of its existence. . . .

POSTSCRIPT

Can the Global Community "Win" the Drug War?

The October 2003 Lisbon International Symposium on Global Drug Policy provided a forum for leading drug policymakers from national governments, senior representatives from various UN agencies, and other experts to address new ideas and innovative solutions. Speakers addressed such varied topics as an international framework for combating drugs, better public health policy, new approaches to the war on drugs, and the variety of new challenges facing the international community. Four key areas of division were spelled out by Martin Jelsma: (1) repression vs. protection; (2) zero tolerance vs. harm reduction; (3) the North-South or donors vs. recipients divide; and (4) demand vs. supply. The failure of nations of the world to reach agreement on these four major areas of contention has resulted, in the judgment of many, in the inability of the global community to address the drug problem successfully. Not enough funds are made available to international agencies like the United Nations, and money that is given is likely to have strings attached to it.

The conference was particularly timely because for the first time in the global war on drugs, policymakers did not debate the issue of whether the current policy was working. Instead, conference participants focused on how to organize a better drug control system. Honorary Secretary-General of Interpol, Raymond Kendall, echoed this view in his closing speech. While Kendall was pleased that debate had shifted to what he called "new levels," the conference did not develop a new plan of action. The issues outlined above were too great to overcome. Nonetheless, progress was made as national examples of successful alternative public policy programs were presented to the delegates.

The United Nations continued the theme of alternative approaches in its *2004 World Drug Report.* Acknowledging that effective strategies are discovered through trial and error, the UN alluded to a number of recent developments that appear to have potential for helping the global war on drugs. Four major ideas were discussed under the rubric of a "holistic approach." The first is "Addressing the drug problem in a broader sustainable development context." On the one hand, the drug problem hinders development in poor countries and compromises peacemaking efforts in countries torn by civil strife. On the other hand, "poverty, strife and feeble governance are fertile ground for drug production, trafficking and abuse." These situations are interconnected and can only be addressed by a comprehensive approach that recognizes the causes as well as the symptoms of the drug problem.

The second idea is "Providing an integrated response to the drugs and crime nexus." The connections between drug trafficking, organized crime, and even the financing of terrorism has meant that those responsible for addressing each of these scourges must work within the same multilateral system rather than in isolation.

The third development is "Addressing the drugs and crime nexus under the new paradigm of human security." Growing out of the 2000 UN Millennium Summit, the Commission on Human Security is developing a new approach to security that combines human development and human rights. The UN report suggests that this could provide a critical link between drugs/crime control and sustainable development.

The fourth development is termed "a more synergistic approach." This simply means that not only must there be an integrated and balanced approach to the war on drugs, but that much more needs to be learned. For example, the structure and dynamics of drug markets at the national, regional, and global levels are a mystery beyond the simple belief that normal supply and demand principles are at work.

The 2004 UN report also called for deeper understanding of and a focus on controlling drug epidemics. The report summed it up this way. "The powerful dynamics created by the combination of the incentives and behavior of a ruthless market with the contagious characteristics of an epidemic explain why drug use can expand so rapidly and become so difficult to stem."

In the selection by Mayor, the basic question of eliminating the supply or drying up the demand is raised. Mayor believes it is critical that a major emphasis be placed on both sides of the equation. His motivation is based on the dual points that eliminating the supply is difficult but that the health issues associated with illegal drug use demand that we educate existing and potential users about the evils of such behavior. Addicts must be treated as patients, not criminals. At the same time, Mayor argues that the most effective way to fight the drug war is to destroy the financial power of organized crime. Both strategies are critically important.

Levine suggests that global drug prohibition and a focus on both punishing the supplier and the user have not worked very well. Instead, these approaches must be reexamined with a view toward addressing the plight of the drug user in a much different way.

A number of publications provide insight into the war on drugs. Twenty-eight speeches from the aforementioned 2003 Lisbon Conference are reproduced in *Global Drug Policy: Building a New Framework* (The Senlis Council, February 2004). A study prepared for the U.S. Congress by the Congressional Research Service describes U.S. international drug policy (Raphael Perl, "Drug Control: International Policy and Approaches," September 8, 2003).

Criticism of the U.S.-dominated global approach can be found in a number of sources. An important one focusing on Latin America is Ted Galen Carpenter's *Bad Neighbor Policy* (2003). *The Economist* suggested in its subtitle to an article on drugs that it was "Time to think again about the rules of engagement in the war on drugs" ("Breaking Convention," vol. 366, issue 8318, April 5, 2003). An article offering another perspective is Adam Isacson's "Washington's 'New War' in Columbia: The War on Drugs Meets the War on Terror" (*NACLA Report on the Americas*, vol. 36, issue 5, March/April 2003).

A recent comprehensive official report is a U.S. Department of State publication, "International Narcotics Control Strategy Report" (March 2006). Its central message is that "international drug control efforts kept the drug trade on the defensive in 2005." Several long-sought drug kingpins were arrested during the year as well.

ISSUE 11

Is the International Community Adequately Prepared to Address Global Health Pandemics?

YES: Global Influenza Programme, from "Responding to the Avian Influenza Pandemic Threat," World Health Organization (2005)

NO: H. T. Goranson, from "A Primer for Pandemics," Global Envision, http://www.globalenvision.org (2005)

ISSUE SUMMARY

YES: The document from the World Health Organization lays out a comprehensive program of action for individual countries, the international community, and WHO to address the next influenza pandemic.

NO: H. T. Goranson, a former top national scientist with the U.S. government, describes the grave dangers posed by global pandemics and highlights flaws in the international community's ability to respond.

Hear the words "global pandemics" and one thinks of the bubonic plague or Black Death of the Middle Ages where an estimated 30 percent of Europe's population died, or the influenza epidemic of 1918 that killed between 25 million and 50 million people worldwide. Both seem (and are) stories from a bygone era, when modern medicine was unknown and people were simply at the mercy of the spreading tendencies of the virulent diseases. The world of medicine is different today, which leads many to assume that somewhere on the shelves of the local pharmacy or the Centers for Disease Control in Atlanta lies a counteragent to whatever killer lurks out there.

The world is far different from that of the fourteenth century or even 1918. Globalization is with us. The world has shrunk, literally and figuratively, as the human race's ability to move people, money, goods, information and also unwanted agents across national boundaries and to the far corners of the globe has increased exponentially. Viruses, germs, parasites, and other virulent disease agents can and do move much more easily than at any time in recorded history.

An article prepared for Risk Management LLC by Anup Shah of www.globalissues.org suggests that the problem is compounded by a number of other factors. One billion people have no access to health systems. Over 10 million people died in a recent year from infectious diseases and a similar number of children under the age of five suffer from malnutrition and other diseases. AIDS/HIV has spread rapidly with 40 million people living with HIV. These conditions help to facilitate the movement of major contagious diseases.

The word "pandemic" is derived from two Greek words, "all" and "people." Thus a global pandemic is an epidemic of some infectious disease that can and is spreading at a rapid rate throughout the world. Throughout history, humankind has fallen victim to many such killers. As early as the Peloponnesian War in fifth-century B.C. Greece, typhoid fever was responsible for the deaths of upwards of 25 percent of combatants and civilians alike, necessitating major changes in military tactics. Imperial Rome felt the wrath of a plague thought to be smallpox, as did the eastern Mediterranean during its political height several centuries later. In the past 100 years, influenza (1918, 1957, and 1968), typhoid, and cholera were major killers. In recent years, other infectious diseases have made front page news: HIV, ebola virus, SARS, and most recently, avian or bird flu. At this moment, the latter flu is striking tremendous fear in the hearts of global travelers and governmental policymakers everywhere.

According to World Health Organization Europe, as many as 175 million to 360 million people could fall victim if the outbreak were severe enough. The bird flu is front page news because more than 150 million birds have died worldwide from one of its strains, H5N1. This strain was first found in humans in 1997, and WHO estimates report that the human fatality rate has been 50 percent, with 69 deaths occurring as of December 2005. One might be prompted to ask: what is the "big deal, only 134 confirmed cases?" It is not quite so simple. Unlike previous pandemics that hit suddenly and without little or any warning, the avian flu is giving us a clear warning. The loss in poultry has been enormous. And with the jump to humans, with an initial high mortality rate, our senses have been awakened to the potential for global human disaster.

But there is good news as well. There does appear to be time to prepare for the worst-case scenario and diminish its likelihood. The flu has the attention of all relevant world health agencies and most national agencies, and steps have been undertaken and/or are currently underway to find a way to combat this contagious disease. While this global issue addresses pandemics in general, we have selected avian flu as a case study of world pandemics and global responses because of its current notoriety with the media and policymakers alike. In the first selection, the World Health Organization lays out a comprehensive program of action for individual countries, the international community, and WHO to address the next influenza pandemic. In the second selection, H. T. Goranson, a former top national scientist with the U.S. government, cautions us that the task is so enormously difficult. He describes the grave dangers posed by global pandemics and highlights flaws in the international community's ability to respond.

YES

Responding to the Avian Influenza Pandemic Threat

Purpose

This document sets out activities that can be undertaken by individual countries, the international community, and WHO to prepare the world for the next influenza pandemic and mitigate its impact once international spread has begun. Recommended activities are specific to the threat posed by the continuing spread of the H5N1 virus. Addressed to policy-makers, the document also describes issues that can guide policy choices in a situation characterized by both urgency and uncertainty. Recommendations are phase-wise in their approach, with levels of alert, and corresponding activities, changing according to epidemiological indicators of increased threat.

In view of the immediacy of the threat, WHO recommends that all countries undertake urgent action to prepare for a pandemic. Advice on doing so is contained in the recently revised *WHO global influenza preparedness plan*[1] and a new *WHO checklist for influenza pandemic preparedness planning.*[2] To further assist in preparedness planning, WHO is developing a model country plan that will give many developing countries a head start in assessing their status of preparedness and identifying priority needs. Support for rehearsing these plans during simulation exercises will also be provided.

Opportunities to Intervene

As the present situation continues to evolve towards a pandemic, countries, the international community, and WHO have several phase-wise opportunities to intervene, moving from a pre-pandemic situation, through emergence of a pandemic virus, to declaration of a pandemic and its subsequent spread. During the present pre-pandemic phase, interventions aim to reduce the risk that a pandemic virus will emerge and gather better disease intelligence, particularly concerning changes in the behaviour of the virus that signal improved transmissibility. The second opportunity to intervene occurs coincident with the first signal that the virus has improved its transmissibility, and aims to change the early history of the pandemic. The final opportunity

From *Communicable Disease Surveillance and Response Global Influenza Programme*, 2005, pp. 1–4, 6–7, 9–10, 12, 14–15. Copyright © 2005 by World Health Organization. Reprinted by permission.

occurs after a pandemic has begun. Interventions at this point aim to reduce morbidity, mortality, and social disruption.

Objectives

The objectives of the strategic actions correspond to the principal opportunities to intervene and are likewise phase-wise.

Phase: pre-pandemic
1. Reduce opportunities for human infection
2. Strengthen the early warning system
Phase: emergence of a pandemic virus
3. Contain or delay spread at the source
Phase: pandemic declared and spreading internationally
4. Reduce morbidity, mortality, and social disruption
5. Conduct research to guide response measures

Strategic Actions

The document describes strategic actions that can be undertaken to capitalize on each opportunity to intervene. Given the many uncertainties about the evolution of the pandemic threat, including the amount of time left to prepare, a wise approach involves a mix of measures that immediately address critical problems with longer-term measures that sustainably improve the world's capacity to protect itself against the recurring pandemic threat.

Background

Influenza pandemics have historically taken the world by surprise, giving health services little time to prepare for the abrupt increases in cases and deaths that characterize these events and make them so disruptive. Vaccines—the most important intervention for reducing morbidity and mortality—were available for the 1957 and 1968 pandemic viruses, but arrived too late to have an impact. As a result, great social and economic disruption, as well as loss of life, accompanied the three pandemics of the previous century.

The present situation is markedly different for several reasons. First, the world has been warned in advance. For more than a year, conditions favouring another pandemic have been unfolding in parts of Asia. Warnings that a pandemic may be imminent have come from both changes in the epidemiology of human and animal disease and an expanding geographical presence of the virus, creating further opportunities for human exposure. While neither the timing nor the severity of the next pandemic can be predicted, evidence that the virus is now endemic in bird populations means that the present level of risk will not be easily diminished.

Second, this advance warning has brought an unprecedented opportunity to prepare for a pandemic and develop ways to mitigate its effects. To date, the main preparedness activities undertaken by countries have concentrated on preparing and rehearsing response plans, developing a pandemic vaccine, and securing supplies of antiviral drugs. Because these activities are costly, wealthy countries are presently the best prepared; countries where H5N1 is endemic—and where a pandemic virus is most likely to emerge—lag far behind. More countries now have pandemic preparedness plans: around one fifth of the world's countries have some form of a response plan, but these vary greatly in comprehensiveness and stage of completion. Access to antiviral drugs and, more importantly, to vaccines remains a major problem because of finite manufacturing capacity as well as costs. Some 23 countries have ordered antiviral drugs for national stockpiles, but the principal manufacturer will not be able to fill all orders for at least another year. Fewer than 10 countries have domestic vaccine companies engaged in work on a pandemic vaccine. A November 2004 WHO consultation reached the stark conclusion that, on present trends, the majority of developing countries would have no access to a vaccine during the first wave of a pandemic and possibly throughout its duration.

Apart from stimulating national preparedness activities, the present situation has opened an unprecedented opportunity for international intervention aimed at delaying the emergence of a pandemic virus or forestalling its international spread. Doing so is in the self-interest of all nations, as such a strategy could gain time to augment vaccine supplies. At present capacity, each day of manufacturing gained can mean an additional 5 million doses of vaccine. International support can also strengthen the early warning system in endemic countries, again benefiting preparedness planning and priority setting in all nations. Finally, international support is needed to ensure that large parts of the world do not experience a pandemic without the protection of a vaccine.

Pandemics are remarkable events in that they affect all parts of the world, regardless of socioeconomic status or standards of health care, hygiene and sanitation. Once international spread begins, each government will understandably make protection of its own population the first priority. The best opportunity for international collaboration—in the interest of all countries—is now, before a pandemic begins.

Situation Assessment

1. The risk of a pandemic is great.

Since late 2003, the world has moved closer to a pandemic than at any time since 1968, when the last of the previous century's three pandemics occurred.

All prerequisites for the start of a pandemic have now been met save one: the establishment of efficient human-to-human transmission. During 2005, ominous changes have been observed in the epidemiology of the disease in animals. Human cases are continuing to occur, and the virus has expanded its geographical range to include new countries, thus increasing the size of the population at risk. Each new human case gives the virus an opportunity to evolve towards a fully transmissible pandemic strain.

2. The risk will persist.

Evidence shows that the H5N1 virus is now endemic in parts of Asia, having established an ecological niche in poultry. The risk of further human cases will persist, as will opportunities for a pandemic virus to emerge. Outbreaks have recurred despite aggressive control measures, including the culling of more than 140 million poultry. Wild migratory birds—historically the host reservoir of all influenza A viruses—are now dying in large numbers from highly pathogenic H5N1. Domestic ducks can excrete large quantities of highly pathogenic virus without showing signs of illness. Their silent role in maintaining transmission further complicates control in poultry and makes human avoidance of risky behaviours more difficult.

3. Evolution of the threat cannot be predicted.

Given the constantly changing nature of influenza viruses, the timing and severity of the next pandemic cannot be predicted. The final step—improved transmissibility among humans—can take place via two principal mechanisms: a reassortment event, in which genetic material is exchanged between human and avian viruses during co-infection of a human or pig, and a more gradual process of adaptive mutation, whereby the capability of these viruses to bind to human cells would increase during subsequent infections of humans. Reassortment could result in a fully transmissible pandemic virus, announced by a sudden surge of cases with explosive spread. Adaptive mutation, expressed initially as small clusters of human cases with evidence of limited transmission, will probably give the world some time to take defensive action. Again, whether such a "grace period" will be granted is unknown.

4. The early warning system is weak.

As the evolution of the threat cannot be predicted, a sensitive early warning system is needed to detect the first sign of changes in the behaviour of the virus. In risk-prone countries, disease information systems and health, veterinary, and laboratory capacities are weak. Most affected countries cannot adequately compensate farmers for culled poultry, thus discouraging the reporting of outbreaks in the rural areas where the vast majority of human cases have occurred. Veterinary extension services frequently fail to reach these areas. Rural poverty perpetuates high-risk behaviours, including the traditional home-slaughter and consumption of diseased birds. Detection of human cases is impeded by patchy surveillance in these areas. Diagnosis of human cases is impeded by weak laboratory support and the complexity and high costs of testing. Few affected countries have the staff and resources needed to thoroughly investigate human cases and, most importantly, to detect and investigate clusters of cases—an essential warning signal. In virtually all affected countries, antiviral drugs are in very short supply.

The dilemma of preparing for a potentially catastrophic but unpredictable event is great for all countries, but most especially so for countries affected by H5N1 outbreaks in animals and humans. These countries, in which rural subsistence farming is a backbone of economic life, have experienced direct and enormous agricultural losses, presently estimated at more than US$ 10 billion. They are being asked to sustain—if not intensify—resource-intensive activities needed to safeguard international public health while struggling to cope with many other competing health and infectious disease priorities.

5. Preventive intervention is possible, but untested.

Should a pandemic virus begin to emerge through the more gradual process of adaptive mutation, early intervention with antiviral drugs, supported by other public health measures, could theoretically prevent the virus from further improving its transmissibility, thus either preventing a pandemic or delaying its international spread. While this strategy has been proposed by many influenza experts, it remains untested; no effort has ever been made to alter the natural course of a pandemic at its source.

6. Reduction of morbidity and mortality during a pandemic will be impeded by inadequate medical supplies.

Vaccination and the use of antiviral drugs are two of the most important response measures for reducing morbidity and mortality during a pandemic. On present trends, neither of these interventions will be available in adequate quantities or equitably distributed at the start of a pandemic and for many months thereafter.

1. Reduce Opportunities for Human Infection

Strategic Actions

- **Support the FAO/OIE control strategy**

  The FAO/OIE technical recommendations describe specific control measures and explain how they should be implemented. The global strategy, developed in collaboration with WHO, takes its urgency from the risk to human health, including that arising from a pandemic, posed by the continuing circulation of the virus in animals. The strategy adopts a progressive approach, with different control options presented in line with different disease profiles, including such factors as poultry densities, farming systems, and whether infections have occurred in commercial farms or small rural holdings.

The proposed initial focus is on Viet Nam, Thailand, Cambodia, and Indonesia, the four countries where human cases of infection with H5N1 avian influenza have been detected.

Clear and workable measures are proposed for different countries and situations within countries. Vaccination is being recommended as an appropriate control measure in some, but not all, epidemiological situations. Other measures set out in the strategy include strict biosecurity at commercial farms, use of compartmentalization and zoning concepts, control of animal and product movements, and a restructuring of the poultry industry in some countries. The strategy notes strong political will to tackle the problem. Nonetheless, time-frames for reaching control objectives are now being measured in years.

In July 2005, OIE member countries approved new standards, recognized by the World Trade Organization, specific to avian influenza and aimed at improving the safety of international trade of poultry and poultry products. The new standards cover methods of surveillance, compulsory international notification of low- and highly-pathogenic strains of avian influenza, the use of vaccination, and food safety of poultry products. Compliance with these standards should be given priority in efforts to strengthen early detection, reporting, and response in countries currently affected by outbreaks of H5N1 avian influenza.

- **Intensify collaboration between the animal and public health sectors**

WHO will appoint dedicated staff to increase the present exchange of information between agricultural and health sectors at the international level. Increased collaboration between the two sectors serves three main purposes: to pinpoint areas of disease activity in animals where vigilance for human cases should be intensified, to ensure that measures for controlling the disease in animals are compatible with reduced opportunities for human exposure, and to ensure that advice to rural communities on protective measures remains in line with the evolving nature of the disease in animals.

WHO will undertake joint action with FAO and OIE to understand the evolution of H5N1 viruses in Asia. Achieving this objective requires acquisition and sharing of a full inventory of H5N1 viruses, from humans, poultry, wild birds, and other animals, and sequences.

WHO will stress the importance of controlling the disease in rural areas. Measures to control the disease in animals of necessity consider how best to regain agricultural productivity and international trade, and this objective is reflected in the FAO/OIE strategy. While elimination of the virus from the commercial poultry sector alone will aid agricultural recovery, it may not significantly reduce opportunities for human exposure, as the vast majority of cases to date have been associated with exposure to small rural flocks. No case has yet been detected among workers in the commercial poultry sector. The FAO/OIE strategy fully recognizes that control of disease in rural "backyard" flocks will be the most difficult challenge; strong support from the health sector, as expressed by WHO, helps gather the political will to meet this challenge. In addition, it is imperative that measures for controlling disease in rural flocks are accompanied by risk communication to farmers and their families.

A joint FAO/OIE/WHO meeting, held in Malaysia in July 2005,[3] addressed the links between animal disease and risks of human exposures and infections, and defined preventive measures that should be jointly introduced by the animal and public health sectors. Priority was given to interventions in the backyard rural farming system and in so-called wet markets where live poultry are sold under crowded and often unsanitary conditions.

WHO, FAO, and OIE have jointly established a Global Early Warning and Response System (GLEWS) for transboundary animal diseases. The new mechanism combines the existing outbreak alert, verification, and response capacities of the three agencies and helps ensure that disease tracking at WHO benefits from the latest information on relevant animal diseases. The system formalizes the sharing of epidemiological information and provides the operational framework for joint field missions to affected areas.

- **Strengthen risk communication to rural residents**

WHO will, through its research networks and in collaboration with FAO and OIE, improve understanding of the links between animal disease, human behaviours, and the risk of acquiring H5N1 infection. This information will be used as the basis for risk communication to rural residents.

Well-known and avoidable behaviours with a high risk of infection continue to occur in rural areas. Ongoing risk communication is needed to alert rural residents to these risks and explain how to avoid them. Better knowledge about the relationship between animal and human disease, obtained by WHO in collaboration with FAO/OIE, can be used to make present risk communication more precise and thus better able to prevent risky behaviours.

- **Improve approaches to environmental detection of the virus**

WHO, FAO, and OIE will facilitate, through their research networks, the rapid development of new methods for detecting the virus in environmental samples. The purpose of these methods is to gain a better understanding of conditions that increase the risk of human infection and therefore favour emergence of a pandemic virus. Such knowledge underpins the success of primary prevention of a pandemic through disease control in animals. It also underpins advice to rural residents on behaviours to avoid. Reliance on routine veterinary surveillance, which is weak in most risk-prone countries, has not produced an adequate understanding of the relationship between animal and human disease. For example, in some cases, outbreaks in poultry are detected only after a human case has been confirmed. In other cases, investigation of human cases has failed to find a link with disease in animals.

2. Strengthen the Early Warning System

Strategic Actions

- **Improve the detection of human cases**

  WHO will provide the training, diagnostic reagents, and administrative support for external verification needed to improve the speed and reliability of case detection. To date, the vast majority of cases have been detected following hospitalization for respiratory illness. Hospitals in affected countries need support in case detection, laboratory confirmation, and reporting. Apart from its role in an early warning system, rapid laboratory confirmation signals the need to isolate patients and manage them according to strict procedures of infection control, and can thus help prevent further cases.

  Diagnostic support continues to be provided by laboratories in the WHO network. However, because the initial symptoms of H5N1 infection mimic those of many diseases common in these countries, accurate case detection requires the testing of large numbers of samples. Improved local capacity is therefore a more rational solution.

  Because of its high pathogenicity, H5N1 can be handled safely only by specially trained staff working in specially equipped laboratories operating at a high level of biosecurity. These facilities do not presently exist in the majority of affected countries. As an alternative, laboratory capacity can be enhanced by strengthening the existing system of national influenza centres or by providing mobile high-containment laboratories. Supportive activities include training in laboratory methods needed for H5N1 diagnosis, distribution of up-to-date diagnostic reagents, and coordination of work between national laboratories and epidemiological institutions.

  An infrastructure needs to be developed to complement national testing with rapid international verification in WHO certified laboratories, especially as each confirmed human case yields information essential to risk assessment. The capacity to do so already exists. WHO offers countries rapid administrative support to ship samples outside affected countries. Such forms of assistance become especially critical when clusters of cases occur and require investigation.

- **Combine detection of new outbreaks in animals with active searches for human cases**

  Using epidemiologists in its country offices and, when necessary, external partners, WHO will ensure that detection of new outbreaks of highly pathogenic H5N1 in poultry is accompanied by active searches for human cases. Surveillance in several countries where H5N1 is considered endemic in birds is inadequate and suspicions are strong that human cases have been missed. Cambodia's four human cases were detected only after patients sought treatment in neighbouring Viet Nam, where physicians are on high alert for cases and familiar with the clinical presentation.

- Support epidemiological investigation

  Reliable risk assessment depends on thorough investigation of sporadic human cases and clusters of cases. Guidelines for outbreak investigation, specific to H5N1 and to the epidemiological situation in different countries, are being developed on an urgent basis for use in training national teams. These guidelines give particular emphasis to the investigation of clusters of cases and determination of whether human-to-human transmission has occurred. Teams assembled from institutions in the WHO Global Outbreak Alert and Response Network (GOARN) can be deployed for rapid on-site investigative support.

- Coordinate clinical research in Asia

  Clinical data on human cases need to be compiled and compared in order to elucidate modes of transmission, identify groups at risk, and find better treatments. Work has begun to establish a network of hospitals, modelled on the WHO global influenza surveillance network, engaged in clinical research on human disease. The network will link together the principal hospitals in Asia that are treating H5N1 patients and conducting clinical research. Technical support will allow rapid exchange of information and sharing of specimens and research results, and encourage the use of standardized protocols for treatment and standardized sampling procedures for investigation.

  Identification of risk groups guides preventive measures and early interventions. Provision of high-quality data on clinical course, outcome, and treatment efficacy meets an obvious and immediate need in countries with human cases. Answers to some key questions—the efficacy of antiviral drugs, optimum dose, and prescribing schedules—could benefit health services elsewhere once a pandemic is under way.

- Strengthen risk assessment

  WHO's daily operations need to be strengthened to ensure constant collection and verification of epidemiological and virological information essential for risk assessment. Ministries of health and research institutions in affected countries need to be more fully engaged in the collection and verification of data. Ministries and institutions in non-affected countries should help assess the significance of these data, and the results should be issued rapidly. These activities, currently coordinated by WHO, need to escalate; influenza viruses can evolve rapidly and in unexpected ways that alter risk assessment, as evidenced by the recent detection of highly pathogenic H5N1 viruses in migratory birds. Functions of the WHO network of laboratories with expertise in the analysis of H5N1 viruses can be improved through tools, such as a genetic database, and a strong collaboration with veterinary laboratory networks to ensure that animal as well as human viruses are kept under constant surveillance.

- **Strengthen existing national influenza centres throughout the risk-prone region**

  Many existing national influenza centres, designated by WHO, already possess considerable infrastructure in the form of equipment and trained personnel. Additional support, particularly in the form of diagnostic reagents, could help strengthen the early warning system in risk-prone countries and their neighbours.

- **Give risk-prone countries an incentive to collaborate internationally**

  The promise of assistance is a strong motivation to report cases and share clinical specimens internationally. A high-level meeting should be convened so that heads of state in industrialized countries and in risk-prone countries can seek solutions and reach agreement on the kinds of support considered most desirable by individual countries.

3. Contain or Delay Spread at the Source

Strategic Actions

- **Establish an international stockpile of antiviral drugs**

  WHO will establish an international stockpile of antiviral drugs for rapid response at the start of a pandemic. The stockpile is a strategic option that serves the interests of the international community as well as those of the initially affected populations. Issues that need to be addressed include logistics associated with deployment and administration, and licensing for use in individual countries. Mechanisms for using an international stockpile need to be defined more precisely in terms of epidemiological triggers for deploying the stockpile and time-frames for emergency delivery and administration. WHO is working closely with groups engaged in mathematical modelling and others to guide the development of early containment strategies.

  While pursuit of this option thus has no guarantee of success, it nonetheless needs to be undertaken as it represents one of the few preventive options for an event with predictably severe consequences for every country in the world. It is also the best guarantee that populations initially affected will have access to drugs for treatment. Should early containment fail to completely halt spread of the virus, a delay in wide international spread would gain time to intensify preparedness. It can be expected that most governments will begin introducing emergency measures only when the threat of a pandemic is certain and immediate. A lead time for doing so of one month or more could allow many health services to build surge capacity and make the necessary conversion from routine to emergency services.

- **Develop mass delivery mechanisms for antiviral drugs**

  Several WHO programmes, such as those for the emergency response to outbreaks of poliomyelitis, measles, epidemic-prone meningitis, and yellow fever, have acquired considerable experience in the urgent mass delivery of vaccines in developing countries. Less experience exists for the mass delivery of antiviral drugs, where administration is complicated by the need for drugs to be taken over several days and the need for different dosing schedules according to therapeutic or prophylactic use. WHO will develop and pilot test delivery mechanisms for antiviral drugs in collaboration with national health authorities and industry. Studies will assess coverage rates that could be achieved, taking into account compliance rates, and ways to support this intervention with other measures, such as area quarantine.

- **Conduct surveillance of antiviral susceptibility**

  Using its existing network of influenza laboratories, WHO will establish a surveillance programme for antiviral susceptibility testing, modelled on a similar programme for anti-tuberculosis drugs. Use of an international stockpile to attempt to halt an outbreak will involve administration of drugs to large numbers of people for several weeks. A mechanism must be in place to monitor any resulting changes in virus susceptibility to these drugs. The development of drug resistance would threaten the effectiveness of national stockpiles of antiviral drugs established for domestic use. The work of WHO collaborating centres for influenza and reference laboratories for H5N1 analysis can be coordinated to include antiviral susceptibility testing.

4. Reduce Morbidity, Mortality and Social Disruption

Strategic Actions

- **Monitor the evolving pandemic in real time**

  Many characteristics of a pandemic that will guide the selection of response measures will become apparent only after the new virus has emerged and begun to cause large numbers of cases. WHO, assisted by virtual networks of experts, will monitor the unfolding epidemiological and clinical behaviour of the new virus in real time. This monitoring will give health authorities answers to key questions about age groups at greatest risk, infectivity of the virus, severity of the disease, attack rates, risk to health care workers, and mortality rates. Such monitoring can also help determine whether severe illness and deaths are caused by primary viral pneumonia or secondary bacterial pneumonia, which responds to antibiotics, and thus guide the emergency provision of supplies. Experts in mathematical modelling will be included in the earliest field assessment teams to make the forecasting of trends as reliable as possible.

- **Introduce non-pharmaceutical interventions**

  Answers to these questions will help officials select measures—closing of schools, quarantine, a ban on mass gatherings, travel restrictions—that match the behaviour of the virus and thus have the greatest chance of reducing the number of cases and delaying geographical spread. WHO has produced guidance on the use of such measures at different stages at the start of a pandemic and after its international spread.

- **Use antiviral drugs to protect priority groups**

  WHO recommends that countries with sufficient resources invest in a stockpile of antiviral drugs for domestic use, particularly at the start of a pandemic when mass vaccination is not an option and priority groups, such as frontline workers, need to be protected.

- **Augment vaccine supplies**

  WHO, in collaboration with industry and regulatory authorities, has introduced fast-track procedures for the development and licensing of a pandemic vaccine. Strategies have also been developed that make the most of scarce vaccine antigen and thus allow more quantities of vaccine to be produced despite the limits of existing plant capacity. Once a pandemic is declared, all manufacturers will switch from production of seasonal vaccines to production of a pandemic vaccine. Countries need to address liability issues that could arise following mass administration of a pandemic vaccine and ensure adequate warehousing, logistics, and complementary supplies, such as syringes.

- **Ensure equitable access to vaccines**

  The present strong interdependence of commerce and trade means that the international community cannot afford to allow large parts of the world to experience a pandemic unprotected by a vaccine. The humanitarian and ethical arguments for providing such protection are readily apparent. As a matter of urgency, WHO must build a political process aimed at finding ways to further augment production capacity dramatically and make vaccines affordable and accessible in the developing world. WHO will also work with donor agencies on the latter issue.

- **Communicate risks to the public**

  As soon as a pandemic is declared, health authorities will need to start a continuous process of risk communication to the public. Many difficult issues—the inevitable spread to all countries, the shortage of vaccines and antiviral drugs, justification for the selection of priority groups for protection—will need to be addressed. Effective risk communication, supported by confidence in government authorities and the reliability of their information, may help mitigate some of the social and economic disruption attributed to an anxious public. Countries are advised to plan in advance. A communication strategy for a pandemic situation should include training in outbreak communication and integration of communicators in senior management teams.

5. Conduct Research to Guide Response Measures

Strategic Actions

- **Assess the epidemiological characteristics of an emerging pandemic**

  At the start of a pandemic, policy-makers will face an immediate need for epidemiological data on the principal age groups affected, modes of transmission, and pathogenicity. Such data will support urgent decisions about target groups for vaccination and receipt of antiviral drugs. They can also be used to support forecasts on local and global patterns of spread as an early warning that helps national authorities intensify preparedness measures. WHO will identify epidemiological centres for collecting these data and establish standardized research protocols.

- **Monitor the effectiveness of health interventions**

  Several non-pharmaceutical interventions have been recommended to reduce local and international spread of a pandemic and lower the rate of transmission. While many of these interventions have proved useful in the prevention and control of other infectious diseases, their effectiveness during a pandemic has never been comprehensively evaluated. More information is needed on their feasibility, effectiveness, and acceptability to populations. WHO will establish study sites and develop study protocols to evaluate these interventions at local, national, and international levels. Comparative data on the effectiveness of different interventions are also important, as several measures are associated with very high levels of social disruption.

- **Evaluate the medical and economic consequences**

  WHO will establish study sites and develop protocols for prospective evaluation of the medical and economic consequences of the pandemic so that future health interventions can be adjusted accordingly. In the past, such evaluations have been conducted only after a pandemic had ended. Their value as a policy guide for the allocation of resources has been flawed because of inadequate data.

Notes

1.　http://whqlibdoc.who.int/hq/2005/WHO_CDS_CSR_GIP_2005.5.pdf.
2.　http://whqlibdoc.who.int/hq/2005/WHO_CDS_CSR_GIP_2005.4.pdf.
3.　http://www.fao.org/ag/againfo/subjects/documents/ai/concmalaysia.pdf.

A Primer for Pandemics

A few times each year, the world is reminded that a pandemic threat is immanent. What can we do to prepare for the next one?

According to Dr. Tim Evans, Assistant Director-General for Evidence and Information for Policy, World Health Organization: "There is a chronic global shortage of health workers, as a result of decades of underinvestment in their education, training, salaries, working environment and management. This has led to a severe lack of key skills, rising levels of career switching and early retirement, as well as national and international migration. In sub-Saharan Africa, where all the issues mentioned above are combined with the HIV/AIDS pandemic, there are an estimated 750,000 health workers in a region that is home to 682 million people. By comparison, the ratio is ten to 15 times higher in OECD countries, whose ageing population is putting a growing strain on an over-stretched workforce. Solutions to this crisis must be worked out at local, national and international levels, and must involve governments, the United Nations, health professionals, non-governmental organizations and community leaders."

"There is no single solution to such a complex problem, but ways forward do exist and must now be implemented. For example, some developed countries have put policies in place to stop active recruitment of health workers from severely understaffed countries. Some developing countries have revised their pay scales and introduced non-monetary incentives to retain their workforce and deploy them in rural areas. Education and training procedures have been tailored to countries' specific needs. Community health workers are helping their communities to prevent and treat key diseases. Action must be taken now for results to show in the coming years."

This article takes a look at global pandemics, and how medical professionals worldwide trained as "detectors" would be the best way to halt the spread of a disease before it became a global threat.

༺◦◉◦༻

A few times each year, the world is reminded that a pandemic threat is immanent. In 2003, it was SARS. Today, it is a potential avian virus similar to the one that killed 30 million people after 1914.

"Bird flu" has already shown that it can jump from fowl to humans, and now even to cats, which indicates that it might be the next global killer. But there are many other potential pandemics, and many are not even viruses. Bacteria, prions, parasites, and even environmental factors could suddenly change in a way that slays us. It is widely predicted that when this happens, the economic and human losses will exceed that of any previous war.

Indeed, it is humbling to remember that some of history's most deadly invasions were carried out by single-cell organisms, such as cholera, bubonic plague, and tuberculosis. Countries with the resources to do so are making resistance plans against pandemics—limited strategies that would protect their own citizens. Most governments are hoping that early detection will make containment possible.

Containment depends heavily on vaccines, but vaccines are only part of the answer. While they are a good defense against many viruses, each vaccine is highly specific to the threat. Viruses are parasites to cells, and each virus attacks a particular type of cell. The virus is shaped so that it can drill into a particular feature of that cell and inject parts of itself inside, confusing the cell into making more viruses and destroying itself in the process. With their very specific forms, the most effective anti-viral vaccines must be designed for a narrow range of factors.

Sometimes the tailored nature of viruses works in our favor. For example, they usually find it difficult to jump between species, because they would have to change their structure. But if large numbers of a host—say, birds—encounter a great number of people, eventually the virus will find a way to prosper in a new type of cell.

Birds are the greatest concern today only because the spread is easy to see. But AIDS jumped from monkeys and several types of flu jumped from swine. Deadly mutations of any kind need to be identified urgently, so that an effective vaccine can be designed before the strain becomes comfortable in the human body. Unfortunately our present methods of detection are not sensitive enough.

This is even more worrying when you realize that scientists should also be monitoring bacteria, prions, and parasites. There are more bacteria than any other life form. Many live harmlessly in our bodies and perform useful functions. They evolve and adapt easily, which means that they learn to sidestep our drugs over time. Bacteria should be checked for two types of mutation: adaptation by a hostile form that enables it to become super-immune to drugs, or a deadly mutant strain that appears in one of the multitude of "safe" bacteria.

Prions are a relatively new discovery. They are made from proteins similar to those that the body uses during healthy operations, which means that they are able to fool the body's tools into making more prions. They have only recently been recognized as the cause of several infectious diseases, including mad cow disease and Creutzfeldt-Jakob Disease, which kill by crowding out healthy brain cells. Many nerve, respiratory and muscle diseases might also be caused by prions.

Finally, parasites, simple animals that infect us, are already classified as pandemics. Malaria afflicts 300 million people and is the world's biggest killer of children. Many parasites are worms: hookworm (800 million people infected), roundworm (1.5 billion), schistosomes (200 million), and the worm that causes Elephantiasis (150 million).

There are also antagonists that are currently ignored. Environmental chemicals and particulates might warrant their own categories. Or consider combinations of problems, such as these chemical infectors mixing with airborne pollens, and apparently pushing up incidences of asthma. New fungal infections are even scarier and might be harder to treat.

The bottom line is that we can't predict where the threat will emerge, so we need a distributed, intelligent detection system. In practical terms, how should it be built?

"Detectors" would have to be expert enough to know when an ordinary-looking symptom is actually an emergency. They would be located everywhere, with an emphasis on vulnerable regions. Initial warning signs of a pandemic are most likely to appear in the developing world, but detection nodes should be positioned in every country, with the least possible expense. This is not as difficult as it sounds. The key is to harness existing infrastructure.

Medical infrastructure exists everywhere, in some form. It also tends to be the least corrupt of institutions in regions where that is a problem. Medical centers and clinics would be expected to investigate the cause of ailments in a large number of their patients, even in cases where the symptoms seem common. A small amount of additional scientific expertise and lab equipment would need to be added to a public health system that serves ordinary needs.

Enhancing existing resources would be effective for two reasons. First, illness is more likely to be reported in a city hospital than at a specialist institute. Second, the investment would boost latent public health in that region. For poor regions, investment in equipment and training would have to come from wealthier counterparts. Rich countries could justify the expense in terms of the savings that would result from early detection of a major threat. Tropical climates and urban slums are humanity's front line against pandemics, and they should be equipped properly.

Public health is an important asset for any nation. With so much at stake, it makes sense to place sentinels near every swamp, city, public market, and farmyard on earth.

POSTSCRIPT

Is the International Community Adequately Prepared to Address Global Health Pandemics?

It is far too easy to adopt one of two extreme positions regarding the potential for a global pandemic such as influenza. The first assumes that since these outbreaks begin somewhere in the poorer regions of the globe far removed geographically from the United States, we in the developed world, particularly in the United States, are not at great risk. One should immediately pause, however, as each winter many Americans find themselves suffering from a much less virulent strain of the flu, which typically has its roots in the same poor regions of the globe. The second position with respect to the future potential for a global pandemic suggests that modern medical science will always find a way to counteract such diseases. As the world's experience with HIV/AIDS has taught us, however, modern viruses and other diseases are increasingly more complicated to address successfully, either in treating the symptoms of the problem or the actual problem itself.

Edwin D. Kilbourne addresses this issue in his analysis of twentieth-century pandemics and lessons learned there from ("Influenza Pandemics of the 20th Century," *Emerging Infectious Diseases*, vol. 12, no. 1, January 2006). Says Kilbourne, "Yes, we can prepare, but with the realization that no amount of hand washing, hand wringing, public education, or gauze masks will do the trick. The keystone of influenza prevention is vaccination." This raises a number of questions, Kilbourne continues. Primary among these are who shall be given the vaccine, the risks of mass administration, the availability of sufficient quantities of a vaccine of "adequate antigenic potency."

Kilbourne suggests that an appropriate strategy was suggested by the World Health Organization as early as 1969 and endorsed repeatedly since then. The approach assumes that a new influenza outbreak will emerge from one of sixteen known subtypes HA in avian or mammalian species. This assumption has not yet resulted in a genetic vaccine that can address these subtypes. As the first reading reveals, however, WHO has developed a comprehensive set of strategic actions for the three phrases of a pandemic: the pre-pandemic phrase, the emergence of a pandemic virus, and the pandemic declared and spreading internationally. As the second reading suggests, however, national governments in the developing world, the likely place of origin of the next global pandemic, suffer from a shortage of skilled medical personnel that would serve as front-line defenders against such infectious diseases.

One piece of good news is that, unlike previous pandemics, the developed world has taken notice of a potential future threat. The U.S. government has pledged close to $4 billion for the current fiscal year, three times its expenditures of five years ago. Private philanthropists led by the Bill and Melinda Gates Foundation have also been active in global health programs.

Literature from both the medical and more general fields has also been more cognizant of the problem and has dramatically focused attention to the potential future problem. One place to begin is a major report from the World Health Organization called *Avian Influenza: Assessing the Pandemic Threat* (2005). It traces the evolution of the outbreaks of the H5N1 avian influenza, lessons from past pandemics, its origins in poultry, and future actions "in the face of an uncertain threat." A second WHO report spells out in greater detail that organization's plan of action, *WHO Global Influenza Preparedness Plan* (2005). One study cautions us about this report, however (Martin Enserink, "New Study Casts Doubt on Plans for Pandemic Containment," Science, vol. 311, no. 5764, February 24, 2006).

A historical overview of pandemics is R. S. Bray's *Armies of Pestilence: The Effects of Pandemics on History* (Lutterworth Press, 1998). Another general source is "Preparing for Pandemic: High Probability of a Flu Pandemic Prompts WHO to Offer Strategies" (*The Futurist*, vol. 40, no. 1, January/February 2006). A source that focuses on the value of global structures in the fight against pandemics is "Global Network Could Avert Pandemics" (J. P. Chretien, J. C. Gaydos, J. L. Malone and D. L. Blazes, *Nature*, vol. 440, 2006).

An excellent Internet website is bmj.com, a comprehensive source of resources by the medical field on global pandemics. One example is "An Iatrogenic Pandemic of Panic" by Luc Bonneux and Wim Van Damme (April 1, 2006).

Finally, a current report on the problem of HIV/AIDS is "The Global Challenge of HIV and AIDS" (Peter R. Lamptey, Jami L. Johnson and Marya Khan, *Population Reference Bureau*, vol. 61, no. 1, March 2006).

ISSUE 12

Has the International Community Designed an Adequate Strategy to Address Human Trafficking?

YES: Janie Chuang, from "Beyond a Snapshot: Preventing Human Trafficking in the Global Economy," *Indiana Journal of Global Legal Studies* (Winter 2006)

NO: Dina Francesca Haynes, from "Used, Abused, Arrested and Deported: Extending Immigration Benefits to Protect the Victims of Trafficking and to Secure the Prosecution of Traffickers," *Human Rights Quarterly* (vol. 26, no. 2, 2004)

ISSUE SUMMARY

YES: Janie Chuang, practitioner-in-residence at the American University Washington College of Law, suggests that governments have been finally motivated to take action against human traffickers as a consequence of the concern over national security implications of the forced human labor movement and the involvement of transnational criminal syndicates.

NO: Dina Francesca Haynes, associate professor of law at the New England School of Law, argues that none of the models underlying domestic legislation to deal with human traffickers is "terribly effective" in addressing the issue effectively.

Human trafficking is defined by the United Nations as "The recruitment, transportation, transfer, harbouring or receipt of persons, by means of the threat or use of force or other forms of coercion, of abduction, of fraud, of deception, of the abuse of power or of a position of vulnerability or of the giving or receiving of payments or benefits to achieve the consent of a person having control over another person, for the purpose of exploitation ("Trafficking in Persons—Global Patterns," United Nations, Office on Drug and Crime, April 2006). Exploitation may take any one of several forms: prostitution, forced labor, slavery, or other forms of servitude. While slavery has been with us since ancient times, the existence of human trafficking across national borders, particularly involving major distances, is a relatively new escalation of a problem that in the past was addressed as a domestic issue, if addressed at all.

Its modern manifestation is eerily similar, as reported by the UN in the above source. People are abducted or "recruited" in the country of origin, transferred through a standard network to another region of the globe, and then exploited in the destination country. If at any point exploitation is interrupted or ceases, victims can be rescued and might receive support from the country of destination. Victims might be repatriated to their country of origin or, less likely, relocated to a third country. Too often, victims are treated as illegal migrants and treated accordingly. The UN estimates that 127 countries act as countries of origin while 137 countries serve as countries of destination. Profits are estimated by the UN to be $7 billion per year, with between 700,00 and four million new victims annually.

When one hears of human trafficking, one usually thinks of sexual exploitation rather than of forced labor. This is not surprising as not only are individual victim stories more compelling, the former type of exploitation represents the more frequent topic of dialogue among policymakers and is also the more frequent occurrence as reported to the UN by a three-to-one margin. With respect to victims, 77 percent are woman, 33 percent children, 48 percent girls, 12 percent boys, and 9 percent males (the sum of percentages is over 100 because one source can indicate more than one victim profile). It is not surprising that most women and female children are exploited sexually, while most male adults and children are subjected to forced labor. Sexual exploitation is more typically found in Central and South Eastern Europe. Former Soviet republics serve as a huge source of origin. Africa ranks high as a region of victim origin as well, although most end up in forced labor rather than in sexual exploitation. Asia is a region of both origin and destination. Countries at the top of the list include Thailand, Japan, India, Taiwan, and Pakistan.

The same UN study found that nationals of Asia and Europe represent the bulk of traffickers. And most traffickers who are arrested are nationals of the country where the arrest occurred. Two principal types of groups characterize the traffickers. The first group was highly structured and organized, following a disciplined hierarchical pattern of control. In addition to trafficking, they were involved in transnational movement of a variety of illegal goods, such as drugs and firearms. Violence was the norm in this group. The second type was a smaller group that was strictly profit-oriented and opportunistic, and appeared to operate under a loose network of associates.

In the first selection, Janie Chuang suggests that governments have adopted a three-pronged approach focusing on prosecuting traffickers, protecting trafficked persons, and preventing trafficking. She argues that there has been greater success with the first two foci, but that the current legal response appears to be having an effect or shows great promise in addressing all three areas. In the second selection, Dina Francesca Haynes suggests that while much rhetoric has occurred, the models underlying domestic legislation are not effective for a variety of reasons.

YES

Janie Chuang

Beyond a Snapshot: Preventing Human Trafficking in the Global Economy

Introduction

Within the last decade, governments have hastened to develop international, regional, and national laws to combat the problem of human trafficking, i.e., the recruitment or movement of persons for forced labor or slavery-like practices. Legal responses to the problem typically adopt a three-prong framework focused on prosecuting traffickers, protecting trafficked persons, and preventing trafficking. In practice, however, these responses emphasize the prosecution of traffickers and, to a lesser extent, the protection of their victims. Most legal frameworks address trafficking as an act (or a series of acts) of violence, with the perpetrators to be punished and the victims to be protected and reintegrated into society. While such responses might account for the consequences of trafficking, they tend to overlook its causes—that is, the broader socioeconomic conditions that feed the problem. Oft-repeated pledges to prevent trafficking by addressing its root causes seldom evolve from rhetoric into reality.

More often than not, trafficking is labor migration gone horribly wrong in our globalized economy. Notwithstanding its general economic benefits, globalization has bred an ever-widening wealth gap between countries, and between rich and poor communities within countries.[1] This dynamic has created a spate of "survival migrants"[2] who seek employment opportunities abroad as a means of survival as jobs disappear in their countries of origin. The desperate need to migrate for work, combined with destination countries tightening their border controls (despite a growing demand for migrant workers), render these migrants highly vulnerable to trafficking. For women in particular, this vulnerability is exacerbated by well-entrenched discriminatory practices that relegate women to employment in informal economic sectors and further limit their avenues for legal migration.

Governments have been deeply reluctant, however, to view trafficking in this broader frame—that is, as a problem of migration, poverty, discrimination, and gender-based violence. They have tended to view trafficking as a "law and order" problem requiring an aggressive criminal justice response. Emerging

From *Indiana Journal of Global Legal Studies*, vol. 13, no. 2, 2004, pp. 137–138, 147–157. Copyright © 2004 by Indiana University Press. Reprinted by permission.

studies reveal the drawbacks of this myopia. Notwithstanding the hundreds of millions of dollars already invested in the criminal justice response to the problem, we have yet to see an appreciable reduction in the absolute numbers of people trafficked worldwide.[3] And even in the rare cases where trafficked persons have received rights protective treatment and aftercare, they nonetheless are left facing the socioeconomic conditions that rendered them vulnerable to abuse in the first instance.

This article explores governments' reluctance to address trafficking in its broader socioeconomic context, and offers both a plea and a proposal for more comprehensive approaches to trafficking. Because close examination of these issues is beyond the scope of this short symposium piece, this article aims only to lay a foundation for further thought and discussion in this area. This article problematizes current approaches to trafficking by refraining the problem of trafficking as a global migratory response to current globalizing socioeconomic trends. It argues that, to be effective, counter-trafficking strategies must also target the underlying conditions that impel people to accept dangerous labor migration assignments in the first place. The article then examines how the international legal response to the problem is, as yet, inadequate to the task of fostering longterm solutions. Moreover, by failing to assess the long-term implications of existing counter-trafficking strategies, these responses risk being not only ineffective, but counterproductive. Observing the need for more focused inquiry into prevention strategies, the article advocates strategic use of the nondiscrimination principle to give more meaningful application to basic economic, social, and cultural rights, the violation of which sustains the trafficking phenomenon.[4]

Given the enduring nature of socioeconomic deprivation in many parts of the world, it is easy to dismiss calls for substantive prevention strategies as too lofty or impracticable. But the reality that millions of lives remain at risk for trafficking demands that we embrace this challenge.

I. Globalization, Migration, and Trafficking

While the problem of human trafficking has captured widespread public attention in recent years, it has mostly been in response to narrow portrayals of impoverished women and girls trafficked into the sex industry by shady figures connected to organized crime.[5] Considerably less attention has been devoted to the widespread practice of the trafficking of women, men, and children into exploitative agricultural work, construction work, domestic work, or other nonsexual labor.[6] Most portrayals—particularly of sex trafficking—depict trafficking as an act (or series of acts) of exploitation and violence, perpetrated by traffickers and suffered by desperate and poverty-stricken victims. While accurate in some respects, such depictions are incomplete. The problem of trafficking begins not with the traffickers themselves, but with the conditions that caused their victims to migrate under circumstances rendering them vulnerable to exploitation. Human trafficking is but "an opportunistic response" to the tension between the economic necessity to migrate, on the one hand, and the politically motivated restrictions on migration, on the other.[7] This section offers a broader view of trafficking as a product of the larger socioeconomic forces that feed the "emigration push" and "immigration pull" toward risky labor migration practices in our globalized economy.

A. Emigration "Push" Factors

Globalization and the opening of national borders have led not only to greater international exchange of capital and goods, but also to increasing labor migration.[8] The wealth disparities created by our globalized economy have fed increased intra- and transnational labor migration as livelihood options disappear in less wealthy countries and communities.[9] As Anne Gallagher explains, trafficking lies at one extreme end of the emigration continuum,[10] where the migration is for survival—that is, escape from economic, political, or social distress—as opposed to opportunity-seeking migration—that is, merely a search for better job opportunities. Contrary to the popular, sensationalized image of trafficked persons as either kidnapped or coerced into leaving their homes, more often than not the initial decision to migrate is a conscious one.[11] Yet, the decision to uproot oneself, leave one's home, and migrate elsewhere cannot be explained as a straightforward "rational choice by persons who assess the costs and benefits of relocating"; rather, an understanding of this decision must account for "macro factors that encourage, induce or often, compel migration."[12] "Push" factors are not created by the traffickers so much as this broader context, i.e., the economic impact of globalization.[13] Traffickers, being opportunity-seeking by nature, simply take advantage of the resulting vulnerabilities to make a profit.

Because women are over-represented among survival migrants, it is not surprising that women comprise the vast majority of trafficked persons. Recent estimates from the U.S. State Department place the figure at 80 percent.[14] This gender disparity is often attributed to the "feminization of poverty" arising from the failure of existing social structures to provide equal and just educational and employment opportunities for women.[15] While women migrate in response to economic hardship, they also migrate to flee gender-based repression.[16] Women will accept dangerous migration arrangements in order to escape the consequences of entrenched discrimination against women, including unjust or unequal employment, gender-based violence, and the lack of access to basic resources for women.[17]

As the former U.N. Special Rapporteur on Violence against Women, Radhika Coomaraswamy, explains, gender discrimination underlying these migratory flows is maintained through the collusion of factors at the market, state, community, and family levels.[18] Women's role in the market tends to be derived from traditional sex roles and division of labor, e.g., housekeeping, childcare, and other unpaid/underpaid subsistence labor. At the community level, women face discrimination through "uneven division of wage labour and salaries, citizenship rights and inheritance rights,"[19] as well as certain religious and customary practices, which, reinforced by state policies, further entrench and validate the discrimination and perpetuate the cycle of oppression of women. At the family level, gender discrimination manifests, for example, in "the preference for male children and [a] culture of male privilege [that] deprives girls and women of access to basic and higher education."[20]

Women's lack of rights and freedoms is further exacerbated by certain (macro-level) globalizing trends that have produced an environment conducive to trafficking. Professor Jean Pyle has identified these trends to include: (1) the shift to "export-oriented" approaches, where the production of essential goods is targeted for external trade rather than countries' own internal markets; (2) the entry of multinational corporations (MNCs) into developing countries and the MNCs' extensive networks of subcontractors; (3) structural adjustment policies (SAPs) mandated by the International Monetary Fund (IMF) or the World Bank (WB) as a condition for loans, requiring governments to open their markets to further financial and trade flows and to undertake austerity measures which fall heavily on the poor, particularly women; and (4) the shift in the structure of power at the international level—that is, the rise in the power of international institutions focused on markets (such as MNCs, the IMF, the WB, and the World Trade Organization (WTO)) relative to those that are more people-centered and concerned with sustainable human development (such as the ILO, many U.N. agencies, and nongovernmental organizations (NGOs)).[21]

These global restructuring trends can have harsh effects on women in developing countries—either fostering exploitative conditions for women working in the formal sector, or pushing women directly into work in the informal sector. To the (limited) extent that women are even permitted to work in the formal economy—such as in small businesses or in agriculture—they are often forced out of business by the cheaper imports that trade liberalization brings.[22] As the manufacturing and service industries have entered developing economies, workers in these countries have joined the "global assembly line"; indeed, many MNCs prefer female workers due to their lower cost and lesser likelihood of resisting adverse working conditions.[23] While MNCs provide a source of jobs, they also create "a pool of low-skilled wage labour exposed to standards of western consumption and representing a potential source of emigration."[24]

Structural adjustment policies add to the pressure on women to migrate in search of work. These policies, which require governments to cut programs and reduce expenditures on social services, cause women to take on additional income-earning activities in order to maintain their families' standards of living, as governments decrease benefits in housing, health care, education, food, and fuel subsidies.[25] This often pushes women to work in the unregulated, informal sectors, thus contributing to the rise of gendered-labor networks—prostitution or sex work, domestic work, and low-wage production work.[26] Women often migrate in search of jobs in these largely unregulated sectors, rendering them all the more vulnerable to traffickers.

Compelled to leave their homes in search of viable economic options, previously invisible, low-wage-earning, migrant women are now playing a critical role in the global economy. Through this dynamic—which Professor Saskia Sassen terms the "feminization of survival"—entire households, communities, and even some governments are increasingly dependent on these women for their economic survival.[27] The changes to the international political economy have caused a number of states in the global south, especially in Asia, which is grappling with foreign debt and rising unemployment, to play a "courtesan's role" to global capital in ways that either directly or indirectly foster these gendered-labor networks.[28] Favored growth strategies include attracting direct foreign investment from MNCs and their subcontracting networks—often sacrificing labor standards to do so—or investing in tourism industries widely associated with recruitment of trafficked females for the entertainment of foreign tourists.[29] Moreover, in an effort to ease their unemployment problems and accumulate foreign currency earnings, deeply indebted countries make use of their comparative advantage in the form of women's surplus labor and encourage their labor force to seek employment in wealthier countries.[30] Through their work and remittances, women enhance the government revenue of deeply indebted countries,[31] helping to "narrow the trade gap, increase foreign currency reserves, facilitate debt servicing, reduce poverty and inequalities in wealth and support sustainable development."[32]

B. Immigration "Pull" Factors

The growth in trafficking reflects not just an increase of "push" factors in the globalized economy, but also the strong "pull" of unmet labor demands in the wealthier destination countries. Most have an aging population, with "[t]he proportion of adults over 60 in high income countries . . . expected to increase from eight per cent to 19 per cent by 2050, while the number of children will drop by one third" due to low fertility rates.[33] The resulting "labour shortages, skills shortages, and increased tax burdens on the working population . . . to support and provide social benefits to the wider population,"[34] means these economies will become increasingly dependent on migrant populations to fill the labor gaps.[35] A number of other factors strengthen the immigration "pull," including, for example, fewer constraints on travel (for example, less restrictions on freedom of movement and cheaper and faster travel opportunities); established migration routes and communities in destination countries plus the active presence of recruiters willing to facilitate jobs or travel; and the promise of higher salaries and standards of living abroad.[36] Advances in information technology, global media, and internet access provide the means to broadcast to even the most isolated communities the promise of better opportunities abroad.[37] This fosters high hopes and expectations of women from poor, unskilled backgrounds who are desperate for employment.[38] The prospect of any job is a strong "pull" factor for survival migrants.

Labor shortages in the informal sector are often filled by migrant workers, who are willing to take the "3-D jobs"—i.e., jobs that are dirty, dangerous, and difficult—rejected by the domestic labor force.[39] The employers' profit potential, particularly in the case of trafficked persons, is much higher than would be the case if local labor were employed. If trafficked persons are paid at all, it is invariably at a lower rate than local workers would require, and the trafficked persons do not receive the costly benefits required in many Western states.[40]

In addition to the cost differential, migrants' "foreignness" appears to be a factor in the demand for migrant workers in the domestic work and commercial sex sectors. As Professors Anderson and O'Connell Davidson report in a recent study of the "demand side" of trafficking, employers favor migrant domestic workers over local domestic workers because of the vulnerability and lack of choice that results from their foreign status.[41] Employers perceive them as more "flexible" and "cooperative" with respect to longer working hours, more vulnerable to "molding" to the requirements of individual households, and less likely to leave their jobs. Moreover, their racial "otherness" makes the hierarchy between employer and employee less socially awkward—it is easier to dress up an exploitative relationship as one of paternalism/maternalism towards the impoverished "other."[42]

Rather than publicly recognize their dependence on migrant labor (skilled and unskilled), destination countries have sought instead to promote increasingly restrictive immigration policies, particularly in the wake of the September 11, 2001, terrorist attacks in the United States. There remains considerable public and political resistance to liberalizing the migration policies of these countries,[43] despite strong demographic and economic evidence that migrants produce more benefit than burden for their host countries.[44] This resistance is linked to popular—yet mistaken—concerns about the negative impact of immigration flows on employment, national security, welfare systems, and national identity.[45] Rather than confront xenophobic reactions to issues of migration, many governments instead have sought electoral or political advantage by promoting increasingly restrictive immigration policies.[46] The tension between economic reality and political expedience thus fosters conditions that enable and promote human trafficking. In reducing the opportunities for regular migration, these policies provide greater opportunities for traffickers, who are "fishing in the stream of migration," to take advantage of the confluence of survival migrants' need for jobs, on the one hand, and the unrelenting market demand for cheap labor, on the other.[47] Indeed, as borders close and migration routes become more dangerous, smuggling costs increase to the point that smugglers turn to trafficking to make a profit.[48]

Situating the trafficking phenomenon in this broader context spotlights how deeply rooted trafficking is in the underlying socioeconomic forces that impel workers to migrate. It also demonstrates how the focus on the back end of the trafficking process—that is, entry of the trafficker and the abuses committed in the course of the trafficking—is but a narrow snapshot of the broader problem of trafficking. Solutions that fail to account for the broader picture can only hope to ameliorate the symptoms, rather than address the cause of the problem.

II. The International Legal Response

Throughout the 1980s and 1990s, human rights advocates worked diligently to draw attention to the problem of trafficking in its broader socioeconomic context.[49] But it was concern over the national security implications of increased labor migration and the involvement of transnational criminal syndicates in the clandestine movement of people that ultimately motivated governments to take action. Viewing trafficking as a border and crime control issue, governments seized the opportunity to develop a new international counter-trafficking law in the form of a trafficking-specific protocol to a new international cooperation treaty to combat transnational crime—the U.N. Convention Against Transnational Organized Crime (Crime Convention).[50] States' eagerness to combat the problem resulted in the conclusion of the U.N. Protocol to Prevent, Suppress and Punish Trafficking in Persons, Especially Women and Children (Palermo Protocol or Protocol) within two years and its entry into force three years later, on December 25, 2003.[51]

The development of the Protocol set the stage for a rapid proliferation of counter-trafficking laws in the past five years. The issue of human trafficking now high on the agenda, the international community has devoted hundreds of millions of dollars in trafficking interventions.[52] Efforts to combat trafficking have proceeded from a narrow view of trafficking as a criminal justice problem, with a clear focus on targeting the traffickers and, to a lesser extent, protecting their victims. Addressing the socioeconomic factors at the root of the problem, by contrast, has largely fallen outside the purview of government action.

A. The Palermo Protocol

The Palermo Protocol is, at base, an international crime control cooperation treaty designed to promote and facilitate States Parties' cooperation in combating trafficking in persons. Together with the Crime Convention, the Protocol establishes concrete measures to improve communication and cooperation between national law enforcement authorities, engage in mutual legal assistance, facilitate extradition proceedings, and establish bilateral and multilateral joint investigative bodies and techniques.[53] While the criminal justice aspects of this framework are a clear priority, the Palermo Protocol also contains measures to protect trafficked persons and to prevent trafficking. Unlike the criminal justice measures, which are couched as hard obligations, these provisions are mostly framed in programmatic, aspirational terms. Thus, "in appropriate cases and to the extent possible under its domestic law," the Protocol requires states to consider implementing measures providing for trafficked persons' physical and psychological recovery and endeavor to provide for their physical safety, among other goals.[54] With respect to "prevention" efforts, states are to endeavor to undertake measures such as information campaigns and social and economic initiatives to prevent trafficking,[55] as well as "to alleviate the factors that make persons . . . vulnerable to trafficking, such as poverty, underdevelopment and lack of equal opportunity," and to discourage demand for trafficking.[56]

Just as the text of the Protocol reflects states' clear prioritization of the criminal justice response, so does that which was excluded from the Protocol. Human rights advocates lobbied to include a provision in the Protocol granting trafficked persons protections against prosecution for status-related offenses, such as illegal migration, undocumented work, and prostitution,[57] citing the well-documented reality that trafficked persons were frequently deported or jailed rather than afforded protection.[58] But states refused to include such a provision for fear that it would lead to the "unwarranted use of the 'trafficking defense' and a resulting weakening of states' ability to control both prostitution and migration flows through the application of criminal sanctions."[59]

States' concern over maintaining strong border controls was also reflected in their efforts to draw a legal distinction between trafficking and migrant smuggling,[60] despite the difficulty in distinguishing between the two in practice. Defined as the illegal movement of persons across borders for profit, "migrant smuggling" technically applies to any trafficked person who begins his/her journey as a smuggled migrant but is ultimately forced into an exploitative labor situation.[61] Consequently, a victim of incomplete trafficking—for example, a victim who is stopped at the border before the end purpose of the movement is realized—could be treated as a smuggled migrant and thus denied the victim status and protections afforded to trafficked persons.[62] As Anne Gallagher concludes, the Protocol drafters' failure to address this issue was "clear evidence of [states'] unwillingness . . . to relinquish any measure of control over the migrant identification process."[63]

States' refusal to adjust their migration control policies is perhaps symptomatic of states' deep reluctance to expand the rights afforded to migrant workers. Tellingly, it took thirteen years for the International Convention on the Protection of the Rights of All Migrant Workers and Members of their Families (the Migrant Workers Convention) to receive enough ratifications to enter into force on July 1, 2003.[64] By contrast, the Palermo Protocol entered into force three years after its adoption.[65] Despite well-documented abuses of migrant workers' rights in countries of destination, these countries discouraged ratification of the instrument on grounds that its provisions—which address the treatment, welfare, and human rights of migrant workers (documented and undocumented) and their families—are too ambitious and detailed to be practicable and realizable.[66] That states would maintain such a restrictive stance even when the violations are egregious enough to constitute trafficking reveals the strong priority placed on the crime and border control aspects of trafficking over concern for the welfare of trafficked persons.

B. Counter-Trafficking Efforts in Practice

In practice, the priorities set forth in the Palermo Protocol are mirrored in counter-trafficking law and policy initiatives undertaken across the globe. As the U.S. State Department's yearly Trafficking in Persons Report (TIP Report) reveals, most countries' counter-trafficking efforts focus on effectuating a strong criminal justice response to the problem.[67] Although there is a growing awareness of a need for stronger protection of trafficked persons' human rights,[68] current models of protection continue "to prioritise the needs of law enforcement over the rights of trafficked persons."[69] Most governments adopt restrictive immigration policies, which, at times, fail to distinguish between smuggling and trafficking and can lead to summary deportation or incarceration of trafficked persons.[70] This not only exposes trafficked persons to further harm, including possible retrafficking, but it deprives them of access to justice and undermines government efforts to prosecute the traffickers.[71] To the extent trafficked persons are afforded an opportunity to remain in the destination countries, their residency status is often conditioned on their willingness to assist in the prosecution of their traffickers, potentially exposing them to further trauma and reprisals from the traffickers.

Even well-intentioned efforts to adopt a more "victim-centered approach"[72] to the problem can promote a narrow conception of trafficking that diverts attention from its broader labor and migration causes and implications. A review of country practices reveals two trends, in particular, that foster this dynamic: (1) the deliberate de-emphasizing of the movement or recruitment element of the trafficking definition; and (2) an over-emphasis on sex trafficking, to the neglect or exclusion of labor trafficking.

Regarding the first trend, the United States, for example, has adopted an interpretation of the trafficking definition that shifts focus away from the movement or recruitment element to the "end purpose" of the trafficking:

> The means by which people are subjected to servitude—their recruitment and the deception and coercion that may cause movement—are important factors but factors that are secondary to their compelled service. It is the state of servitude that is key to defining trafficking. . . . The movement of [a] person to [a] new location is not what constitutes trafficking; the force, fraud or coercion exercised on that person by another to perform or remain in service to the master is the defining element of trafficking in modern usage.[73]

Granted, de-emphasizing the recruitment or movement aspect of the definition perhaps helps draw much-needed attention to the broader problem of forced labor. But it also has the detrimental effect of diverting attention from the fact that trafficking is a crime committed during migration and against migrants. It also departs from the international legal definition of trafficking, of which movement or recruitment of the person is a defining element:

> [Trafficking is defined as] (a) . . . the recruitment, transportation, transfer, harbouring or receipt of persons, by means of the threat or use of force or other forms of coercion, of abduction, of fraud, of deception, of the abuse of power or of a position of vulnerability or of the giving or receiving of payments or benefits to achieve the consent of a person having control over another person, for the purpose of exploitation. Exploitation shall include, at a minimum, the exploitation of the prostitution of others or other forms of sexual exploitation, forced labour or services, slavery or practices similar to slavery, servitude or the removal of organs; (b) the consent of a victim of trafficking to the intended exploitation set forth in subparagraph (a) shall be irrelevant where any of the means set forth in subparagraph (a) have been used.[74]

The migration element of the definition speaks to the particular vulnerability that migrants face as a result of living and working in an unfamiliar milieu, where language and cultural barriers can prevent the migrant from accessing assistance.[75] De-emphasizing the migration aspect of trafficking thus overlooks a substantial source of vulnerability. It also narrows the focus of state responsibility to the confines of that which has taken place within its borders—that is, the explorative end purpose of the facilitated movement. Moreover, it conveniently sets to the side thorny questions regarding how to address a victim's (often undocumented) immigration status—an issue of immediate and pressing concern to trafficked persons, who often fear return to their home communities. In sum, this formulation glosses over any responsibility on the part of the state for fostering emigration push or immigration pull factors discussed in Part I, above.

As for the second trend, despite the fact that the international legal definition of trafficking encompasses trafficking for nonsexual as well as sexual purposes, many states—including, until recently, the United States[76]—have focused their efforts on trafficking for sexual purposes.[77] Significantly less attention has been devoted to "labor trafficking" or trafficking for nonsexual purposes, despite recent estimates that this practice accounts for at least one-third of all trafficking cases.[78] The moral outrage that images of women trapped in "sexual slavery" so easily provoke has been a galvanizing force behind global efforts to combat trafficking. Sex trafficking and its associated sex crimes also fall neatly within the purview of a criminal justice response. By contrast, labor trafficking, though hardly benign, is perhaps less likely to engender a criminal justice response given our arguably greater moral tolerance for explorative labor conditions. An overemphasis on sex trafficking thus not only risks overlooking a significant portion of the trafficked population, but it diverts attention away from states' responsibility to promote safe labor conditions.

If protection of the victims is of secondary concern to states, then prevention of trafficking (at least, in the long term) is practically an afterthought. Despite the Protocol's requirement that states should take measures to alleviate the root causes of trafficking, such as "poverty, underdevelopment and lack of equal opportunity,"[79] in practice, "prevention" efforts focus on short-term strategies such as public awareness campaigns regarding the risks of migration. For instance, in her ground-breaking study assessing prevention efforts in southeastern Europe (SEE),[80] Barbara Limanowska found a tendency to adopt "repressive" prevention strategies that "focus on suppressing the negative (or perceived as negative) phenomena related to trafficking, such as [undocumented migration] . . . illegal and forced labor, prostitution, child labor or organized crime."[81] Common strategies include bar raids, computerized border checks and databases that register the names of undocumented migrants, and public awareness campaigns that broadcast to the general public the risks of trafficking.[82] While efforts to prevent re-trafficking of victims are more victim-focused—providing housing, social services, and legal and medical assistance to victims to assist in reintegration into their home communities—these are only provided on a short-term basis.[83] As Limanowska has concluded with respect to SEE, "[t]here is no comprehensive long-term prevention strategy for the region, nor any clear understanding of what such a strategy should include."[84] Although prevention strategies from other regions of the world have yet to be assessed in as comprehensive a fashion,[85] a review of the country practices in other regions of the world reveals a similar focus on repressive approaches to prevention.[86]

Preliminary evaluation of these strategies indicates that they are ineffective, if not counterproductive. Rather than deterring risky migration, large-scale public awareness campaigns have been dismissed by their target audiences as anti-migration measures resulting from "the manipulation of the anti-trafficking agenda by rich countries that want to keep the poor away from their territory."[87] Efforts to "reintegrate" trafficked persons into their home communities cannot overcome the grim reality that the underlying social conditions that led to their trafficking—such as poverty and unemployment—still exist. Indeed, the myopic failure to recognize, much less address, the root causes of trafficking can actually increase vulnerability to trafficking. For example, as Limanowska reports with respect to SEE, the failure to link domestic violence to trafficking at the policy level has led to the creation of separate shelters for trafficked persons and victims of domestic violence, with the former underutilized and the latter underfunded and overcrowded.[88] Rather than recognizing domestic violence as a possible early warning sign of trafficking, the closing of domestic violence shelters has gone unaddressed, thus increasing the vulnerability to trafficking of an already at-risk population.

States' resistance to addressing the broader social problems that feed human trafficking is, in some respects, unsurprising. Treating trafficking as a criminal justice issue is far less resource-intensive and politically risky than developing long-term strategies to address the labor migration aspects of the problem. Moreover, addressing the socioeconomic root causes of trafficking means confronting vexing questions concerning the measure and content of states' obligations to achieve "progressive realization" of the social, economic, and cultural rights half of the human rights corpus.[89] A long-term strategy would thus require attention to deeper, systemic problems that states have proven highly reluctant to confront—for example, the economic need to migrate and the politically motivated restrictions against doing so, not to mention the cycle of poverty, discrimination, and violence that causes these migratory flows. As discussed below, however, such a strategy is critical to the success of global efforts to eliminate trafficking.

III. Prevention as Necessary Core of Counter-Trafficking Strategy

There is no doubt that a strong criminal justice response is a critical component of any effective global counter-trafficking strategy. Absent meaningful victim protection and long-term prevention measures, however, it is, at best, a temporary solution to a chronic and potentially growing problem.[90] Stopping the vicious cycle of trafficking demands a strategy that frames the problem within its broader socioeconomic context and takes seriously the project of targeting the root causes of this complex problem. As with any call to confront the world's ubiquitous social problems, it is an ambitious task, but one for which a few modest steps could help transform the rhetorical commitment to prevention into a substantive one. Two such measures are proposed and briefly described here.

The first proposed step is to undertake rigorous and independent assessment of the potential long-term effects of existing counter-trafficking strategies.

This speaks to the need to ensure that existing counter-trafficking measures do not operate at cross-purposes with the goal of long-term prevention. In their haste to adopt counter-trafficking policies and legislation, governments have largely taken on faith that these strategies are effective with little or no basis in objective evaluations of their outcomes.[91] The sobering results of the few assessments that have been conducted—such as Limanowska's SEE study and even the United States' self-assessment[92]—illustrate the critical need for further evaluation. With data from at least five years of state practice since the adoption of the Protocol by the U.N. General Assembly, there is now a basis for some preliminary evaluations.

The second is to use international human rights law to provide a conceptual framework for addressing the root causes of trafficking. Framing the project of alleviating the root causes of trafficking as a human rights issue would encourage more proactive efforts to address these problems rather than the traditional assumption that such issues are solely within the province of broader development policy. The Palermo Protocol obliges states to "take or strengthen measures . . . to alleviate the factors that make persons . . . vulnerable to trafficking, such as poverty, underdevelopment and lack of equal opportunity."[93] While development policy can provide detailed prescriptions for action on the ground, international human rights law offers an important normative framework within which these strategies can be constructed. Most significantly, a human rights framework offers legal and political space for the disenfranchised to begin to claim these needs as rights, and thereby bring the scope of state responsibility into sharper focus.

A. Assessment of Existing Counter-Trafficking Strategies

A 2003 expert report to the U.N. Commission on the Status of Women concluded that, despite ten years of counter-trafficking laws and policies in the Balkans region, there was no evidence of a significant decrease in trafficking or increase in the number of assisted victims or number of traffickers punished.[94] Considering the hundreds of millions of dollars spent on counter-trafficking programs around the world—the United States contributed $96 million in 2004 alone[95]—and the vast numbers of lives affected by trafficking, this conclusion should give us pause. Regrettably, however, as the International Organization for Migration (IOM) recently reported, "there has been relatively little independent evaluation of counter-trafficking policies and programmes to assess the real impact and effectiveness of different interventions."[96]

The few assessments that have been conducted thus far demonstrate why further evaluation of state practices is vital. Studies such as those conducted by Limanowska not only provide critical, pragmatic insight into best (and worst) practices, but they also expose weaknesses and inaccuracies in the ways in which the problem is conceptualized. For instance, Limanowska's findings concerning the ineffectiveness of large-scale public awareness campaigns in the SEE region underscore how these efforts fail to appreciate fully the migrant perspective. That the target audiences of some of these campaigns so readily dismiss them as rich countries' anti-migration propaganda[97]-despite recognizing the accuracy of the risks portrayed—illustrates the depths of the migrants' need to migrate and the great risks they are willing to assume to do so. This is similarly demonstrated in the fact that the vast majority of calls to helplines created to reach victims of trafficking were "preventive and informative"—that is, to seek information regarding migration for work abroad.[98] Limanowska's evaluation of these and other counter-trafficking initiatives thus underscores governments' chronic failure to appreciate fully the power of the socioeconomic forces underlying migratory flows.

Another area where preliminary studies of programs have called into question the wisdom of existing counter-trafficking strategies relates to efforts to target the demand side of trafficking. Most of these programs are punitive in nature—that is, designed to clamp down on consumer demand, particularly with respect to the commercial sex industry. But as Anderson and O'Connell Davidson demonstrated in their pioneering study of the demand side of trafficking,[99] there are no easy solutions to reducing demand for trafficked labor—sexual or nonsexual. On the one hand, clamping down on demand for street prostitution may actually strengthen demand on other segments of the sex industry where trafficked labor can be an issue, such as pornography, escort agency prostitution, lap- and table-dance clubs, etc.[100] On the other hand, regulating the sex or domestic work sectors "does nothing, in itself, to counteract racism, xenophobia and prejudice against migrants and ethnic minority groups" who tend to comprise the trafficked end of these labor markets and could actually reinforce existing racial, ethnic, and national hierarchies in these sectors.[101] Accordingly, Anderson and O'Connell Davidson suggest that policy makers "pay much closer attention to the unintended and negative consequences of legislating prostitution . . . or of regulating . . . domestic work."[102] Policy makers instead ought to consider concentrating efforts on educational and preventive work targeting the social construction of demand—that is, the social norms that permit exploitation of vulnerable labor.

In addition to evaluating specific counter-trafficking programs and policies, governments should endeavor to assess their overall priorities vis-a-vis the types of programs they pursue—that is, whether oriented toward short-term or long-term results. The SEE experience reveals that funding for programs tends to be channeled toward anti-migration projects reflecting the interests of countries of destination, or in the alternative, "charity work" focused on direct assistance to victims.[103] This has had the unfortunate effect of diverting money away from programs focused on development, equality, and human rights, which hold greater promise of long-lasting change.[104] Trafficking research suffers from the same shortsightedness. Most of the research in the trafficking field is "action-oriented" or designed to prepare for specific counter-trafficking interventions on the ground, typically conducted within a six- to nine-month time frame. "There has been less funding for long-term research [into] the causes of trafficking and the best ways to prevent and combat it, or [into] the impacts of different interventions and policy responses."[105]

The importance of rigorous and independent assessment of existing counter-trafficking programs and research cannot be underestimated. Obtaining meaningful results requires a deeper understanding of the problem and the operational value of the proposed solutions than currently exists today.

B. Addressing Root Causes through a Human Rights Lens

Although there is a general understanding that trafficking has its root causes in poverty, unemployment, discrimination, and violence against women, no large-scale counter-trafficking program has been implemented to address these underlying problems.[106] Even at the level of legal analysis, there is a persistent failure to analyze how international human rights law could be used to address the root causes of the problem. While resource limitations might necessarily slow the implementation of programs targeting root causes on the ground, no such barrier exists to articulating a legal framework to address root causes. Emerging norms and analysis in the field of women's human rights, specifically, and economic, social, and cultural rights, generally, provide a basis upon which such a framework might be developed. Utilizing the principle of nondiscrimination is one potential avenue, as described briefly below.

When one considers trafficking in its broader socioeconomic context, it is not difficult to connect the root causes of trafficking to violations of economic, social, and cultural rights. These include violations of such rights as the right of opportunity to gain a living by work one freely chooses or accepts, the right to just and favorable conditions of work, the right to an adequate standard of living, and the right to education.[107] Race- and gender-based discrimination in the recognition and application of these rights are also critical factors rendering women particularly vulnerable to trafficking.[108] Many of the rights implicated in the root causes of trafficking are the subject of states' obligations under the International Covenant on Economic, Social, and Cultural Rights (ICESCR). With women arguably encountering the most severe deprivations in the area of economic, social, and cultural life,[109] the Convention on the Elimination of All Forms of Discrimination against Women (CEDAW) also plays a critical role in safeguarding these rights vis-a-vis women.

As readily identifiable as these rights violations are, however, legal analyses of trafficking have persistently neglected the economic, social, and cultural rights implications of trafficking. This likely has to do with the fact that, despite being touted as indivisible, interdependent, interrelated, and of equal importance for human dignity,[110] the norm development, monitoring, and implementation of economic, social, and cultural rights—half of the human rights corpus—has fallen far behind that of civil and political rights. The traditional view of economic, social, and cultural rights as merely "programmatic" or "aspirational" in nature—in contrast to the apparently immediately realizable civil and political rights—has fed their marginalization in human rights discourse. Vexing questions and enduring debates over the justiciability of economic, social, and cultural rights—or their capacity to be subject to formal third-party adjudication with remedies for noncompliance[111]—are another likely cause of this relative neglect.

Evolving jurisprudence regarding economic, social, and cultural rights, generally, and their application to women, specifically, nonetheless provides a basis for at least conceptualizing a legal framework to address the root causes of trafficking. The traditional assumption that economic, social, and cultural rights are inherently aspirational, necessarily resource-intensive, and therefore not immediately realizable, has now been cast into doubt.[112] By distinguishing between the types or levels of obligations human rights impose on States Parties—to respect, to protect, and to fulfill—commentators have demonstrated how certain aspects of economic, social, and cultural rights can be of immediate effect.[113] Many of these rights can be safeguarded by virtue of states' noninterference with the freedom and use of resources possessed by individuals. Accordingly, the Committee on Economic, Social and Cultural Rights (the treaty body charged with monitoring state compliance with the ICESCR) has made clear that states have an immediate obligation to ensure that ICESCR rights be exercised without discrimination.[114] Thus, states are obliged to abolish any laws, policies, or practices that affect enjoyment of these rights and, moreover, to take action to prevent discrimination by private persons and bodies in any field of public life.

Interpreted to have broad application under international human rights law, the nondiscrimination principle is particularly well-suited to a human rights analysis of the broad range of root causes of trafficking—poverty, unequal educational and employment opportunities, and violence against women, among others. Under the International Covenant for Civil and Political Rights, states are obliged not only to refrain from discriminatory practices, but also to adopt punitive measures to make equality and nondiscrimination a concrete reality.[115] As General Comment 18, issued by the Human Rights Committee, makes clear, the prohibition on discrimination in law or in fact applies "in any field regulated and protected by public authorities," and thus encompasses economic, social, and cultural rights.[116] In practice, the nondiscrimination principle has been applied to prohibit gender-based differential treatment in the allocation of social benefits, such as unemployment benefits.[117] It has also provided a framework for addressing gender-based violence, "or violence that is directed against a woman because she is a woman or that affects women disproportionately."[118] Poverty is another root cause of trafficking to which the nondiscrimination principle can be applied, as "poverty not only arises from a lack of resources—it may also arise from a lack of access to resources, information, opportunities, power, and mobility. . . . [D]iscrimination may cause poverty, just as poverty may cause discrimination."[119]

As discussed above in Part I, discrimination against women with respect to educational and employment opportunities, the disproportionate burden economic restructuring places on women, the feminization of migration due to violence against women, and the feminization of poverty, among other factors, render women particularly vulnerable to trafficking. "Gender-based discrimination [often] intersects with discriminations based on other forms of 'otherness,' such as race, ethnicity, [and] religion," among others.[120] The nondiscrimination principle, particularly as articulated, interpreted, and applied by treaty bodies such as the Committee on the Elimination of Discrimination against Women (Women's Committee) and the Committee on the Elimination of Racial Discrimination, thus offers a useful framework for addressing the root causes of trafficking.

Moreover, the recent entry into force of the CEDAW Optional Protocol contributes to the practical appeal of a nondiscrimination approach to root causes. The Optional Protocol provides individuals alleging violations of their CEDAW rights the opportunity to pursue complaints against States Parties to the Optional Protocol, and for the Women's Committee to conduct inquiries into allegations of systematic and gross violations of those rights.[121] Using the discrimination framework thus affords rare access to an enforcement mechanism otherwise unavailable for violations of economic, social, and cultural rights.

Conclusion

Situated within its broader frame, the problem of human trafficking demands that efforts to combat this international crime and human rights violation take seriously the need to address its root causes. Over a decade of global counter-trafficking initiatives adopting a "law and order" approach to the problem has yielded questionable, if not disappointing, results. The international community is coming to the growing realization that treating trafficking predominantly, if not solely, as a border and crime control issue is but to respond only to a snapshot view of a much larger problem. There is no question that confronting the poverty, unemployment, discrimination, and gender-based violence, among other factors, that increase an individual's vulnerability to trafficking is a tremendous task that demands creative and long-term strategic thinking. This article has provided a cursory view of two possible approaches by which we might begin to undertake this project. Far more analysis and deeper understanding of the trafficking problem are necessary prerequisites of the project, as is dispossessing ourselves of the traditional view that realization of economic, social, and cultural rights can wait. As daunting of a task as this may be, it is a necessary one if global efforts to eliminate trafficking are to succeed.

Notes

1. See United Nations High Commissioner for Human Rights, Report of the United Nations High Commissioner, [paragraph] 6, delivered to the Economic and Social Council, U.N. Doc. E/1999/96 (July 29, 1999) (noting that while globalization has had its benefits, there is a "clear trend towards a smaller percentage of the population receiving a greater share of wealth, while the poorest simultaneously lose ground"); see generally Executive Summary to U.N. Econ. & Soc. Council [ECOSOC], Comm'n on Human Rights, Integration of the Human Rights of Women and the Gender Perspective: Violence Against Women, at 4 U.N. Doc. E/CN.4/2000/68 (Feb. 29, 2000) (prepared by Radhika Coomaraswamy) [hereinafter Coomaraswamy Report].

2. BIMAL GHOSH, HUDDLED MASSES AND UNCERTAIN SHORES: INSIGHTS INTO IRREGULAR MICRATION 35 (1998).

3. The number of people trafficked remains staggering. The International Labour Organization (ILO) estimates that 2.5 million people are trafficked at any point in time, generating $32 billion in profits for organized crime. INTERNATIONAL LABOUR OFFICE, A GLOBAL ALLIANCE AGAINST FORCED LABOUR: GLOBAL REPORT UNDER THE FOLLOW-UP TO THE ILO

DECLARATION ON FUNDAMENTAL PRINCIPLES AND RIGHTS AT WORK 46, 55 (2005) [hereinafter ILO GLOBAL REPORT].

4. Michael J. Dennis & David P. Stewart, Justiciability of Economic, Social, and Cultural Rights: Should There Be an International Complaints Mechanism to Adjudicate the Rights to Food, Water; Housing, and Health?, 98 AM. J. INT'L L. 462,464 (2004).

5. See, e.g., Peter Landesman, The Girls Next Door, N.Y. TIMES MAG., Jan. 25, 2004, at 30; Nicholas D. Kristof, Girls for Sale, N.Y. TIMES, Jan. 17, 2004, at A 15; Nicholas D. Kristof, Bargaining for Freedom, N.Y. TIMES, Jan. 21, 2004, at A27; Nicholas D. Kristof, Loss of Innocence, N.Y. TIMES, Jan. 28, 2004, at A25; Nicholas D. Kristof, Stopping the Traffickers, N.Y. TIMES, Jan. 31, 2004, at Al7.

6. David A. Feingold, Think Again: Human Trafficking, FOREIGN POL'Y, Sept.-Oct. 2005, at 26.

7. ILO GLOBAL REPORT, supra note 3, at 46.

8. COMM. ON FEMINISM AND INT'L LAW, INT'L LAW ASS'N, WOMEN AND MIGRATION: INTERIM REPORT ON TRAFFICKING IN WOMEN 2 (2004), available at http://www.ila-hq.org/pdf/Feminism%20&%20International%20Law/ Report%202004.pdf.

9. See generally GHOSH, SUpm note 2 (distinguishing between survival migration and opportunity seeking migration); MIKE KAYE, ANTI-SLAVERY INT'L, THE MIGRATION-TRAFFICKING NEXUS: COMBATING TRAFFICKING THROUGH THE PROTECTION OF MIGRANTS' HUMAN RIGHTS 13 (2003), available at http://www.antislavery.org/homepage/resources/the%20migration%20 trafficking%20nexus%202003.pdf.

10. ANNE GALLAGHER ET AL., CONSIDERATION OF THE ISSUE OF TRAFFICKING: BACKGROUND PAPER 16-17 (2002) (citing GHOSH, supra note 2, at 35), available at http://www.nhri.net/pdf/ACJ%20Trafficking%20Background%20 Paper.pdf.

11. Feingold, supra note 6, at 28.

12. Patrick A. Taran, Human Rights of Migrants: Challenges of the New Decade, INT'L MIGRATION, Vol. 38, No. 6 (Special Issue 2), Feb. 2001, at 12.

13. See Saskia Sassen, Women's Burden: Counter-Geographies of Globalization and the Feminization of Survival, 71 NORDIC J. INT'L L. 255, 257 (2002).

14. U.S. DEP'T OF STATE, TRAFFICKING IN PERSONS REPORT 6 (2005) [hereinafter 2005 TIP REPORT], available at http://www.state.gov/documents/organization/ 47255.pdf.

15. Coomaraswamy Report, supra note 1, [paragraph] 58.

16. Id. [paragraph][paragraph] 54-60.

17. See id. [paragraph] 60.

18. Id. [paragraph] 57.

19. Id.

20. Id.

21. See Jean L. Pyle, How Globalization Fosters Gendered Labor Networks and Trafficking 23 (Nov 13-15, 2002) (unpublished manuscript), available at http:// www.hawaii.edu/global/projects_activities/Trafficking/Pyle.doc.

22. Id. at 5.

23. Id.

24. CHRISTINA BOSWELL & JEFF CRISP, POVERTY, INTERNATIONAL MIGRATION AND ASYLUM 6 (United Nations Univ., World Inst. for Dev. Econ. Research, Policy Brief No. 8, 2003).

25. See Sassen, supra note 13, at 263.

26. See U.N. Dip. for the Advancement of Women, The New Borderlanders: Enabling Mobile Women and Girls for Safe Migration and Citizenship Rights, at 5-6, U.N. Doc. CM/MMW/2003/ CRE3 (Jan. 14, 2004) (prepared by Jyoti Sanghera).

27. Sassen, supra note 13, at 258.

28. Vidyamali Samarasinghe, Confionting Globalization in Anti-Trafficking Strategies in Asia, BROWN J. WORLD AFF., Summer-Fall 2003, at 91, 94 (citing JIM MITTLEMAN, THE GLOBALIZATION SYNDROME: TRANSFORMATION AND RESISTANCE 15 (2000)).

29. Id. at 92, 94.

30. Id. at 95.

31. Sassen, supra note 13, at 258. Thus, according to Professor Sassen, "[t]he growing immiseration of governments and whole economies in the global south has promoted and enabled the proliferation of survival and profit-making activities that involve migration and trafficking of women." Id. at 255 (from the Abstract). According to the International Organization for Migration, remittances through official channels totaled $93 billion in 2003, INTERNATIONAL ORGANIZATION FOR MIGRATION, WORLD MIGRATION 2005: COSTS AND BENEFITS OF INTERNATIONAL MIGRATION 491 (2005), approached $100 billion in 2004, id. at 124, and now seriously rival development aid in many countries; unofficial remittances are likely to be two to three times that figure. For example, in El Salvador, "remittances accounted for more than 80 per cent of the total financial inflows in 2000, with overseas development assistance and foreign direct investment accounting for less than 20 per cent." Kaye, supra note 9, at 14.

32. KAYE, supra note 9, at 14 (spending remittances on locally produced goods and services can have a multiplier effect by simulating demand).

33. Id. at 13.

34. Id.

35. "In order to stabilise the size of the working population in the 15 EU member states, there needs to be a net inflow of some 68 million foreign workers and professionals between 2003 and 2050." Id. (citing INT'L ORG. FOR MIGRATION) WORLD MIGRATION 2003, at 245 (2003)).

36. Id. at 11; accord BOSWELL & CRISP, supra note 24, at 10.

37. Samarasinghe, supra note 28, at 96-97.

38. See BOSWELL & CRISP, supra note 24, at 6.

39. See ILO GLOBAL REPORT, supra note 3, at 46.

40. See Taran, supra note 12, at 15-16.

41. BRIDGET ANDERSON & JULIA O'CONNELL DAVIDSON, IS TRAFFICKING IN HUMAN BEINGS DEMAND DRIVEN?: A MULTI-COUNTRY PILOT STUDY 29-32 (Int'l Org. for Migration, IOM Migration Research Series No. 15, 2003).

42. Id. at 32.

43. BOSWELL & CRISP, supra note 24, at 1.

44. INT'L ORG. FOR MIGRATION, supra note 31, at 170, 188-89 (noting that a recent U.K. study calculated that in 1999-2000, migrants contributed $4 billion more in taxes than they received in benefits, and a U.S. study estimated that national income had expanded $8 billion in 1997 because of immigration).

45. BOSWELL & CRISP, supra note 24, at 21-22.

46. KAYE, SUPRA note 9, at 13.

47. Helen Thomas, Fishing in the Stream of Migration, ADB REV. (ASIAN DEV. BANK, MANILA, PHIL.), February 2004, at 16, 16-17, available at http://www.adb.org/Documents/Periodicals/ADB_Review/2004/vo136_1/fishing.asp.

48. Migrant smuggling entails payment by a third party to facilitate the movement of the migrant. See Protocol Against the Smuggling of Migrants by Land, Sea and Air, Supplementing the United Nations Convention Against Transnational Organized Crime, G.A. Res. 55/25, annex III, pmbl. & art. 3, U.N. Doc. A/RES/55/25 (Nov. 2, 2000) [hereinafter Migrant Smuggling Protocol]. In addition to whatever profit is to be made from the facilitated migration, traffickers can also profit from the revenue generated from the exploitative end purpose of the movement—e.g., the forced labor or slavery-like practice.

49. See generally HUMAN RIGHTS WATCH, A MODERN FORM or SLAVERY: TRAFFICKING OF BURMESE WOMEN AND GIRLS INTO BROTHELS IN THAILAND (1993), available at http://www.hrw.org/reports/1993/thailand/; HUMAN RIGHTS WATCH, RAPE FOPPROFIT: TRAFFICKING OF NEPALI GIRLS AND WOMEN TO INDIAN BROTHELS (1995), available at http://www.hrw.org/reports/pdfs/c/crd/india957.pdf (report by Human Rights Watch demonstrating that unequal access to education and employment opportunities, among other factors, fed the feminization of poverty and migration and increased women's vulnerability to traffickers).

50. United Nations Convention Against Transnational Organized Crime, G.A. Res. 55/25, annex I, U.N. Doc. A/RES/55/25 (Nov. 2, 2000) [hereinafter Crime Convention].

51. Protocol to Prevent, Suppress and Punish Trafficking in Persons, Especially Women and Children, G.A. Res. 55/25, annex II, U.N. Doc. A/RES/55/25 (Nov. 2, 2000) [hereinafter Palermo Protocol].

52. See 2005 TIP REPORT, supra note 14, at 245 (reporting that the United States alone has invested $295 million in counter-trafficking efforts over the last four fiscal years).

53. Crime Convention, supra note 50, arts. 16, 18, 19, 27, 28.

54. Palermo Protocol, supra note 51, arts. 6-8.

55. Id. art. 9, [paragraph] 2.

56. Id. art. 9, [paragraph][paragraph] 4-5.

57. See Position Paper on the Draft Protocol to Prevent, Suppress and Punish Trafficking in Women and Children, submitted by the Special Rapporteur on Violence Against Women, Report of the Ad Hoc Committee on the Elaboration of a Convention Against Transnational Organized Crime on its Fourth Session, Held in Vienna June 28 to July 9, 1999, U.N. Doc. A/AC.254/CRP.13 (May 20, 1999) [hereinafter "Coomaraswamy Position Paper"].

58. See, e.g., Coomaraswamy Report, supra note 1, [paragraph] 44.

59. See Anne Gallagher, Human Rights and the New UN Protocols on Trafficking and Migrant Smuggling: A Preliminary Analysis, 23 HuM. RTs. Q. 975, 991 (2001).

60. See Migrant Smuggling Protocol, supra note 48 (migrant smuggling is the subject of one of the other two protocols to the Crime Convention).

61. Smuggling is defined as "the procurement, in order to obtain, directly or indirectly, a financial or other material benefit, of the illegal entry of a person into a State Party of which the person is not a national or a permanent resident." Id. art. 3.

62. "Smuggled migrants are assumed to be acting voluntarily," and are thus afforded less protection under international law. Anne Gallagher, Trafficking, Smuggling and Human Rights: Tricks and Treaties, 12 FORCED MIGRATION REV. 25, 26 (2002).

63. Gallagher, supra note 59, at 1001.

64. For a current list of ratifications, see http://0-untreaty.un.org.unistar.uni.edu:80/ ENGLISH/bible/englishinternetbible/partl/chapterlV/treaty25.asp.

65. For a current list of ratifications, see http://www.unodc.org/unodc/en/crime_cicp_ signatures trafficking.html.

66. Taran, supra note 12, at 18-22.

67. 2005 TIP REPORT, SUPRA note 14, at 34.

68. The United States, for example, is increasingly recognizing how the failure to protect trafficked persons' human rights compromises efforts to prosecute traffickers. For example, noting the significant disparity between the numbers of people trafficked to the United States (14,500–17,500 each year) and the numbers of those who have reported the abuse to law enforcement (757 as of November 2003), the U.S. Department of Justice has made concerted efforts to collaborate more effectively with NGOs and consider more victim-centered approaches to prosecution. DEP'T OF JUSTICE, ASSESSMENT OF U.S. ACTIVITIES TO COMBAT TRAFFICKING IN PERSONS 5, 22, 26-27 (2004).

69. ELAINE PEARSON, ANTI-SLAVERY INT'L, HUMAN TRAFFIC, HUMAN RIGHTS: REDEFINING VICTIM PROTECTION 4 (2002).

70. See, e.g., U.S. DEP'T OF STATE, TRAFFICKING IN PERSONS REPORT 148, 165, 185 (2004) (citing the Italian, Portuguese, and British governments' failure to distinguish between trafficking and smuggling). The 2004 Trafficking in Persons Report also described how trafficked persons in the Czech Republic "were treated as illegal immigrants and expressed fear of testifying due to safety concerns," id. at 134, and in Morocco, were "jailed and/or detained for violating immigration or other laws [and were] not provided adequate legal representation." Id. at 199. In the 2005 Trafficking in Persons Report, Italy, the United Kingdom, and Portugal persisted in their failure to distinguish between trafficking and smuggling or illegal immigration. 2005 TIP REPORT, supra note 14, at 130, 181, 221. France has apparently adopted a practice of "arresting, jailing, and fining trafficking victims as a means of discouraging the operation of trafficking networks and to gain information to pursue cases against traffickers," which, as the U.S. State Department notes, "harms trafficking victims and allows for [their] deportation . . . regardless of possible threats [in their country of origin]." Id. at 106.

71. PEARSON, supra note 69, at 2.

72. See 2005 TIP RETORT, supra note 14, at 5 (referring to the "three P's" of prosecution, protection, and prevention, noting that "a victim-centered approach to trafficking requires us equally to address the 'three R's'—rescue, rehabilitation and reintegration').

73. Id. at 9-10.

74. Palermo Protocol, supra note 51, art. 3.

75. See Coomaraswamy Position Paper, supra note 57, at 3.

76. In the 2005 TIP REPORT, the United States expanded its coverage of trafficking for nonsexual purposes. 2005 TIP REPORT, supra note 14, at 1. This expansion was undoubtedly in response to years of NGO protests over the United States' focus on sex trafficking.

77. Frank Laczko, Introduction, INT'L MICRATION, Jan. 2005, at 5, 9 (introduction to a special issue entitled "Data and Research on Human Trafficking: A Global Survey," noting that research on trafficking has focused on the sex trafficking of women and children, neglecting other forms of trafficking). Liz Kelly, "You Can Find Anything You Want": A Critical Reflection on Research on Trafficking in Persons Within and into Europe, INT'L MIGRATION, Jan. 2005, at 235, 239 (article in a special issue entitled "Data and Research on Human

Trafficking: A Global Survey," noting how most of the content and data in the TIP Reports issued by the U.S. State Department for years 2002–2004 was "confined to sexual exploitation").

78. ILO GLOBAL REVORT, Supra note 3, at 46.

79. Palermo Protocol, supra note 51, art. 9, [paragraph] 4.

80. BARBARA LIMANOWSKA, TRAFFICKING IN HUMAN BEINGS IN SOUTH EAST-ERN EUROPE xiii (2005) [hereinafter SEE REPORT] (assessing prevention strategies in Albania, Bosnia and Herzegovina, Bulgaria, Croatia, the former Yugoslav Republic of Macedonia, Moldova, Romania, Servia and Montenegro, and Kosovo).

81. Id. at 2.

82. See generally SEE REPORT, Supra note 80.

83. Id. at 36–37.

84. Id. at xiii.

85. The SEE REPORT is one of the few to undertake an assessment of prevention programs. As the IOM found in its survey of data and research on trafficking, there is a lack of information regarding trafficking programs in many regions of the world, including the Middle East, the Americas, and Africa. See Laczko, supra note 77, at 7.

86. See generally 2005 TIP REPORT, Supra note 14.

87. SEE REVORT, supra note 80, at 22.

88. Id. at 20-21.

89. International Covenant on Economic, Social and Cultural Rights, G.A. Res. 2200, U.N. GAOR, 21st Sess., Supp. No. 16, at 49, U.N. Doc. A/6316 (Dec. 16, 1966) [hereinafter ICESCR].

90. The current reality is that the number of traffickers arrested is low compared to the efforts expended to capture them, and of those who are tried, few are convicted and even fewer serve sentences. Am. Soc'y of Int'l Law, Trafficking in Humans: Proceedings of the 99th Annual Meeting (forthcoming 2006)]hereinafter 2005 ASIL Human Trafficking Panel] (noting that "virtually no kingpins are brought to justice, and criminal networks remain largely undisturbed," and that sentences are relatively minor and often not served) (draft on file with author). It appears to be a socioeconomic reality that there will be others to take advantage of the substantial profit-making potential to be had wherever there is both economic necessity to migrate, yet shrinking avenues for legal migration. See Sassen, supra note 13, at 266-70.

91. See Laczko, supra note 77, at 9.

92. See generally SEE REVORT, supra note 80; DEP'T OF JUSTICE, ASSESSMENT or U.S. ACTIVITIES TO COMBAT TRAFFICKING IN PERSONS (2004).

93. Palermo Protocol, supra note 51, art. 9, [paragraph] 4.

94. U.N. Comm'n on the Status of Women, Women's Human Rights and Elimination of All Forms of Violence Against Women and Girls as Defined in the Beijing Platform of Action and the Outcome Documents of the Twenty-third Special Session of the General Assembly, at 47, U.N. Doc. E/CN.6/ 2003/12 (Mar. 13, 2003) (prepared by Barbara Limanowska). This report was based on information from the Balkans region. Id. at 1 n.2. See also 2005 ASIL Human Trafficking Panel, supra note 90, at 2-3 (noting the disparity between the proliferation of new counter-trafficking legal and institutional mechanisms, on the one hand, and the achievement of meaningful results, on the other).

95. 2005 TIP REPORT, supra note 14, at 1.

96. Laczko, supra note 77, at 9.

97. See supra text accompanying note 87.

98. SEE REPORT, Supra note 80, at 32-33. On a positive note, however, this experience also suggests the potential preventive role that helplines can play.

99. See generally ANDERSON & O'CONNELL DAVIDSON, supra note 41.

100. Id. at 43. Reports of the Swedish experience illustrate this point. "[W]hen Sweden introduced laws in 1999 to criminalize men who purchase sex, while decriminalizing [the prostitutes/sex workers], the incidence of female sex trafficking dropped. . . . [W]hile the demand for prostitution decreased in Sweden, it increased in neighboring countries. The male clients simply went [elsewhere to satisfy their desires]." Samarasinghe, supra note 28, at 102.

101. ANDERSON & O'CONNELL DAVIDSON, supra note 41, at 44.

102. Id. at 46, 47.

103. SEE RERORT, supra note 80, at 54.

104. Id.

105. Laczko, supra note 77, at 9.

106. See 2005 ASIL Human Trafficking Panel, supra note 90, at 3.

107. ICESCR, supra note 89, arts. 6, 7, 11, 13.

108. See generally Convention on the Elimination of All Forms of Discrimination Against Women, G.A. Res. 34/180, U.N. Doc. A/RES/34/180 (Dec. 18, 1979), reprinted in 19 I.L.M. 33 (1980).

109. Katarina Frostell & Martin Scheinin, Women, in ECONOMIC, SOCIAL AND CULTURAL RIGHTS 331,331 (Asbjorn Eide et al. eds., 2d. rev. ed. 2001).

110. World Conference on Human Rights, June 14–25, 1993, Vienna Declaration and Programme of Action, art. 1, [paragraph] 5, U.N. Doc. A/CONE.157/23 (July 12, 1993), reprinted in 32 I.L.M. 1663 (1993).

111. See Martin Scheinin, Economic and Social Rights as Legal Rights, in ECONOMIC, SOCIAL AND CULTURAL RIGHTS, Supra note 109, at 29, 29; Dennis & Stewart, supra note 4, at 463.

112. See, e.g., Asbjorn Eide, Economic, Social and Cultural Rights as Human Rights, in ECONOMIC, SOCIAL AND CULTURAL RIGHTS, supra note 109, at 9, 23-25.

113. See U.N. Comm'n on Econ., Soc. and Cultural Rights, General Comment 3, annex III, [paragraph] 10, U.N. Doc. E/1991/23 (Dec. 14, 1990) [hereinafter General Comment 3]. "Progressive realization" cannot be used as a "pretext for non-compliance." The Maaotricht Guidelines on Violations of Economic, Social and Cultural Rights, reprinted in 20 HUM. RTS. Q. 691,694 (1998).

114. General Comment 3, supra note 113, [paragraph] 1.

115. See International Covenant on Civil and Political Rights, G.A. Res. 2200, art. 26, U.N. GAOR, 21st Sess., Supp. No. 16, at 52, U.N. Doc. A/6316 (Dec. 16, 1966).

116. See Frostell & Scheinin, supra note 109, at 334 (citing General Comment 18).

117. Id. at 334 & n. 15.

118. U.N. Comm. on the Elimination of All Foms of Discrimination Against Women, General Recommendation 19, U.N. Doc. HRI/GEN/Rev. 3 (noting that "[p]overty and unemployment increase opportunities for trafficking in women").

119. UNITED NATIONS, OFFICE OF THE HIGH COMM'R FOR HUMAN RIGHTS, HUMAN RIGHTS AND POVERTY REDUCTION: A CONCEPTUAL FRAMEWORK 17 (2004).

120. Coomaraswamy Report, supra note 1, [paragraph] 55.

121. See Optional Protocol to the Convention on the Elimination of All Forms of Discrimination Against Women, G.A. Res. 54/4, annex, U.N. Doc. A/54/49 (Oct. 6, 1999).

Dina Francesca Haynes  **NO**

Used, Abused, Arrested and Deported: Extending Immigration Benefits to Protect the Victims of Trafficking and to Secure the Prosecution of Traffickers

Prologue

Madeleina was a slight, delicate-looking sixteen-year-old girl from Moldova. She left Moldova in 1998, when her sister's husband convinced her and another girl to go with his friend who promised to find them hostess jobs in Italy. She was given a fake passport, and after about a week of traveling, found herself locked in a brothel in what she later discovered was the Republika Srpska, Bosnia and Herzegovina (Bosnia).[1] A woman interpreting for the brothel owner told her that she had been sold to him to be his "wife." The brothel owner forced Madeleina to have sex with him and his friends and told her that she could begin working off her debt to him immediately. He told her that she already owed him more than $2000 for her purchase price and working papers.

Madeleina had no money and no friends. She could not speak the local language and the owner threatened her regularly, beating her and telling her that police would arrest her if she tried to leave. There were at least eleven other girls and women at this brothel, all foreigners. Most of them were from Moldova or Romania, and the brothel owner tried to keep them separated as much as possible to prevent their collusion and escape. The owner sometimes forced them to take drugs to keep them more compliant, the cost of which was added to their debt. The brothel owner kept Madeleina for about five months, forcing her to have sex with as many as twenty men a day. She thought that some of the customers at the brothel were local police. She also knew that Russian and either American, Canadian, or British men, and she thinks Italian, had visited her and had sex with her, in addition to local men.

When police raided that brothel, she was taken by car to Arizona Market, near Brcko, where cars, goods, and women are sold. Two foreign men purchased her; she thinks they were Swiss and American peacekeepers. These two

Haynes, Dina Francesca. "Used, Arrested and Deported: Extending Immigration Benefits to Protect the Victims of Trafficking and to Secure the Prosecution of Traffickers". *Human Rights Quarterly* 26:2 (2004), 222–237. © The John Hopkins University Press. Reprinted with permission of the John Hopkins University Press.

men put her in a car and took her to an apartment in Tuzla where they kept her locked up and came to visit her every day or two, often with friends, and forced her to have sex with them. Over the course of these months, Madeleina had begun to teach herself some of the Serbian language.

One day, after no one had visited her for several days and she was running out of food, the landlord of the apartment opened the door and told her to get out. It was winter, and with no warm clothes Madeleina went out to find the local police, not because she believed the police would help her, but because she knew she would freeze to death with no place to go.

The local police promptly jailed her for prostitution. A Human Rights Officer with the Organization for Security and Co-operation in Europe (OSCE) intervened, and Madeleina was transported to a makeshift shelter in Sarajevo.[2] International and local nongovernmental organizations were then just establishing the shelter.

I. Introduction

Trafficking in human beings is an extremely lucrative business, with profits estimated at $7 billion per year[3] and a seemingly endless supply of persons to traffic, estimated at between 700,000 and four million new victims per year.[4] Trafficked persons, typically women and children, can be sold and resold, and even forced to pay back their purchasers for the costs incurred in their transport and purchase.[5] In fact, the United States Central Intelligence Agency estimates that traffickers earn $250,000 for each trafficked woman.[6] Economic instability, social dislocation, and gender inequality in transitioning countries foster conditions ripe for trafficking.

Trafficking in human beings involves moving persons for any type of forced or coerced labor, for the profit of the trafficker.[7] Several countries are finally adopting domestic legislation to criminalize trafficking in human beings, although many continue to punish the victims of trafficking, charging them with prostitution, possession of fraudulent documents, or working without authorization.[8] Many international organizations and consortiums of grassroots anti-trafficking organizations have also put forward models for combating trafficking.

None of these models is yet terribly effective, for a variety of reasons. At the forefront of these reasons is the fact that several countries have yet to adopt anti-trafficking laws.[9] Second, of those that have, many completely fail to implement those laws even after undertaking domestic and international obligations.[10] A third major reason is that some governments have failed to incorporate the advice of grassroots and international anti-trafficking organizations that have worked for years drafting recommended legislation based upon their observations in the field.[11]

A particular contemporary problem is trafficking for the sexual exploitation of women[12] in and from Central and Southeastern Europe.[13] Currently, Central and Southeastern Europe are the primary sources from which women are drawn into global sex traffic through Europe,[14] and some countries in this region are actively engaged in developing anti-trafficking initiatives pursuant to their obligations as signatories to the 2000 Protocols to the UN Convention on Transnational Organized Crime.[15] In addition, some countries in the Balkans have the added presence of international peacekeepers and humanitarian workers, which in many respects exacerbates the problem.[16]

This paper will, in Part II, discuss the recent increase in trafficking. Part II will explore how and why governments have failed to effectively address the problem, despite being aware of its existence for decades. Part IV illustrates that two dominant anti-trafficking models have emerged in recent years, one of which is oriented towards prosecution of traffickers while the other emphasizes victim protection. Part V proposes a specific combination of the best of the two models, recommending several additional elements to create a new model that will more effectively combat trafficking, highlighting immigration benefits, and responds to anticipated arguments against such an expansion.

The principal recommendation of this article is that the best of the "jail the offender" and "protect the victim" models should be combined. The new model should incorporate advice from grassroots organizations that work directly with trafficked persons, in order to craft anti-trafficking programs that promote protection of victims. This new model should include immigration protection, should hit traffickers where it hurts, and should prioritize full implementation.

II. The Recent Rise of Trafficking in Human Beings

The horrific practice of trafficking in human beings has long been a serious problem throughout the world, but in the last fifteen years trafficking from European countries has been on the rise. Trafficking in Europe has been fueled by the social dislocations, increasing pockets of poverty, gender imbalance, bureaucratic chaos, and legislative vacuums resulting from the collapse of communism.[17]

Women already disenfranchised within their communities are most often those who fall prey to traffickers: ostracized minorities, women without employment or future economic prospects, and girls without family members to look out for them or who have fallen outside of the educational system.[18] These girls and women are lured by traffickers into leaving their countries, believing that they will work in the West as dancers, hostesses, or nannies, and instead find themselves forced to have sex for the profit of the men and women who purchased them.[19]

In order to secure their silence and compliance, traffickers threaten, beat, rape, drug, and deprive their victims of legitimate immigration or work documents. Women are forced to sell themselves in brothels, often receiving several clients per day.[20] They rarely see any wages for their work; in fact, most victims are kept in indentured servitude and told that they owe their traffickers or the brothel owners for their own purchase price and for the price of procuring working papers and travel documents.[21]

The rings of traffickers are often vast, extremely well connected to police and government officials, well hidden, and reach across borders and continents.[22] Traffickers in human beings are also known to traffic in weapons and drugs, and to use trafficking in human beings to bring in initial cash flow to support the riskier traffic in drugs and arms.[23] Human beings, being reusable commodities that can be sold and resold, are both more lucrative[24] and less risky to traffic than drugs and arms, in that traffickers of human beings are rarely prosecuted for this particular offense.[25]

While between 700,000 and four million women are trafficked each year,[26] only a fraction of those are known to have received assistance in order to escape trafficking.[27] Many are re-victimized by being deported from the countries in which they are found,[28] sanctioned by law when attempting to return to their countries of origin,[29] and ostracized within their communities and families.[30]

Governments appear to have recognized the importance of the issue, many having ratified international instruments established to eradicate trafficking in human beings. Nevertheless, trafficking is neither slowing, nor is the prosecution of traffickers or the protection of their victims becoming any more certain.

III. Governmental Failures to Confront the Issue

As early as 1904, concern over "white slavery," in which European women were exported to the colonies, prompted the adoption of the International Agreement for the Suppression of White Slave Traffic, addressing the fraudulent or abusive recruitment of women for prostitution in another country.[31] The issue was addressed again in 1933 with the International Convention on the Suppression of the Traffic in Women of Full Age, by which parties agreed to punish those who procured prostitutes or ran brothels.[32] In 1949, the United Nations adopted the Convention for the Suppression of the Traffic in Persons and of the Exploitation of the Prostitution of Others.[33] Until 2000, the only other international treaty to address trafficking was the 1979 UN Convention on the Elimination of All Forms of Discrimination Against Women (CEDAW), which required states to take all measures to suppress both trafficking and "exploitation of prostitution," meaning forced prostitution.[34]

Beginning in the late 1980s, the European Union and the United Nations began addressing the issue repeatedly, yet little progress was made and the collapse of communism flooded trafficked persons throughout Europe. With trafficking recognized as a distinct problem since 1903, with the ratification of four treaties by many nations, and with trafficking recently and dramatically on the rise, why has so little progress been made?

A. Some Politicians Use Trafficking to Direct Attention to Unrelated Political Agendas

Trafficking is a low priority for many governments who pay lip service to solving the problem only to harness more support for other political objectives. Because of the visceral reaction trafficking elicits with the public, it has recently been used by politicians and governments to bolster other political agendas, such as curtailing illegal migration, fighting prostitution, and even combating terrorism.

Some governments pretend to care about trafficking when the real objective is controlling unwanted migration.[35] Trafficking in human beings is a very serious topic in its own right, but the gravity and emotional impact of the topic unfortunately render it vulnerable to political manipulation. With illegal migration, smuggling, terrorism, and prostitution now on many political agendas, the pledge to combat trafficking is misused as justification for "clamping down" on these other threats that also have immigration implications.[36] Authorities have remained cynical and hardened to the plight of victims who are easier to treat as prostitutes or illegal immigrants.[37]

In fact, some countries seem to view the existence of trafficked women within their sovereign borders as evidence of a breach of security or the failure of their domestic immigration mechanisms, and accordingly attempt to address trafficking through simple reconfiguration of their border control mechanisms.[38] Traffickers are often extremely savvy transnational organized criminals, while their victims are most often women and children already victimized by economic, political, or social conditions in their home countries. Viewing trafficking as an immigration issue overly simplifies the complexity of preparing effective anti-trafficking measures.

As this section will demonstrate, politicians and governments have blurred the distinctions between illegal migration, trafficking, and smuggling, taking advantage of the current world fear of terrorism committed by legal and illegal immigrants, to restrict immigration and freedom of movement further. They have purposely co-mingled anti-trafficking initiatives with anti-prostitution initiatives. They have tried to further curtail migration by blurring the distinction between trafficking and smuggling. Finally, it is my opinion that some governments are motivated not by a keen belief in the necessity of curtailing trafficking, but by a desire to secure international financial assistance or enter the European Union.

1. Prostitution

Prostitution and trafficking are not one and the same, yet some would treat them as such.[39] Prostitution involves persons willingly engaging in sex work. Although there may be a gray area involving different degrees of consent, choice, and free will, trafficking goes well outside of this gray area. While a valid argument could be made that gender imbalances in economic or social factors drive a woman to consent to such labor as her chosen profession, thus effectively removing her "will,"[40] trafficking involves clear deprivation of choice at some stage, either through fraud, deception, force, coercion, or threats.

Whether a trafficked woman was initially willing or unwilling when she entered into sex work should make no legal difference when the outcome is enslavement or forced servitude; a person cannot consent to enslavement or forced labor of any kind.[41] While some trafficked persons may be willing to work in the sex industry, they do not anticipate being forced to pay off large forcibly imposed debts, being kept against their will, having their travel documents taken from them, or being raped, beaten, and sold like chattel.[42]

Nevertheless, within the community of NGOs, international organizations, governments, and working groups laboring to define and combat trafficking, the issue of prostitution regularly enters the deliberation. As recently as 2001, for example, some persons working for the United Nations Mission in Bosnia and Herzegovina (UNMIBH) and partner organizations tasked with assisting the Bosnian government to eradicate trafficking refused to provide trafficking protection assistance to women who at any point willingly engaged in prostitution.[43]

The Organized Crime Convention has encouraged countries to focus on coercion and use of force in identifying whether a woman is a victim of trafficking, rather than on whether she has ever engaged in prostitution. Nevertheless, the US government agency tasked with distributing funding to international trafficking initiatives recently determined that it would refuse to fight trafficking where doing so might appear to treat prostitution as a legitimate activity.[44] Thus, trafficking is politicized, a volatile topic easily used to affix other political agendas. Even while most experts working in anti-trafficking initiatives agree that trafficking and prostitution are separate issues, to be handled separately as a matter of law, the United States took a step backwards in attempting to tackle prostitution under the guise of combating trafficking.

2. Smuggling

Politicians have also attempted to link smuggling and trafficking in order to achieve tightened border controls. While most governments acknowledge that smuggling and trafficking are two distinct crimes, they nonetheless use trafficking statistics and horrific trafficking stories to justify tightened border controls when the primary goal is not the elimination of trafficking, but the reduction of illegal migration, some of which occurs via smugglers, and perhaps preventing terrorism.

The United States Department of State, for instance, opened the Migrant Smuggling and Trafficking in Persons Coordination Center in December 2000, even while acknowledging, "at their core . . . these related problems are distinct."[45] The US government nevertheless justified combining the two issues by pointing out that "these related problems result in massive human tragedy and affect our national security, primarily with respect to crime, health and welfare, and border control."[46] By way of another example, the Canadian government supported a study jointly reviewing both smuggling and trafficking, even while pointing out the legal distinctions between the two.[47] The study was justified under the premise that "as human smuggling and trafficking are increasing, the tightening of border controls has taken on a new urgency from the fear of terrorism in the West, as well as restrictive measures placed on irregular migratory movements."[48]

Smuggling involves delivering persons to the country they wish to enter, initiated by the potential migrant. Smuggling often takes place under horrible and possibly life threatening conditions, but smuggled persons are left to their own devices upon delivery. Smuggling is not as lucrative for the perpetrators, as smugglers usually make only a short-term profit on the act of moving a person, while traffickers regard people as highly profitable, reusable, re-sellable, and expendable commodities.[49]

In order for anti-trafficking initiatives to be effective, politicians must make the eradication of trafficking and the protection of trafficked persons into a prioritized goal, distinct from the elimination of smuggling or the tightening of border controls.

3. Some Governments are Motivated by a Desire to Meet EU Entrance Requirements or to Obtain Financial Assistance

Not surprisingly, the European Union and the United States, among other institutions and governments, are conditioning financial assistance[50] and entry into the European Union on the country's willingness to develop legislation curtailing trafficking within and across its borders[51]. Countries set to enter the European Union in 2004[52] are eager to pass legislation recommended by the European Union and the Council of Europe (CoE), and join working groups that address stemming the flow of trafficking and smuggling.[53]

Passing recommended legislation and making real efforts to stem the flow of trafficking, however, are often two different things. When countries simply adopt legislation in order to secure entry into the European Union or to meet financial assistance requirements, there is no real ownership or commitment to eradicating trafficking. The legislation, no matter how meticulously in conformity with international standards, will not be fully or adequately implemented at the local level without serious political will.

B. Governments Ignore Obvious Problems with Anti-Trafficking Initiatives

Many countries have now finally adopted some domestic legislation addressing trafficking, and most have eradicated earlier laws that punished trafficked persons for immigration or prostitution offenses.[54] This section points out reasons no current laws are very effective in the fight to eradicate trafficking.

By no means, however, have all countries adopted laws to specifically target trafficking.[55]

1. Governments Fail to Prioritize the Implementation of Anti-Trafficking Laws

A piece of legislation is useful to trafficked persons and threatening to violators only if it is implemented and known by the traffickers to be fully in force. No matter how great the economic or political pressure applied by the European Union or the United States to encourage countries to introduce legislation to prosecute traffickers, no incentive can create the political will to *implement* legislation if such will or ability does not exist or is not prioritized.[56]

In Bosnia, for example, UNMIBH reported that of sixty-three cases brought against traffickers in 2000, only three were successfully prosecuted.[57] Of those three, the defendants were *all* tried on charges related to prostitution, not trafficking. According to the HRW Report, all of the thirty-six cases brought involved charges related to prostitution and not trafficking—not just the three successful ones.[58] In one of the three cases, three trafficked women and two brothel owners were arrested in a raid. Although the defendants admitted that they had purchased the women for prices ranging between $592 and $1162, the court convicted the three women for prostitution and dropped the charges against the male defendants.[59]

Coordination among responsible agencies to implement the law is often flawed in the best of circumstances, further obstructing implementation.[60] Meetings are held at the highest levels and those in attendance come away full of self-congratulations that plans are being made and laws adopted. Yet out in the community, brothels are raided and no screening is done for victims of trafficking; victims identify themselves to police and face prosecution;[61] traffickers supply false passports to border police,[62] and the girls and traffickers are waived through.

For example, during the author's tenure in Belgrade, Serbia, and Montenegro, a brothel was raided and trafficked women were placed in jail, rather than the new shelter for trafficked persons, on the very same day that a high-level regional meeting took place in Belgrade between ministries and Stability Pact, UN, and OSCE officials to discuss follow up victim protection mechanisms for the new shelter. There seemed to be no communication between those making the decisions to adopt new laws and practices and those carrying them out in the field, and there was an inability or unwillingness to train these low-level government employees.

2. Governments Fail to Penalize or Even Acknowledge the Complicity of Peacekeepers and International Workers in Trafficking

Despite a growing awareness that peacekeeping forces and humanitarian workers regularly and knowingly obtain the services of trafficked women and sometimes even engage in or aid and abet trafficking, governments have failed to publicly address this issue. Trafficked women in Bosnia, for instance, report that approximately 30 percent of their clients are internationals.[63] Countries that had never before been countries of destination began receiving trafficked women when peacekeepers and international aid workers moved into Bosnia, Croatia, and Kosovo.[64] Neighboring countries quickly became countries of transit and origin. While the use of trafficked women by international workers might constitute only a fraction of the total number of trafficked women and the fraction of those trafficked by international workers is even less, the participation of international humanitarian workers and peacekeeping forces in trafficking conveys a powerful symbolic message to local authorities and traffickers. The message is this: governments working to "democratize" developing countries do not really care about eradicating trafficking.

For years international organizations operating in the Balkans have been unwilling to determine how they can best prevent their employees from frequenting brothels known to harbor trafficked women. In recent years, when it has become clear that most brothels in the Balkans, for instance, do contain trafficked women,[65] these international organizations have still failed to enforce internal rules or laws against frequenting brothels.[66]

Ninety percent of foreign sex workers in the Balkans are estimated to be trafficked, although less than 35 percent are identified and deemed eligible to receive protection assistance, and less than 7 percent actually do receive long-term support.[67] It is therefore well known among those charged with teaching

Bosnians how to better enforce their laws, e.g. peacekeepers, the International Police Task Force [IPTF][68], and international humanitarian workers, that by visiting a prostitute, one stands a good chance of visiting a trafficked woman.[69] One would think, therefore, that workers paid by the foreign ministries whose goals are combating trafficking and promoting safety and democracy would be strictly forbidden to visit brothels, but they are not. In fact, sometimes they receive no punishment whatsoever even when caught engaging in such activity.[70] How can a victim of trafficking be expected to escape her captor and seek safety with the very men paying her captors for her services?

Some international organizations such as the OSCE and some branches of the United Nations have recently developed "Codes of Conduct" which implicitly forbid their personnel from seeing prostitutes by exhorting that they not "engage in any activity unbecoming of a mission member," subsequent to widely-publicized scandals involving international troops engaged in trafficking.[71] Nevertheless, several recent articles indicate that local police, international peacekeepers, and humanitarian aid workers continue to be major users of brothels in the Balkans in particular.[72] Developing and enforcing prohibitions against this practice are crucial, because the international police, peacekeepers, and humanitarian workers are the very persons whose duty it is to work with local authorities to eradicate trafficking in this part of the world, and the victims are supposed to be looking to international police and peacekeepers for protection.[73]

Notes

1. Bosnia is currently divided into two entities and a district: the Republika Srpska, the Federation, and Brcko District.

2. As related to the author during her work with the OSCE in Bosnia. For similar stories, see Human Rights Watch, Hopes Betrayed: Trafficking of Women and Girls to Post-Conflict Bosnia and Herzegovina for Forced Prostitution (2002) [hereinafter HRW Report]; John McGhie, *Bosnia–Arizona Market: Women for Sale* (UK Channel 4 News television broadcast, 8 June 2000), *available at* fpmail.friends-partners.org/pipermail/stop-traffic/2000-August/000113; William J. Kole & Aida Cerkez-Robinson, *U.N. Police Accused of Involvement in Prostitution in Bosnia,* Assoc. Press, 28 June 2001; Colum Lynch, *U.N. Halted Probe of Officers' Alleged Role in Sex Trafficking,* Wash. Post, 27 Dec. 2001, at A17, *available at* www.washingtonpost.com/wp-dyn/articles/A28267-2001Dec26.

3. United Nations Children's Fund (UNICEF) et al., Trafficking in Human Beings in Southeastern Europe xiii (2002) [hereinafter Joint Report on Trafficking] (stating that trafficking in human beings is the third most lucrative organized crime activity after, and often conjoined with, trafficking in arms and drugs). *See also* Gillian Caldwell et al., Crime and Servitude: An Exposé of the Traffic in Women for Prostitution from the Newly Independent States 14 (1997), *available at* www.qweb.kvinnoforum.se/misc/crimeru.rtf (citing 1988 German police estimates that "traffickers earned US \$35–50 million annually in interest on loans to foreign women and girls entering Germany to work as prostitutes").

4. U.S. Dep't of State Office to Monitor and Combat Trafficking in Persons, Trafficking in Persons Report 1 (2002).

5. *The Sex Trade: Trafficking of Women and Children in Europe and the United States: Hearing before the Commission on Security and Cooperation in Europe,*

106th Cong., 1st Sess. 22 (1999) (testimony of Laura Lederer) [hereinafter, The Lederer Report] (stating that women trafficked into North America are sold for as much as $16,000 to each new brothel owner, and have to pay or work off a debt of $20,000 to $40,000); *see also*, Jennifer Lord, *EU Expansion Could Fuel Human Trafficking*, United Press Int'l, 9 Nov. 2002, *available at* cayman netnews.com/Archive/Archive%20Articles/November%202002/Issue%20286%20 Wed/EU%20Expansion.html.

6.   Caldwell, *supra* note 3, at 10.

7.   While there are a multitude of definitions of trafficking, the most widely used definition derives from the current legal standard bearer, the Protocol to Prevent, Suppress and Punish Trafficking in Persons, Especially Women and Children, Supplementing the United Nations Convention Against Transnational Organized Crime, *adopted* 15 Nov. 2000, G.A. Res. A/55/25, Annex II, 55 U.N. GAOR Supp. (No. 49), at 60, U.N. Doc. A/45/49 (Vol. I) (2001) (*entered into force* 25 Dec. 2003) [hereinafter Trafficking Protocol]. Article 3 of The Protocol defines trafficking as:

> the recruitment, transportation, transfer, harbouring or receipt of persons, by means of the threat or use of force or other forms of coercion, of abduction, of fraud, of deception, of the abuse of power or of a position of vulnerability or of the giving or receiving of payments or benefits to achieve the consent of a person having control over another person, for the purpose of exploitation.

See United Nations, Office of Drugs and Crime *available at* www.unodc.org/ unodc/en/trafficking_human_beings. Solely for the purposes of narrowing discussion, this article will emphasize trafficking for sex work. This narrow focus should not be viewed as support for a definition of trafficking that bifurcates trafficking that results in sex work from other forms of trafficking (such as indentured domestic service, forced labor, forced marriage, subjugation in making pornography, etc.). All trafficking in human beings is a violation of human rights in that it involves affronts to human dignity and arguably constitutes a form of slavery.

8.   *See infra* text and accompanying notes pt. IV(A).

9.   In South Eastern Europe, for instance, Croatia, Serbia, and Montenegro, have no distinct criminal offense for trafficking, despite being known countries of origin, transit, or destination, although a law is under consideration in Serbia. For review of laws related to trafficking in these countries, *see* Kristi Severance, ABA: Central European an Eurasian Law Initiative Survey of Legislative Frameworks for Combating Traficking in Persons (2003) [hereinafter ABA CEELI Report] *available at* www.abanet.org/ceeli/publications/conceptpapers/ humantrafficking/home. *See infra* note 109. In March 2003, the Office of the High Representative imposed a law criminalizing trafficking as a distinct offense, as the Bosnian authorities had failed to do. As yet, however, no traffickers have been charged under this new law.

10.   *See infra* text pt. III(B)(1).

11.   *See infra* text pt. IV(B)(1)(b).

12.   For the purposes of simplicity, the paper will refer to women in particular, and use the feminine pronouns when referring to victims of trafficking, as the majority of victims of trafficking for sexual exploitation are women and girls.

13.   Since the early 1990s countries in political and economic transition in Central, Eastern, and South Eastern Europe and the Former Soviet Union have not only become main countries of origin for trafficked persons, but also of transit and destination. See Office for Democratic Institutions and Human Rights,

Organization for Security and Co-operation in Europe, Reference Guide for Anti-Trafficking Legislative Review 20 (2001) [hereinafter OSCE Reference Guide]. South Eastern European countries offer the unique combination of being countries deeply mired in trafficking, and simultaneously interested in entering the European Union (EU). As such, they are in the process of bringing their legislation and administrative bodies into compliance with European standards, and are particularly useful for viewing the process of developing anti-trafficking initiatives. Cyprus, the Czech Republic, Estonia, Hungary, Latvia, Lithuania, Malta, Poland, the Slovak Republic, and Slovenia, and are set to join the EU on 2004, while Romania, Bulgaria, and Turkey all have active applications for EU membership. See European Union Website, Candidate, Countries, *available at* europa.eu.int/comm/enlargement/candidate.

14. Central and Eastern Europe have surpassed Asia and Latin America as countries of origin since the breakdown of the Soviet Union in 1989. See OSCE Reference Guide, *supra* note 13, at 7.

15. U.N. Convention Against Transnational Organized Crime, G.A. Res. 55/25, Annex I, U.N. GAOR 55th Sess., Supp. No. 49, at 44, U.N Doc. A/45/49 (Vol. 1) (2000), *entered into force* 29 Sept. 2003 [hereinafter Organized Crime Convention]. Serbia, Montenegro, and Bosnia have ratified the Organized Crime Convention, and its Protocols. All other South Eastern European countries are parties and it remains unclear as to how they will implement their commitments.

16. *See infra* text pt. III(B)(2). International Administration is still in effect in Kosovo (through the U.N. Mission in Kosovo, pursuant to U.N. Resolution 1244 (S.C. Res 1244, U.N. SCOR, 4011th mtg., S/RES/1244 (1999)), and partially in Bosnia and Herzegovina.

17. *See* Jenna Shearer Demir, the Trafficking of Women for Sexual Exploitation: A Gender-Based and Well-founded Fear of Persecution? 4–5 (2003), *available at* www.unhcr.ch/cgi-bin/texis/vtx/home/opendoc.pdf?tbl=RESEARCH&id=3e71f 84c4&page=publ (arguing that women disproportionately suffer the effects of an economic upheaval); Sergei Blagov, *Equal Opportunities Remain a Pipedream*, Asia Times Online, 10 Mar. 2000, *available at* www.atimes.com/c-asia/BC10Ag01.html (stating that "[s]ome 70 percent of Russian unemployed with college degrees are women. In some regions, women make up almost 90 percent of the unemployed.")

18. Based on the author's discussion with anti-trafficking NGOs and UN officials in Bosnia and Serbia, and on direct discussion with trafficking victims.

19. *Id.*

20. HRW Report, *supra* note 2, at 18.

21. *Id.* at 4, 11.

22. *Report of the Special Rapporteur on Violence Against Women, its Causes and Consequences, Ms. Radhika Coomaraswamy,* U.N. ESCOR, Comm'n on Hum. Rts., 53rd Sess., U.N. Doc. E/CN.4/1997/47 (1997) § IV (expressing concern about government complicity in trafficking).

23. *See* Amy O'Neill Richard, U.S. Dep't of State Center for the Study of Intelligence: Internaitonal Trafficking in Women to the United States: A Contemporary Manifesttion of Slavery and Organized Crime 1 (1999), *available at* www.odci.gov/csi/monograph/women/trafficking.pdf [hereinafter CSI Report]. *See also* IOM, Applied Research and Data Collection on a Study of Trafficking in Women and Children for Sexual Exploitation to through and from the Balkan Region 7 (2001) [hereinafter IOM Report].

24. *See* CSI Report, *supra* note 23, at 19–20.

25. *See infra* text pt. IV (C)(2)(a).

26. U.S. Dep't of State, Office to Monitor and Combat Trafficking in Persons, Trafficking in Persons Report 1, *supra* note 4, at 1. The numbers for South Eastern Europe in particular are difficult to specify. For example, one Swedish NGO estimates that "500,000 women . . . are trafficked each year into Western Europe alone. A large proportion of these come from the former Soviet Union countries." Joint Report on Trafficking, *supra* note 3, at 4. IOM estimates that in 1997, "175,000 women and girls were trafficked from Central and Eastern Europe and the Newly Independent States." *Id.* As of 2002, IOM estimates that 120,000 women and children are trafficked into the EU each year, mostly through the Balkans, and that 10,000 are working in Bosnia alone, mostly from Moldova, Romania and the Ukraine. *Id.*

27. Joint Report on Trafficking, *supra* note 3, at xv (only 7 percent of the foreign migrant sex workers known to be victims of trafficking receive any long term assistance and support).

28. HRW Report, *supra* note 2, at 38.

29. Global Alliance Against Traffic in Women et al., Human Rights Standards for the Treatment of Trafficked Persons 13, 15 (1999), *available at* www.hrlawgroup.org/resources/content/IHRLGTraffickin_tsStandards.pdf. Countries from which trafficked persons originate are referred to as countries of origin. Countries through which victims are trafficked are called countries of transit, and destination countries are those in which victims ultimately find themselves engaged in sex work.

30. *Id.* at 13.

31. International Agreement for the Suppression of White Slave Traffic, 1 U.N.T.S. 83 (*signed* 18 May 1904) (*entered into force* 18 July 1905) (amended by the Protocol signed at Lake Success, New York, 4 May 1949). The Agreement was ratified by Belgium, Denmark, France, Germany, Italy, the Netherlands, Portugal Russia, Spain, Sweden, and Norway, Switzerland, and the United Kingdom and consented to by their respective colonies, and dealt with European women being exported to the colonies for prostitution, sometimes forcibly.

32. International Convention for the Suppression of the Traffic in Women of Full Age, Concluded at Geneva 11 Oct. 1933, as amended by the Protocol signed at Lake Success, New York, on 12 Nov. 1947, *registered* 24 Apr. 1950, No. 772.

33. Convention for the Suppression of the Traffic in Persons and of the Exploitation of the Prostitution of Others, *opened for signature* 21 Mar. 1950, 96 U.N.T.S. 271 (*entered into force* 25 July 1951). Parties agreed to "punish any person who, to gratify the passions of another: (1) Procures, entices or leads away, for purposes of prostitution, another person, even with the consent of that person; (2) Exploits the prostitution of another person, even with the consent of that person." *Id.* art. 1.

34. Convention on the Elimination of All Forms of Discrimination Against Women, *adopted* 18 Dec. 1979, G.A. Res. 34/180, U.N. GAOR, 34th Sess., Supp. No. 46, U.N. Doc. A/34/46 (1980) (*entered into force* 3 Sept. 1981), 1249 U.N.T.S. 13, *reprinted in* 19 I.L.M. 33 (1980).

35. *See* CSI Report, *supra* note 23, at 31 (stating that "[d]efinitional difficulties still persist regarding trafficking in women. . . . Distinctions regarding trafficking in women, alien smuggling, and irregular migration are sometimes blurred with INS [former US immigration department] predisposed to jump to the conclusion that most cases involving illegal workers are alien smuggling instead of trafficking cases").

36. *See,* e.g., Richard Monk, Organization for Security and Co-operation in Europe Mission to the Fry: Study on Policing in the Federal Republic of Yugoslavia 21 (2001), *available at* www.osce.org/yugoslavia/documents/reports/files/report-policing-e.pdf [hereinafter Monk Report] (Commenting: "Additionally, these statistics [on successful anti-trafficking ventures] are used for various political

purposes—for example, prevention of trafficking is used as an argument for refus- ing young women entry to a country or for refusing to issue them a visa, and then, in the police statistics, these cases are relabeled as successful cases of rescu- ing 'victims of trafficking.'").

37. *See,* e.g., CSI Report, *supra* note 23, at 35 (US government officials cited as hold- ing the opinion that trafficking victims are part of the conspiracy and therefore view them as accomplices).

38. "More often than not, anti-trafficking laws, be it domestic or international, tend to be conceived and are employed as border-control and immigration mechanisms," Agnes Khoo, *Trafficking and Human Rights: Some Observations and Questions,* 12 Asia Pacific-Forum on Law, Women and Development 3 (Dec. 1999), *available at* www.apwld.org/vol123-02.htm.

39. In explaining its priorities for 2003, the Stability Pact of South-Eastern Europe stated: "Attention will be drawn to maintain the differentiation between victims of human trafficking and prostitutes, which is currently becoming blurred, to the detriment of effective and targeted victim protection." Special Co-ordinator of the Stability Pact for South Eatern Europe Task Force on Trafficking in Human Beings, Anti-Trafficking Policy-Outline for 2003 [hereinafter SP Trafficking Task Force Priorities], *available at* www.stabilitypact.org/trafficking/info.html#four. For more discussion on the Stability Pact, see discussion *infra* pt. IV(D)(3).

40. NGO Consultation with the UN/IGO's on Trafficking in Persons, Prostitution and the Global Sex Industry, Trafficking and the Global Sex Industry: The Need for a Human Rights Framework, 21–22 (1999), Room XII Palais des Nations, Geneva, Switzerland [Panel A and Panel B] (some IGO's arguing that all prostitution is forced prostitution and calling for its abolition, with others arguing for a distinction between voluntary and forced prostitution in order to focus on preventing the worst forms of exploitation of prostitutes).

41. *See,* e.g., CSI Report, *supra* note 23, at vi ("The Thirteenth Amendment outlaw- ing slavery prohibits an individual from selling himself or herself into bond- age, and Western legal tradition prohibits contracts consenting in advance to assaults and other criminal wrongs."). This argument is further developed in pt. V(A)(1).

42. *See* HRW Report, *supra* note 2, at 15–20 (detailing common treatment and expectations of trafficked women).

43. *Id.* at 13. This practice of excluding prostitutes from victim protection results from criteria set by donor agencies rather than international law; *see e.g., infra* note 44 and accompanying text.

44. In its report entitled "Trafficking in Persons, The USAID Strategy for Response," designed to implement several provisions within the Trafficking Victim's Protec- tion Act (TVPA), the US Agency for International Development (USAID) states that it will only work with [e.g. fund] local NGOs "committed . . . to combat traffick- ing *and prostitution,*" [emphasis added], explaining that: "organizations advocat- ing prostitution as an employment choice or which advocate or support the legalization of prostitution are not appropriate partners," US Agency for Interna- tional Development, Trafficking in Persons: The USAID Strategy for Response (Feb. 2003), *available at* www.usaid.gov/wid/ pubs/pd-abx-358-final.pdf.

45. U.S. Dep't of State International Information Programs, Fact Sheet: Migrant Smuggling and Trafficking in Persons (2000), *available at* www.usembassy.it/ file2000_12/alia/a0121523. htm.

46. *Id.*

47. *See* Jacqueline Oxman-Martinez, Human Smuggling and Traficking: Achieving the Goals of the UN Protocols? 1 (2003), *available at* www.maxwell.syr.edu/ campbell/XBorder/OxmanMartinez%20oped.pdf.

48. *Id.* at 1.

49. In the last decade, Southeast Asia alone has produced nearly three times as many victims of trafficking than produced during the entire history of slavery from Africa. Melanie Nezer, *Trafficking in Women and Children: "A Contemporary Manifestation of Slavery,"* 21 Refugee Reports 1, 3 (2000) (400 years of slavery from Africa produced 11.5 million victims; victims of trafficking in the 1990s within and from Southeast Asia are estimated to be more than 30 million).

50. The United States Trafficking Victims Protection Act of 2000, Pub. L. No. 106-386, 114 Stat. 1464 (2000) [hereinafter TVPA], for instance, requires an annual submission to Congress by the Department of State on the status of trafficking in each country. Financial assistance is tied directly to the level of each country's compliance with US directives. U.S. Dep't of State Office to Monitor and Combat Trafficking in Persons, Trafficking in Persons Report 10 (2002). (Beginning in 2003, those countries ranked lowest in this report "will be subject to certain sanctions, principally termination of non-humanitarian, non-trade-related assistance. Consistent with the Act, such countries also would face U.S. opposition to assistance . . . from international financial institutions.")

51. In the case of the European Union, entry into the Union is conditioned upon compliance with general respect for human rights and compliance with human rights standards.

52. For list of applicant countries to the European, *see supra* note 13.

53. In the author's experience working with ministries of justice, interior, and human rights in Bosnia, Croatia, and Serbia and Montenegro, high level government authorities were typically keen to attend high level working groups addressing the drafting of trafficking legislation, but were much harder to pin down when it came to establishing work plans to train field level government authorities.

54. *See infra* text pt. IV(A). For example, in Israel as recently as 1998, a victim's best hope was to have the brothel or massage parlor she worked in raided by police. She would then be taken to prison, not a shelter or detention center, and offered two options: be deported and have criminal prostitution charges dropped, or file a complaint against her trafficker or those holding her in involuntary servitude. If she chose to file charges, however, she would remain in prison until a trial was held. Not surprisingly, no women between 1994 and 1998 chose to testify against their traffickers in Israel. Most traffickers were well aware that the laws favored them, if only because the women they trafficked were illegally in the country and were engaging in criminal activity. Michael Specter, *Traffickers' New Cargo: Naïve Slavic Women,* N.Y. Times, 11 Jan. 1998, at A1.

55. Serbia, Montenegro, and Croatia, for example, have no distinct criminal offense for trafficking. *See* generally ABA CEELI Report, *supra* note 9, for updates on domestic trafficking legislation. Although Bosnia's law criminalizing trafficking was imposed in March 2003, it has yet to yield a prosecution. *See infra* note 109.

56. One way to encourage implementation of anti-trafficking laws is for the European Union and United States to condition their assistance on implementation, rather than on simple passage of anti-trafficking laws, a recommendation made in this paper, and finally acknowledged in the 2003 Trafficking in Persons Report, U.S. Dep't of State Office to Monitor and Combat Trafficking in Persons, Trafficking in Persons Report 2 (2003) [hereinafter 2003 Traficking in Persons Report], *available at* www.state.gov/documents/organization/21555.pdf.

57. HRW Report, *supra* note 2, at 36.

58. *Id.*

59. *Id.*

60. CSI Report, *supra* note 23, at 31. Questions about whether the United States can be considered an example of the "best of circumstances" aside, the CSI Report states that at least in 1999, prior to passage of the TVPA, "information sharing among the various entities remain[ed] imperfect. Several Department of Justice [DOJ] offices look at the trafficking issue through the prism of their particular offices' interest, be it eliminating civil rights violations, tackling organized crime, or protecting minors. Even within the [DOJ], information is not always shared." *See also* Monk Report, *supra* note 36, at 76. Although Serbia and Montenegro are actively participating in high level working groups to combat trafficking, including suggesting progressive programs for victim protection, the police force is incapable of coping with the scale of the phenomenon:

> Apart from within the border police departments, there is poor awareness and interest generally on the part of police and the public about the subject [of trafficking], and the prevailing disregard for gender equality contributes to indifference about the plight of victims. . . . Because of the lack of reciprocal agreements with neighboring States, the incompatibility of laws, the absence of [domestic] laws which enable successful prosecutions to be brought against the traffickers and pimps and the lack of [domestic] legal authority to produce evidence obtained by the internal use of technical and surveillance aids, victim's cases are generally viewed as time and energy consuming and inevitably unproductive. The very fact that victim's statements, both verbal and written, will be in a foreign language further reduces responsiveness.

61. HRW Report, *supra* note 2, at 61.

62. *See id.* at 16.

63. *Id.* at 11. *See also,* 2003 Traffickign in Persons Report, *supra* note 56, at 35 (acknowledging that the international civilian and military personnel have contributed to trafficking in Bosnia).

64. *See, e.g.,* HRW Report, *supra* note 2, at 4, 11. ("According to [IGOs and NGOs] trafficking first began to appear [in Bosnia] in 1995," and "[l]ocal NGOs believe that the presence of thousands of expatriate civilians and soldiers has been a significant motivating factor for traffickers to Bosnia and Herzegovina.")

65. *See id.* at 4 (227 of the nightclubs in Bosnia are suspected of harboring trafficked women).

66. *Id.* at 46–60.

67. *See* Joint Report on Trafficking, *supra* note 3, at xv.

68. In January 2003, the duties of the IPTF were assumed by the European Union, and are now referred to as "European Union Police Mission."

69. In Serbia for example, of 600 women questioned by police during brothel raids between January 2000 and July 2001, 300 were determined to be victims of trafficking. *See id.,* at 78.

70. HRW Report, *supra* note 2, at 62–67.

71. The author, a member of the OSCE Mission to Bosnia, signed such a Code of Conduct.

72. *See, e.g.,* McGhie, *supra* note 2; Kole, *supra* note 2; Lynch, *supra* note 2; Daniel McGrory, *Woman Sacked for reveling UN Links with Sex Trade,* The Times Online (London), 7 Aug. 2002; Robert Capps, *Crime Without Punishment,* SALON.COM, 27 Jun. 2002, *available at* www.salon.com/news/feature/2002/

06/27/military; Robert Capps, Outside the Law, SALON.COM, 26 June. 2002, *available at* www.salon.com/news/feature/2002/06/26/bosnia/index_np; *US Scandal, Prostitution, Pimping and Trafficking,* Bosnia Daily, Daily e-newspaper, 25 Jul. 2001, No. 42, at 1 (on file with author).

73.   UNHCHR recently addressed this issue openly in its guideline covering "Obligations of peacekeepers, civilian police and humanitarian and diplomatic personnel," asking states to consider "[e]nsuring that staff employed in the context of peacekeeping, peace-building, civilian policing, humanitarian and diplomatic missions do not engage in trafficking and related exploitation or use the services of persons in relation to which there are reasonable grounds to suspect that they may have been trafficked." *Recommended Principles and Guidelines on Human Rights and Human Trafficking, U.N. High Commissioner for Human Rights,* E/2002/68/Add.1, Guideline 10, ¶ 3 [hereinafter *Recommended Principles and Guidelines*]. *See infra* text pt. IV(C)(2).

POSTSCRIPT

Has the International Community Designed an Adequate Strategy to Address Human Trafficking?

Human trafficking has been part of the global landscape for centuries. What is different today are the magnitude and scope of the trafficking and the extent to which organized crime is involved in facilitating such nefarious activity. And yet the global community is still only in the position of trying to identify the nature and extent of the problem, let alone ascertaining how to deal with it. In April 2006, the UN Office on Drugs and Crime released its most recent report on the human trafficking problem. Titled *Trafficking in Persons: Global Patterns* (United Nations Office on Drugs and Crime, April 2006), its message was clear. The starting point for addressing the problem is the implementation of the Protocol to Prevent, Suppress and Punish Trafficking in Persons, especially Women and Children. National governments are called upon to take a leading role in: (1) the prevention of trafficking, (2) prosecution of violators, and (3) protection of victims.

Consider the task of prevention. Nations are expected to establish comprehensive policies and programs to prevent and combat trafficking, including research, information and media campaigns. Nations must attempt to alleviate the vulnerability of people, especially women and children. They must create steps to discourage demand for victims. Nations must also prevent transportation opportunities for traffickers. Finally, they must exchange information and increase cooperation among border control agencies.

The UN report also suggests several steps with respect to prosecution. The first step is to "ensure the integrity and security of travel and identity documents" and thus prevent their misuse. Domestic laws must be enacted making human trafficking a criminal offense, and these laws must apply to victims of both genders and all ages. Penalties must be adequate to the crime. Finally, victims must be protected and possibly compensated.

The third role outlined in the UN report focuses on protection of victims. Specifically, victims must be able to achieve "physical, psychological and social recovery." The physical safety of victims is also paramount. The final step relates to the future home of victims, whether they want to remain in the location where found or whether they wish to return home.

The essence of the report suggests the changing character of global issues in this age of globalization. No longer can nation-states solve problems alone. Moreover, no longer can they simply create a new international organization to address the problem. The issue is simply too complex. An array of interlocking structures, agreements, international law, and national initiatives (what was termed an international regime earlier) is needed and is well on the way to being created. As with other issues in this volume, modern technology, combined with the process of globalization, demands such a strategy if the international community is going to successfully address the evils of human trafficking.

The UN report cited above represents an excellent starting point for additional readings about human trafficking. A second comprehensive source is "Victims of Trafficking for Forced Prostitution: Protection Mechanisms and the Right to Remain in the Destination Countries" (*Global Migration Perspectives*, Global Commission on International Migration, no. 2, July 2004).

A 2005 book edited by Kamala Kempadoo, *Trafficking and Prostitution Reconsidered: New Perspectives on Migration, Sex Work, and Human Rights* (Paradigm Publishers) examines the contemporary situation in Asia.

A number of articles provide useful insights into aspects of the problem. Ilana Kramer's "Modern-Day Sex Slavery" (*Lilith*, vol. 31, no. 1, spring 2006) describes the problem as it relates to Israel. John R. Miller's "Modern-Day Slavery" (*Sheriff*, vol. 58, no. 2, March/April 2006) describes trafficking problems within the United States.

Finally, a comprehensive look at the new agreements can be found in Anne Gallagher, "Human Rights and the New UN Protocols on Trafficking and Migrant Smuggling: A Preliminary Analysis" (*Human Rights Quarterly*, vol. 23, 2001).

A number of websites are also useful. Among these are http://www.wnhcr.ch (the UN High Commissioner for Refugees); http://www.unicef.org (International Child Development Center); http://www.unifem.org (United Nations Development Fund for Women); http://www.ecre.org/research/smuggle.shtml (European Council on Refugees and Exiles); http://www.bayswan.org?FoundTraf.html (Foundation against Trafficking in Women); http://www.trafficked-women.org (Coalition to Abolish Slavery and Trafficking); http://www.uri.edu/artsci/wms/hughes/pubvio.htm (Coalition Against Trafficking in Women); www.gaatw.org (Global Alliance Against Traffic in Women); www.ecpatw.org (End Child Prostitution and Trafficking); and http://www.antislavery.org (Antislavery International).

ISSUE 13

Is Globalization a Positive Development for the World Community?

YES: Johan Norberg, from "How Globalization Conquers Poverty," The Cato Institute (September 9, 2003)

NO: J. Scott Tynes, from "Globalization Harms Developing Nations' Cultures," in Berna Miller and James D. Torr, eds., *Developing Nations* (Greenhaven Press, 2003)

ISSUE SUMMARY

YES: Norberg argues that throughout history the expansion of trade through the capitalist system has created wealth in nations. He argues that developing countries need only overcome their own shortcomings and adopt this model within a globalizing world to take advantage of this reality.

NO: Tynes contends that globalization not only creates economic poverty among developing states, but first world countries as well through mass media control, corporate action, and policies that exacerbate this problem and create deeper gaps between rich and poor.

Globalization is a phenomenon and a revolution. It is sweeping the world with increasing speed and changing the global landscape into something new and different. Yet, like all such trends, its meaning, development, and impact puzzle many. We talk about globalization and experience its effects, but few of us really understand the forces that are at work in the global political economy.

When people use their cell phones, log onto the Internet, view events from around the world on live television, and experience varying cultures in their own backyards, they begin to believe that this process of globalization is a good thing that will bring a variety of new and sophisticated changes to people's lives. Many aspects of this technological revolution bring fun, ease, and sophistication to people's daily lives. Yet the anti–World Trade Organization (WTO) protests in Seattle, Washington, in 1999 and Washington, D.C., in 2000 are graphic illustrations of the fact that not everyone believes that globalization is a good thing. Many Americans who have felt left out of the global economic boom, as well as Latin Americans, Africans, and Asians who feel that their job skills and abilities are being exploited by multinational corporations (MNCs) in a global division of labor, believe that this system does not meet their needs. Local cultures that believe that Wal-Mart and McDonald's bring cultural change and harm rather than inexpensive products and convenience criticize the process. In this way, globalization, like all revolutionary forces, polarizes people, alters the fabric of their lives, and creates rifts within and between people.

Many in the West, along with the prominent and elite—among MNCs, educators, and policymakers—seem to have embraced globalization. They argue that it helps to streamline economic systems, disciplines labor and management, brings forth new technologies and ideas, and fuels economic growth. They point to the relative prosperity of many Western countries and argue that this is proof of globalization's positive effects. They see little of the problems that critics identify. In fact, those who recognize some structural problems in the system argue that despite these issues, globalization is like an inevitable tide of history, unfortunate for some but unyielding and impossible to change. Any problems that are created by this trend, they say, can be solved.

Many poor and middle-class workers, as well as hundreds of millions of people across the developing world, view globalization as an economic and cultural wave that tears at the fabric of centuries-old societies. They see jobs emerging and disappearing in a matter of months, people moving across the landscape in record numbers, elites amassing huge fortunes while local cultures and traditions are swept away, and local youth being seduced by promises of American material wealth and distanced from their own cultural roots. These critics look past the allure of globalization and focus on the disquieting impact of rapid and system-wide change.

The irony of such a far-ranging and rapid historical process such as globalization is that both proponents and critics may be right. The realities of globalization are both intriguing and alarming. As technology and the global infrastructure expand, ideas, methods, and services are developed and disseminated to greater and greater numbers of people. As a result, societies and values are altered, some for the better and others for the worse.

In the selections that follow, the authors explore the positive and negative impacts of globalization and reach different conclusions. Norberg views globalization within a larch sweep of history viewing wealth creation, expansion of services, and development as inherently positive and thus central to the beneficial dimensions of globalization as a world phenomenon. He cites macro-economic data to show that growth and wealth creation across the world as a result of globalization will alleviate global poverty. He makes this assertion with the understanding that such forces take time to develop. Tynes argues that globalization is nothing more than a western construct designed to exploit and transfer wealth from one part of the world to another and thus is inherently negative. He, too, places globalization within a historical context but his is imperialistic and based on an exploitive view of capitalism. Thus, for Tynes, globalization is nothing more than a system for extracting wealth and consequently can never be a force for positive development.

YES

Johan Norberg

How Globalization Conquers Poverty

In 1870, Sweden was poorer than Congo is today. People lived twenty years shorter than they do in developing countries today, and infant mortality was twice as high as in the average developing country. My forefathers were literally starving.

But reforms for liberalization at home and free trade abroad changed all of this. A trade agreement with England and France in 1865 made it possible for Swedes to specialize. We couldn't produce food well, but we could produce steel and timber, and sell it abroad. For the money we made, we could buy food.

In 1870, the industrial revolution began in Sweden. New companies exported to countries across the world, and production grew rapidly. The competition forced our companies to become more efficient, and old industries were closed so that we could meet new demands, such as better clothes, sanitation, health care and education.

By 1950—when the Swedish welfare state was no more than a glint in the social democrat's eye—the Swedish economy had quadrupled. Infant mortality had been reduced by 85 per cent and life expectancy had increased by a miraculous 25 years. We were on our way to abolishing poverty. We had globalized.

Even more interesting is that Sweden grew at a much faster rate than the developed countries it traded with. The wages in Sweden grew from 33 per cent of the average wage in the US in 1870 to 56 per cent in the early 1900s, even though American wages soared at the same time.

This shouldn't surprise anyone. Economic models predict that poor countries should have higher growth rates than affluent ones. They have more latent resources to harness, and they can benefit from the existence of wealthier nations to which they export goods and from which they import capital and more advanced technology, whereas affluent countries have already captured many of those gains.

It's a clear-cut case. Except for one small problem. This relationship does not exist.

Most poor countries grow more slowly than the industrialized countries. The reason is simple: most developing countries cannot make use of these international opportunities. And the two most significant reasons for this are manmade: domestic and external obstacles. Domestic barriers such as a lack of the rule of law, a stable climate for investment, and the protection of property rights. External barriers such as rich country protectionism in goods of particular importance to the third world—textiles and agriculture—that (according to UNCTAD) deprives developing countries of nearly $700 billion in export income a year—almost 14 times more than they receive in foreign aid.

But when we look at the poor countries with good institutions, and which are open to trade, we see that they are making rapid progress, much faster than the wealthy countries. A classic study by Jeffrey Sachs and Andrew Warner of 117 countries in the 1970s and 1980s showed that open-developing countries had an annual growth rate of 4.5 percent, compared with 0.7 percent in closed-developing countries and 2.3 percent in open industrialized countries. A recent World Bank report concluded that 24 developing countries with a total population of 3 billion are integrating into the global economy more quickly than ever. Their growth per capita has also increased from 1 per cent in the 1960s to 5 per cent in the 1990s (compared to a rich country growth of 1.9 per cent). At the present rate, the average citizen in these developing countries will see her income doubled in less than 15 years.

This points to the conclusion that globalization, the increase in international trade, communications and investments, is the most efficient means in history of extending international opportunity. The anti-globalists are correct when they claim that large parts of the world are left out, especially Sub-Saharan Africa. But that also happens to be the least liberal part of the world, with the most controls and regulations, and the weakest tradition of property rights. When anti-globalists blame globalization for African misery, it rings just as bizarre as the North Korean officials who once explained to a visiting Mongolian politician that the average North Korean is unhappy and miserable because he is sad about American imperialism.

On the whole, official statistics from governments, the UN and the World Bank all point in the direction that mankind has never before seen such a dramatic improvement of the human condition as we've seen in the last three decades. We have heard the opposite view repeated so many times, that we take it for granted, without examining the evidence.

During the last 30 years, chronic hunger and the extent of child labour in the developing countries have been cut in half. In the last half century, life expectancy has gone up from 46 to 64 years and infant mortality has been reduced from 18 to 8 per cent. These indicators are much better in the developing world today than they were in the richest countries a hundred years ago.

In a generation, the average income in developing countries has doubled. As the United Nations Development Programme has observed, in the last 50 years global poverty has declined more than in the 500 years before that. The number of absolute poor—people with less than $1/day—has according to the World Bank been reduced by 200 million in the last two decades, even though world population grew by about 1.5 billion during the same time. [Update: 400 million]

Even those encouraging findings, however, probably overestimate world poverty, because the World Bank uses survey data as the basis for its assessments. This data is notoriously unreliable. It suggests that South Korean is richer than the Swedes and British, for example, and that Ethiopia is richer than India.

Furthermore, surveys capture less and less of an individual's income. The average poor person at exactly the same level of poverty in surveys in 1987 and 1998 had in reality seen her income increase by 17 per cent. Former World Bank economist Surjit S Bhalla recently published his own calculations supplementing survey results with national accounts data (in the book *Imagine There's No Country,* Institute for International Economics, 2002). Bhalla found that UN's goal of lowering world poverty to below 15 percent by 2015 has already been achieved and surpassed. Absolute poverty had actually fallen from a level of 44 percent in 1980 to 13 percent in 2000.

Bhalla also shows that the GDP per capita of the developing countries taken as a whole (not as individual countries) grew by 3.1 percent 1980–2000, compared to the industrialized world's 1.6 percent. These countries are now repeating the Swedish experience from the late 19th century, only faster. From 1780, it took England almost 60 years to double its wealth. A hundred years later, Sweden did it in about 40 years, and another century later it took South Korea just a bit more than 10 years.

The world has never been a better place to live in than it is today. Poverty has never been this low, and living standards so high. And the era of globalization has created the setting for an even faster growth of opportunities and wealth creation.

Hold on to your hat.

J. Scott Tynes  **NO**

Globalization Harms Developing Nations' Cultures

Part I: Understanding Globalization and Cultural Imperialism

The concept of globalization and the new religion of the masses, capitalism, needs to come under even greater scrutiny. The developed world beats a drum of such deafening proportion, the cries and needs of smaller niche participants are being drowned out.

At present the idea and implementation of "globalization" is largely Western in its scope. It remains a Westernizing process integrally tied to the ideals of capitalism, serving to broaden the gap Western nations have enjoyed economically, politically and militarily.

It is the inherent interest of globalization to bring forward values and ideals largely Western in nature because they are understood and thereby seen as superior. They provide a sense of control, a sense of the known. This is not to suggest these ideals are necessarily in direct opposition to any Eastern or Occidental economic design. But it does posit an idea no less disturbing. These ideals are so readily available, so pervasive, they have begun to supercede the native values of developing nations.

Imperialism in the old world sense was an economic, political or military method used to undermine the sovereignty of a nation. The new imperialism is a cultural imperialism, using the machinery of the rapidly expanding communications network to inundate nations with Western ideology.

To suggest "imperialism" is to suggest a deliberate attempt by Western capitalist countries to dominate the mass media of other nations and affect the ways they think and live their lives. How deliberate an attempt is being made is not clear. But the advantages of such domination are easily identifiable. The creation of markets for Western products and the opening of labor markets with workers eager to earn money soon put right back into the system is of primary importance.

Western economies entered a period of change from labor to service and information economies over the last half-century. As such the maintenance and development of information pathways and communications technologies has become vital to economic growth. A need for control and domination of world communication and data flows has emerged. Having this control ensures continued economic expansion and the continual growth now referred to as globalization.

It is not too far a stretch to see how old imperialist desires can evolve and integrate into capitalist interests to control a variety of communication and data flow technologies, the laws that govern them and even the bandwidth that they occupy.

The emergence of transnational corporations (TNCs) was the clarion burst announcing the arrival of a new way of handling global business. Business would no longer be limited by geographic boundaries. Every corner of the earth could become a potential market; every corner could provide labor, so long as control over key aspects of the development could be maintained. Such control allows for a directed diversification, the "globalization" of production and advertising.

This change is bringing about a gradual erosion of the sovereignty of many nations. Overseeing bodies such as the World Bank and the International Monetary Fund (IMF) were brought in to monitor and control this development and shape it to fit Western standards of success and growth; sometimes at great disadvantage to their protégés.

A study by the Canadian Department of Communications revealed "the information revolution may accelerate the erosion of national sovereignty by further increasing the dominance of multi-national corporations in the world economy."

Globalization leaves developing nations as niche participants, requiring them to find some part of the production process they can perform better or cheaper than any other nation. It forces them to do that exclusively, often ignoring the importance of internal needs so they can remain a part of the global marketplace.

Western Control of Communications and Data Flows

Real concerns begin to emerge when direct connections suggesting any "imperialist" attitudes are examined. Currently there are five areas of control, which all bring forth a rather ominous conclusion. Efforts to control communications and data flows suggest, whether deliberately planned or not, globalization is resulting in, at the very least, a cultural imperialism.

The majority of information transmitted between nations is that of transnational corporations (TNCs), coordinating and maintaining control over various branches of their organization. The data they are relaying involves information directly affecting a nation's economy: Raw material supplies, inventory, pricing, labor policy, tax matters, currency holdings and investment plans. These are vital national interests. TNCs end up making decisions affecting these nations from distant locales all the while maintaining their own priorities, totally without the oversight of the national authority.

Remote sensing, where orbiting satellites are capable of mapping out terrain to reveal hidden resources and potential physical trouble spots, is currently employed by military and corporate interests to maintain global dominance and dependency. Resources such as agricultural and oceanic phenomena revealed with this technology can greatly aid developing nations. It could enable them to greatly accelerate their development, taking advantage of exclusive data instead of wasting resources on dead ends.

Database construction and access is controlled, again, for corporate and military gain. The majority of databases in existence are for the purpose of maintaining and supporting the needs of TNCs. These databases contain customer names, payroll information, property ownership, credit rating, and health information. This is vital information key to interests in globalist expansion. Understanding and controlling this information provides an upper hand in the control of markets and the focus of advertising.

Direct Broadcast Satellites (DBS), seen on the sides of most every home, provide just what they advertise: Direct access with little or no interference from a national authority. Images and programming are beamed directly circumventing any control or supervision, any national interest. Despite the international community's cries for prior consent to programming before broadcast, the US has replied that such consent restricts the "free flow of information" and as such violates the US right to free speech.

Western governments even extend control over slots in space for satellites to orbit. This allows TNCs primary, nearly exclusive, access to markets, allowing for more rapidly developed communication amongst the various global branches of these corporations. Nations shut out of the skies are not only limited by restricted access but also subjugated by the data these satellites communicate.

This control creates markets for products by exportation of commercials and the portrayal of the "good life" as seen through Western television programs and films. Once the desire is created within these new markets, a "globalization" door opens for potential labor markets. Once people want products they have seen, once they feel they "need" these products to be happy, they are willing to work to satisfy their desire.

Advertising and the development of a globalized structure such as unified currencies can bode ill for developing countries if they operate unaware. Western advertising currently helps to maintain monopolies, it forms and guides tastes, and shifts consumer demand to largely Western interests. Western control can come through these less direct means or through financial controls handed down from overseeing bodies. Regardless of how it arrives, the effect is to put the control of a nation's economy and the culture of its people in the hands of those whose only real interest is profit.

Part II: The "Free" Flow of Information

Concern is growing over US government protection of its abilities to reach untapped markets with targeted advertising under claims of "free information flows."

The World Press Freedom Committee (WPFC) seemed to concur that such actions were legitimate saying, "We believe that the free flow of information is essential for mutual understanding and world peace." The WPFC sees a correlation between advertising and freedom of the press, suggesting without the income generated through advertising a free press could not exist. Thus suggesting without the advertising of large transnational corporations (TNCs) a free press cannot exist and therefore those who resist this free flow of advertising/information are resisting the free press and thereby resisting freedom.

When globalization is examined in the light of the five previously discussed more passive efforts, some degree of cultural imperialism can be divined. But a more active form exists in activities such as this, termed "soft power." Soft power strategies are designed to control specific information causing people of other resistant nations to act and think as the provider of this information. By infiltrating a nation through communication pathways a controlling nation can defeat resistance to globalization desires from the inside.

From a political perspective, using propaganda to stabilize ethnic unrest and promote democracy is a positive thing, but it also gives birth to a cultural development for the acceptance of capitalism; something perhaps not desired by those being stabilized and developed—at the very least this development may not be in their best interests.

Soft power proves to be cultural imperialism with a twist. Soft power, unlike the previous five methodologies, is a blatant attempt at control. Such attempts often yield results opposite of those desired. Soft power is cultural imperialism naked and unabashed. As such it rarely works.

Any effects, whether active as in the case of soft power, or a more passive means such as control over communication and data flows, are in the end filtered by the cultural capacity of the receiver. The bias of both the generator of such a message and the bias of the receiver must be considered.

But the bias of an individual is really too vague to completely explain the resistance or lack of resistance a particular nation has to cultural attacks. The most lucid explanation of what actually controls a culture's capacity to embrace or reject outside input is found in an examination of the number and complexity of the culture's organizations and institutions.

In instances where these mediators "places of worship, cultural icons, leaders, schools" have a great hold on the culture of the nation through their complexity and history, the input of any imperialist machinery will be rejected outright. When the Shah of Iran made attempts to modernize, and hence Westernize, his nation by filling the airwaves with the glories of the west, his intentions to suppress extremists failed miserably as the people of Iran wholly rejected Western values that so differed from Islamic tradition.

These are traditions and beliefs of such complexity they present formidable resistance to the Western way. This makes the people much less susceptible to "soft power" or other forms of cultural imperialism because a structure is in place. They have an alternative. They are not forced to accept the way of the West as the only means for "salvation."

There are other instances where Western culture is accepted but modified by the receiver. South America has seen instances where on the surface US commercialism is rejected. But digging deeper it is apparent this commercialism/capitalism/globalism has been taken and modified. American lifestyles seen in film and television are not directly accepted but rather transformed into "Brazilianized faces" and thereby deconstructing and reconstructing a message to more aptly fit a cultural norm.

The Harms of Cultural Imperialism

But at the other extreme are examples where cultural capacities are such that nations sacrifice their own best interest to be part of a globalist system. In Mexico a shift began in the Mexican diet resulting from the penetration of 130 TNCs into the Mexican market. These corporations poured millions of dollars into advertising and promotion for their products and increased the consumption of products that cost twenty times more than comparable native foods.

Yet the foods they promoted were much less nutritious than the native staples. But they made the TNCs money. Their profits increased by close to 27 percent. In essence the Mexican people stopped eating the cheaper and healthier foods of their tradition and their ability and opted for poor, processed, more expensive foodstuffs advertised and promoted by TNCs.

In India many left traditional occupations for more lucrative niche positions allowing them to buy the wares of the west. The television bombarded them with what they "need" and many within India began to accept they actually needed these products. For example, a decade ago India had only three or four brands of soap. Through television advertising and a variety of road shows "a sort of infomercial on wheels" there are now several hundred types of soap; more than nearly any other nation in the world, western or not. India moved from a near dearth of soap to one of the largest consumers in the world through a gradual cultural penetration.

The real need may not exist but the desire created and the acceptance of the advertised can, and many suggest did, change Indian society for the worse. Western ideals can be more "enlightened," but they are not necessarily so and certainly not superior by their nature. When they begin to replace traditional values and ideals that are key to the establishment of a unique sense of cultural identity a very significant danger exists.

Nearly all these unique identities are pared with unique non-Western cultures whose value may not be fully understood by those promoting such development. And most recently many of these cultures have been Islamic prompting fears of the wars of "us" versus "them." The "free" flow of information's real cost might be in the homogenization of global cultures and in the end the loss of a diversity key to the development of man.

Part III: Cultural Imperialism: Do All the Pieces Fit?

It should be more and more apparent a degree of cultural imperialism is inherent in the structure of globalization and capitalism. But by its very nature it is more and more difficult to identify any fingerprints that would point to a specific organizer of any effort to implant capitalism in the third and rapidly shrinking second world.

The reality is the US government is not capable of organizing and implementing a cultural imperialist policy. One hand rarely knows what the other is doing in the US government. To suggest an organized effort by the US to be cultural imperialists is ludicrous in light of its difficulties in implementing economic policies at home.

But US and other Western governments are participants and enablers through the economic relationships they have with transnational corporations (TNCs). What we are witnessing now is not governmental domination, not a governmental imperialism like the imperialism of old, but instead an imperialism that has the markings of transnational corporate cultural domination, albeit American/Western in nature.

Again, though, to suggest the TNCs of the world have organized and focused on a cultural imperialist agenda and are utilizing their influence over the nations of the world to this end is a bit premature, ringing a bit of conspiracy theories of *Illuminati* and the like that have recently made the rounds in movie theaters.

TNCs presently have a largely free reign across the globe and do have a motive to expose and exploit the markets available to them. But TNCs do not operate from an organized cultural imperialist position. They simply cannot. To do so would be to operate in a vacuum away from the critical eyes of the rest of the world. But it is also not in their best interest to do so.

TNC motives are pure and simple: profit and markets. When those motives coincide with that of another TNC, the result may be a 'cultural imperialist action.' For example, if two corporations see that combining forces and influencing the government of a developing nation might work to their mutual benefit, they will do so. TNCs usually do not operate in this spirit. They compete with one another. Sharing profits is not part of their capitalist makeup.

Cultural imperialism exists; but when the curtain is raised, and it is examined in detail, it is a perhaps less sensational form than initially imagined. There exists little real organization and focus in creating these imperialist intrusions other than simple profit motives.

TNCs are interested in the bottom line and Western nations are dependent on the success of TNCs to grease the gears that keep their nation afloat. This has become the nature of the modern nation state. Economic power has replaced military power as the most important, most wieldy, form of control and expansion. Countries are less interested in capturing new geographic territory, and more interested in economic dominance that in the end builds their own coffers. Successful economies mean a country can survive and thrive.

It is a complex interrelation that results in legislation and actions that would sometimes seem to point to coordinated efforts at capitalist global domination. TNC advisors to nations on UN councils controlling satellite positioning make many very nervous.

There has begun a great deal of speculation and fear about a unified globe. The idea of a single currency such as the euro and singular bodies to control information flows and economic agendas makes many, quite justly, anxious.

It should be clear that removing impediments to economic expansion benefits the large before the small. Opening the doors of small and developing nations means TNCs can waltz in with their tremendous resources and dominate the economy. It is a scary and very real thought.

No Single Culprit

But rather than finding a singular organized culprit for cultural domination such as TNCs or Western governments responsible, it is more likely the villain is capitalism itself. It is the true motivator of all those involved. It would perhaps serve the agendas of many afraid of a conspiracy to raise a curtain and find a single group to point a finger at. Some have seen the G8 [Canada, France, Germany, Italy, Japan, Russia, the United Kingdom, and the United States] or WTO [World Trade Organization] as the wizards running the show. But the implementation and organization at this point does not suggest this. The pieces are there, but they have yet to be fully assembled.

Some consideration of the cultural capacities of the nations involved is also a point of note. Cultural imperialism will not always be completely successful. Its results may vary from complete failure in nations with complex organizations and institutions to adaptations of capitalism in nations that are less complex and organized. It might also result in the complete abandonment of those institutions and organizations in nations desperate to have the 'good life' of the West.

The import of Western advertising and programming into the former communist nation of Albania and the introduction of a prime piece of Western capitalism, the pyramid scheme, led to the destruction of that nation. Albanians living for years beside Western nations saw the 'good life' transmitted across the border to them. As they lived their lives under the oppression of communism desires grew, fed by the advertising and programming of western television and radio. When given the chance to have it all, albeit through risky pyramid schemes, they risked it all. Unfortunately they lost. But no single figure or group can be blamed.

Cultural imperialism is perhaps best defined as capitalism's attempts at domination. Capitalism has become the new opiate of many, many people. But it must be recognized that capitalism is an autonomous entity, part of Western society but not specifically controlled by any single party or organized group.

It exists because people want—they want newer, they want better and they want more. Those that are able to control their wants, or those whose wants don't fit the mainstream mold are able to sidestep the impact.

As long as a 'want' exists on a global level, capitalism and its brother, globalization, will have at its core methods of transmission resulting in cultural imperialism. Advertising from Western TNCs will continue to inundate the globe, databases will continue to be used primarily for TNC benefit, satellite positioning and the information they provide will remain in the hands of the already powerful, and old men in grass huts will continue to watch reruns of *I Love Lucy* as they drift off into capitalist dreams of the good life, wearing Levi's, drinking Coke and driving a brand new Mercedes Benz.

POSTSCRIPT

Is Globalization a Positive Development for the World Community?

It is hard to argue that this kind of revolution is all positive or all negative. Many will find the allure of technological growth and expansion too much to resist. They will adopt values and ethics that seem compatible with a materialistic Western culture. And they will embrace speed over substance, technical expertise over knowledge, and wealth over fulfillment.

Others will reject this revolution. They will find its promotion of materialism and Western cultural values abhorrent and against their own sense of humanity and being. They will seek enrichment in tradition and values rooted in their cultural pasts. This resistance will take many forms. It will be political and social, and it will involve actions ranging from protests and voting to division and violence.

Trying to determine whether a force as dominant and all-encompassing as globalization is positive or negative is like determining whether the environment is harsh or beautiful. It is both. One can say that in the short term, globalization will be destabilizing for many millions of people because the changes that it brings will cause some fundamental shifts in beliefs, values, and ideas. Once that period is past, it is conceivable that a more stable environment will result as people come to grips with globalization and either learn to embrace it, cope with it, or keep it at bay.

The literature on globalization is growing rapidly, much of it centering on defining its parameters and evaluating its impact. One important work that presents a deeper understanding of globalization is Friedman's best-seller *The Lexus and the Olive Tree* (Farrar, Straus & Giroux, 1999), which provides a positive view of the globalization movement. Two counterperspectives are William Greider's *One World Ready or Not: The Manic Logic of Global Capitalism* (Simon & Schuster, 1997) and David Korten's *When Corporations Rule the World* (Kumarian Press, 1996). Also see Jan Nederveen Pieterse, ed., *Global Futures: Shaping Globalization* (Zed Books, 2001). A key proponent of globalization is of course Thomas Friedman. His two books, *The Lexus and the Olive Tree* (Farrar, Straus, and Giroux, 2000) and *The World is Flat* illustrate this perspective.

The literature that will help us to understand the full scope of globalization has not yet been written. Also, the determination of globalization's positive or negative impact on the international system has yet to be decided. For certain, globalization will bring profound changes that will cause people from America to Zimbabwe to rethink assumptions and beliefs about how the world works. And equally certain is the realization that globalization will empower some people but that it will also leave others out, helping to maintain and perhaps exacerbate the divisions that already exist in global society.

ISSUE 14

Is the World a Victim of American Cultural Imperialism?

YES: George Monbiot, from "Thanks to Corporations Instead of Democracy We Get *Baywatch*," *The Guardian* (September 13, 2005)

NO: Philippe Legrain, from "In Defense of Globalization," *The International Economy* (Summer 2003)

ISSUE SUMMARY

YES: George Monbiot argues that U.S. media control and the desire for profits have not created a political or social awakening around the globe but rather a quest for profits that spreads American culture in its basest forms without any inherent benefits. Thus, by implication, the world is suffering under the yoke of American cultural imperialism.

NO: Philippe Legrain is a British economist who presents two views of cultural imperialism and argues that the notion of American cultural imperialism "is a myth" and that the spreading of cultures through globalization is a positive, not negative, development.

In 1989 the Berlin Wall collapsed. Two years later the Soviet Union ceased to exist. With this relatively peaceful and monumental series of events, the cold war ended, and with it one of the most contentious and conflict-ridden periods in global history. It is easy to argue that in the wake of those events the United States is in ascendancy. The United States and its Western allies won the cold war, defeating communism politically and philosophically. Since 1990 democracies have emerged and largely flourished as never before across the world stage. According to a recent study, over 120 of the world's 190 nations now have a functioning form of democracy. Western companies, values, and ideas now sweep across the globe via airwaves, computer networks, and fiber-optic cables that bring symbols of U.S. culture and values (such as Michael Jordan and McDonald's) into villages and schools and cities around the world.

If American culture is embodied in the products sold by many multinational corporations (MNCs), such as McDonald's, Ford, IBM, The Gap, and others, then the American cultural values and ideas that are embedded in these products are being bought and sold in record numbers around the world. Globalization largely driven by MNCs and their control of technology brings with it values and ideas that are largely American in origin and expression. Values such as speed and ease of use, a strong emphasis on leisure time over work time, and a desire for increasing material wealth and comfort dominate the advertising practices of these companies. For citizens of the United States, this seems a natural part of the landscape. They do not question it; in fact, many Americans enjoy seeing signs of "home" on street corners abroad: a McDonald's in Tokyo, a Sylvester Stallone movie in Djakarta, or a Gap shirt on a student in Nairobi, for example.

While comforting to Westerners, this trend is disquieting to the hundreds of millions of people around the world who wish to partake of the globalizing system without abandoning their own cultural values. Many people around the globe wish to engage in economic exchange and develop politically but do not want to abandon their own cultures amidst the wave of values embedded in Western products. This tension is most pronounced in its effect on the youth around the world. Millions of impressionable young people in the cities and villages of the developing world wish to emulate the American icons that they see on soft drink cans or in movie theaters. They attempt to adopt U.S. manners, language, and modes of dress, often in opposition to their parents and local culture. These young people are becoming Americanized and, in the process, creating huge generational rifts within their own societies. Some of the seeds of these rifts and cultural schisms can be seen in the actions of the young Arab men who joined Al Qaeda and participated in the terrorist attacks of September 11, 2001.

George Monbiot presents the notion that in a world where corporate media is multinational, driven by profit and oblivious to issues of social conscious or standards, cultural imperialism abounds in a variety of forms. He implies that while issues of democracy and freedom and social awareness could be promoted they are not and instead the world is flooded with programming that promotes a materialistic culture above other more crucial values.

YES

George Monbiot

Thanks to Corporations Instead of Democracy We Get *Baywatch*

"Several of this cursed brood, getting hold of the branches behind, leaped up into the tree, whence they began to discharge their excrements on my head." Thus Gulliver describes his first encounter with the Yahoos. Something similar seems to have happened to democracy. In April, Shi Tao, a journalist working for a Chinese newspaper, was sentenced to 10 years in prison for "providing state secrets to foreign entities." He had passed details of a censorship order to the Asia Democracy Forum and the website Democracy News. The pressure group Reporters Without Borders (RSF) was mystified by the ease with which Mr. Tao had been caught. He had sent the message through an anonymous Yahoo! account. But the police had gone straight to his offices and picked him up. How did they know who he was?

Last week RSF obtained a translation of the verdict, and there they found the answer. Mr. Tao's account information was "furnished by Yahoo Holdings." Yahoo!, the document says, gave the government his telephone number and the address of his office. So much for the promise that the Internet would liberate the oppressed. This theory was most clearly formulated in 1999 by the *New York Times* columnist Thomas Friedman. In his book *The Lexus and the Olive Tree,* Friedman argues that two great democratizing forces—global communications and global finance—will sweep away any regime which is not open, transparent and democratic.

"Thanks to satellite dishes, the Internet and television," he asserts, "we can now see through, hear through and look through almost every conceivable wall. . . . No one owns the Internet, it is totally decentralised, no one can turn it off. . . . China's going to have a free press. . . . Oh, China's leaders don't know it yet, but they are being pushed straight in that direction." The same thing, he claims, is happening all over the world. In Iran he saw people ogling "Baywatch" on illegal satellite dishes. As a result, he claims, "within a few years, every citizen of the world will be able to comparison shop between his own . . . government and the one next door."

He is partly right. The Internet at least has helped to promote revolutions of varying degrees of authenticity in Serbia, Ukraine, Georgia, Kyrgyzstan, Lebanon, Argentina and Bolivia. But the flaw in Friedman's theory is that he forgets the intermediaries. The technology which runs the internet did not sprout from the ground. It is provided by people with a commercial interest in its development. Their interest will favour freedom in some places and control in others. And they can and do turn it off.

In 2002 Yahoo! signed the Chinese government's pledge of "self-regulation": it promised not to allow "pernicious information that may jeopardise state security" to be posted. Last year Google published a statement admitting that it would not be showing links to material banned by the authorities on computers stationed in China. If Chinese users of Microsoft's Internet service MSN try to send a message containing the words "democracy," "liberty" or "human rights," they are warned that "This message includes forbidden language. Please delete the prohibited expression."

A study earlier this year by a group of scholars called the OpenNet Initiative revealed what no one had thought possible: that the Chinese government is succeeding in censoring the Net. Its most powerful tool is its control of the routers—the devices through which data is moved from one place to another. With the right filtering systems, these routers can block messages containing forbidden words. Human-rights groups allege that western corporations—in particular Cisco Systems—have provided the technology and the expertise. Cisco is repeatedly cited by Thomas Friedman as one of the facilitators of his global revolution. "We had the dream that the Internet would free the world, that all the dictatorships would collapse," says Julien Pain of Reporters Without Borders. "We see it was just a dream."

Friedman was not the first person to promote these dreams. In 1993 Rupert Murdoch boasted that satellite television was "an unambiguous threat to totalitarian regimes everywhere." *The Economist* had already made the same claim on its cover: "Dictators beware!" The Chinese went berserk, and Murdoch, in response, ensured that the threat did not materialise. In 1994 he dropped BBC world news from his Star satellite feeds after it broadcast an unflattering portrait of Mao Zedong. In 1997 he ordered his publishing house HarperCollins to drop a book by Chris Patten, the former governor of Hong Kong. He slagged off the Dalai Lama and his son James attacked the dissident cult Falun Gong. His grovelling paid off, and in 2002 he was able to start broadcasting into Guangdong. "We won't do programmes that are offensive in China," Murdoch's spokesman Wang Yukui admitted. "If you call this self-censorship then of course we're doing a kind of self-censorship."

I think, if they were as honest as Mr Wang, everyone who works for Rupert Murdoch, or for the corporate media anywhere in the world, would recognise these restraints. To own a national newspaper or a television or radio station you need to be a multimillionaire. What multimillionaires want is what everybody wants: a better world for people like themselves. The job of their journalists is to make it happen. As Piers Morgan, the former editor of the *Mirror,* confessed, "I've made it a strict rule in life to ingratiate myself with billionaires." They will stay in their jobs for as long as they continue to interpret the interests of the proprietorial class correctly.

What the owners don't enforce, the advertisers do. Over the past few months, AdAge.com reveals, both Morgan Stanley and BP have instructed newspapers and magazines that they must remove their adverts from any edition containing "objectionable editorial coverage." Car, airline and tobacco companies have been doing the same thing. Most publications can't afford to lose these accounts; they lose the offending articles instead. Why are the papers full of glowing profiles of the advertising boss Martin Sorrell? Because they're terrified of him.

So instead of democracy we get *Baywatch*. They are not the same thing. Aspirational TV might stimulate an appetite for more money or more plastic surgery, and this in turn might encourage people to look, for better or worse, to the political systems that deliver them, but it is just as likely to be counter-democratic. As a result of pressure from both ratings and advertisers, for example, between 1993 and 2003 environmental programmes were cleared from the schedules on BBC TV, ITV and Channel 4. Though three or four documentaries have slipped out since then, the ban has not yet been wholly lifted. To those of us who have been banging our heads against this wall, it feels like censorship.

Indispensable as the Internet has become, political debate is still dominated by the mainstream media: a story on the Net changes nothing until it finds its way into the newspapers or on to TV. What this means is that while the better networking Friedman celebrates can assist a democratic transition, the democracy it leaves us with is filtered and controlled. Someone else owns the routers.

Philippe Legrain

In Defense of Globalization

Fears that globalization is imposing a deadening cultural uniformity are as ubiquitous as Coca-Cola, McDonald's, and Mickey Mouse. Many people dread that local cultures and national identifies are dissolving into a crass all-American consumerism. That cultural imperialism is said to impose American values as well as products, promote the commercial at the expense of the authentic, and substitute shallow gratification for deeper satisfaction.

Thomas Friedman, columnist for the *New York Times* and author of *The Lexus and the Olive Tree,* believes that globalization is "globalizing American culture and American cultural icons." Naomi Klein, a Canadian journalist and author of *No Logo,* argues that "Despite the embrace of polyethnic imagery, market-driven globalization doesn't want diversity; quite the opposite. Its enemies are national habits, local brands, and distinctive regional tastes."

But it is a myth that globalization involves the imposition of Americanized uniformity, rather than an explosion of cultural exchange. And although—as with any change—it can have downsides, this cross-fertilization is overwhelmingly a force for good.

The beauty of globalization is that it can free people from the tyranny of geography. Just because someone was born in France does not mean they can only aspire to speak French, eat French food, read French books, and so on. That we are increasingly free to choose our cultural experiences enriches our lives immeasurably. We could not always enjoy the best the world has to offer.

Globalization not only increases individual freedom, but also revitalizes cultures and cultural artifacts through foreign influences, technologies, and markets. Many of the best things come from cultures mixing: Paul Gauguin painting in Polynesia, the African rhythms in rock 'n' roll, the great British curry. Admire the many-colored faces of France's World Cup-winning soccer team, the ferment of ideas that came from Eastern Europe's Jewish diaspora, and the cosmopolitan cities of London and New York.

Fears about an Americanized uniformity are overblown. For a start, many "American" products are not as all-American as they seem; MTV in Asia promotes Thai pop stars and plays rock music sung in Mandarin. Nor are American products all-conquering. Coke accounts for less than two of the 64 fluid ounces that the typical person drinks a day. France imported a mere

From *The International Economy* by Philippe Legrain, vol. 17, no. 3, Summer 2003, pp. 62–65.

$620 million in food from the United States in 2000, while exporting to America three times that. Worldwide, pizzas are more popular than burgers and Chinese restaurants sprout up everywhere.

In fashion, the ne plus ultra is Italian or French. Nike shoes are given a run for their money by Germany's Adidas, Britain's Reebok, and Italy's Fila. American pop stars do not have the stage to themselves. According to the IFPI, the record-industry bible, local acts accounted for 68 percent of music sales in 2000, up from 58 percent in 1991. And although nearly three-quarters of television drama exported worldwide comes from the United States, most countries' favorite shows are homegrown.

Nor are Americans the only players in the global media industry. Of the seven market leaders, one is German, one French, and one Japanese. What they distribute comes from all quarters: Germany's Bertelsmann publishes books by American writers; America's News Corporation broadcasts Asian news; Japan's Sony sells Brazilian music.

In some ways, America is an outlier, not a global leader. Baseball and American football have not traveled well; most prefer soccer. Most of the world has adopted the (French) metric system; America persists with antiquated British Imperial measurements. Most developed countries have become intensely secular, but many Americans burn with fundamentalist fervor—like Muslims in the Middle East.

Admittedly, Hollywood dominates the global movie market and swamps local products in most countries. American fare accounts for more than half the market in Japan and nearly two-thirds in Europe. Yet Hollywood is less American than it seems. Top actors and directors are often from outside America. Some studios are foreign-owned. To some extent, Hollywood is a global industry that just happens to be in America. Rather than exporting Americana, it serves up pap to appeal to a global audience.

Hollywood's dominance is in part due to economics: Movies cost a lot to make and so need a big audience to be profitable; Hollywood has used America's huge and relatively uniform domestic market as a platform to expand overseas. So there could be a case for stuffing subsidies into a rival European film industry, just as Airbus was created to challenge Boeing's near-monopoly. But France's subsidies have created a vicious circle whereby European film producers fail in global markets because they serve domestic demand and the wishes of politicians and cinematic bureaucrats.

Another American export is also conquering the globe: English. By 2050, it is reckoned, half the world will be more or less proficient in it. A common global language would certainly be a big plus—for businessmen, scientists, and tourists—but a single one seems far less desirable. Language is often at the heart of national culture, yet English may usurp other languages not because it is what people prefer to speak, but because, like Microsoft software, there are compelling advantages to using it if everyone else does.

But although many languages are becoming extinct, English is rarely to blame. People are learning English as well as—not instead of—their native tongue, and often many more languages besides. Where local languages are dying, it is typically national rivals that are stamping them out. So although, within the United States, English is displacing American Indian tongues, it is not doing away with Swahili or Norwegian.

Even though American consumer culture is widespread, its significance is often exaggerated. You can choose to drink Coke and eat at McDonald's without becoming American in any meaningful sense. One newspaper photo of Taliban fighters in Afghanistan showed them toting Kalashnikovs—as well as a sports bag with Nike's trademark swoosh. People's culture—in the sense of their shared ideas, beliefs, knowledge, inherited traditions, and art—may scarcely be eroded by mere commercial artifacts that, despite all the furious branding, embody at best flimsy values.

The really profound cultural changes have little to do with Coca-Cola. Western ideas about liberalism and science are taking root almost everywhere, while Europe and North America are becoming multicultural societies through immigration, mainly from developing countries. Technology is reshaping culture: Just think of the Internet. Individual choice is fragmenting the imposed uniformity of national cultures. New hybrid cultures are emerging, and regional ones re-emerging. National identity is not disappearing, but the bonds of nationality are loosening.

Cross-border cultural exchange increases diversity within societies—but at the expense of making them more alike. People everywhere have more choice, but they often choose similar things. That worries cultural pessimists, even though the right to choose to be the same is an essential part of freedom.

Cross-cultural exchange can spread greater diversity as well as greater similarity: more gourmet restaurants as well as more McDonald's outlets. And just as a big city can support a wider spread of restaurants than a small town, so a global market for cultural products allows a wider range of artists to thrive. If all the new customers are ignorant, a wider market may drive down the quality of cultural products: Think of tourist souvenirs. But as long as some customers are well informed (or have "good taste"), a general "dumbing down" is unlikely. Hobbyists, fans, artistic pride, and professional critics also help maintain (and raise) standards.

A bigger worry is that greater individual freedom may undermine national identity. The French fret that by individually choosing to watch Hollywood films they might unwittingly lose their collective Frenchness. Yet such fears are overdone. Natural cultures are much stronger than people seem to think. They can embrace some foreign influences and resist others. Foreign influences can rapidly become domesticated, changing national culture, but not destroying it. Clearly, though, there is a limit to how many foreign influences a culture can absorb before being swamped. Traditional cultures in the developing world that have until now evolved (or failed to evolve) in isolation may be particularly vulnerable.

In *The Silent Takeover*, Noreena Hertz describes the supposed spiritual Eden that was the isolated kingdom of Bhutan in the Himalayas as being defiled by such awful imports as basketball and Spice Girls T-shirts. But is that such a bad thing? It is odd, to put it mildly, that many on the left support multiculturalism in the West but advocate cultural purity in the developing world—an attitude they would tar as fascist if proposed for the United States. Hertz appears to want people outside the industrialized West preserved in unchanging but supposedly pure poverty. Yet the Westerners who want this supposed paradise preserved in aspic rarely feel like settling there. Nor do most people in developing countries want to lead an "authentic" unspoiled life of isolated poverty.

In truth, cultural pessimists are typically not attached to diversity per se but to designated manifestations of diversity, determined by their preferences. Cultural pessimists want to freeze things as they were. But if diversity at any point in time is desirable, why isn't diversity across time? Certainly, it is often a shame if ancient cultural traditions are lost. We should do our best to preserve them and keep them alive where possible. Foreigners can often help, by providing the new customers and technologies that have enabled reggae music, Haitian art, and Persian carpet making, for instance, to thrive and reach new markets. But people cannot be made to live in a museum. We in the West are forever casting off old customs when we feel they are no longer relevant. Nobody argues that Americans should ban nightclubs to force people back to line dancing. People in poor countries have a right to change, too.

Moreover, some losses of diversity are a good thing. Who laments that the world is now almost universally rid of slavery? More generally, Western ideas are reshaping the way people everywhere view themselves and the world. Like nationalism and socialism before it, liberalism is a European philosophy that has swept the world. Even people who resist liberal ideas, in the name of religion (Islamic and Christian fundamentalists), group identity (communitarians), authoritarianism (advocates of "Asian values") or tradition (cultural conservatives), now define themselves partly by their opposition to them.

Faith in science and technology is even more widespread. Even those who hate the West make use of its technologies. Osama bin Laden plots terrorism on a cellphone and crashes planes into skyscrapers. Antiglobalization protesters organize by e-mail and over the Internet. China no longer turns its nose up at Western technology: It tries to beat the West at its own game.

Yet globalization is not a one-way street. Although Europe's former colonial powers have left their stamp on much of the world, the recent flow of migration has been in the opposite direction. There are Algerian suburbs in Paris, but not French ones in Algiers. Whereas Muslims are a growing minority in Europe, Christians are a disappearing one in the Middle East.

Foreigners are changing America even as they adopt its ways. A million or so immigrants arrive each year, most of them Latino or Asian. Since 1990, the number of foreign-born American residents has risen by 6 million to just over 25 million, the biggest immigration wave since the turn of the 20th century. English may be all-conquering outside America, but in some parts of the United States, it is now second to Spanish.

The upshot is that national cultures are fragmenting into a kaleidoscope of different ones. New hybrid cultures are emerging. In "Amexica" people speak Spanglish. Regional cultures are reviving. The Scots and Welsh break with British monoculture. Estonia is reborn from the Soviet Union. Voices that were silent dare to speak again.

Individuals are forming new communities, linked by shared interests and passions, that cut across national borders. Friendships with foreigners met on holiday. Scientists sharing ideas over the Internet. Environmentalists campaigning together using e-mail. Greater individualism does not spell the end of community. The new communities are simply chosen rather than coerced, unlike the older ones that communitarians hark back to.

So is national identity dead? Hardly. People who speak the same language, were born and live near each other, face similar problems, have a common experience, and vote in the same elections still have plenty in common. For all our awareness of the world as a single place, we are not citizens of the world but citizens of a state. But if people now wear the bonds of nationality more loosely, is that such a bad thing? People may lament the passing of old ways. Indeed, many of the worries about globalization echo age-old fears about decline, a lost golden age, and so on. But by and large, people choose the new ways because they are more relevant to their current needs and offer new opportunities.

The truth is that we increasingly define ourselves rather than let others define us. Being British or American does not define who you are: It is part of who you are. You can like foreign things and still have strong bonds to your fellow citizens. As Mario Vargas Llosa, the Peruvian author, has written: "Seeking to impose a cultural identity on a people is equivalent to locking them in a prison and denying them the most precious of liberties—that of choosing what, how, and who they want to be."

POSTSCRIPT

Is the World a Victim of American Cultural Imperialism?

Globalization is a process of technological change and economic expansion under largely capitalist principles. The key actors driving the globalization process are multinational corporations like McDonald's, Coca-Cola, Nike, and Exxon Mobil. These companies are rooted in the American-Western cultural experience, and their premise is based on a materialistic world culture that is striving for greater and greater wealth. That value system is Western and American in origin and evolution. It is therefore logical to assume that as globalization goes, so goes American culture.

Evidence of "American" culture can be seen across the planet: kids in Djakarta or Lagos wearing Michael Jordan jerseys and Nike shoes, for example, and millions of young men and women from Cairo to Lima listening to Michael Jackson records. Symbols of American culture abound in almost every corner of the world, and most of that is associated with economics and the presence of multinational corporations.

As the youth of the world are seduced into an American cultural form and way of life, other cultures are often eclipsed. They lose traction and fade with generational change. Many would argue that this loss is unfortunate, but others would counter that it is part of the historical sweep of life. Social historians suggest that the cultures of Rome, Carthage, Phoenicia, and the Aztecs, while still influential, were eclipsed by a variety of forces that were dominant and historically rooted. While tragic, it was inevitable in the eyes of some social historians.

Regardless of whether this eclipse is positive or negative, the issue of cultural imperialism remains. Larger and more intrusive networks of communication, trade, and economic exchange bring values. In this world of value collision comes choices and change. Unfortunately, millions will find themselves drawn toward a lifestyle of materialism that carries with it a host of value choices. The losers in this clash are local cultures and traditions that, as so often is the case among the young, are easily jettisoned and discarded. It remains to be seen whether or not they will survive the onslaught.

Works on this subject include Benjamin Barber, "Democracy at Risk: American Culture in a Global Culture," *World Policy Journal* (Summer 1998); "Globalism's Discontents," *The American Prospect* (January 1, 2002); Seymour Martin Lipset, *American Exceptionalism: A Double-Edged Sword* (W. W. Norton, 1996); and Richard Barnet and John Cavanagh, *Global Dreams: Imperial Corporations and the New World Order* (Simon & Schuster, 1995).

ISSUE 15

Do Global Financial Institutions and Multinational Corporations Exploit the Developing World?

YES: The Global Exchange, from "The Top Ten Reasons to Oppose the IMF" (2005)

NO: Flemming Larsen, from "The IMF's Views and Actions in Dealing with Its Poorest Member Countries," Introductory Remarks at World Council of Churches—World Bank—IMF Meeting (September 11, 2003)

ISSUE SUMMARY

YES: The Global Exchange, an international group with an egalitarian agenda, argues that the IMF is a Western construct designed purely to exploit the developing world and promote corporate interests against those of the people.

NO: Flemming Larsen, director of the IMF office in Europe, contends that the IMF exists solely to promote the expansion of development and wealth throughout the world using capitalist strategies of loans and assistance designed to make countries economically sustainable within the global system.

The global economy is one in which capitalist principles of free trade, credit accessibility, free market, and comparative advantage reign supreme. States with alternative models, whether they are communist or socialist, still must operate within this sea of capitalist exchange.

Aside from states, the dominant actors in this world are IFIs and MNCs. These actors possess the bulk of the world's liquidity (i.e., cash, technology, transportations networks, and access to markets). Working with and among states, these entities are able to dominate trade routes, technology, access to markets, production capability, and access to funds necessary to jump-start trade, investment, and growth. Third World states must deal with these actors if they hope to produce market and sell products on the world stage.

This economic reality places Third World states in a precarious position. They must deal with actors who possess a clear bias for capitalist development and who also have vested economic interests in a variety of activities, including loaning funds, investing money, and removing barriers to trade and local economic policies. Is this the recipe for an exploitive relationship to one of mutual benefit?

The historical evidence suggests both situations have occurred in earnest. Third World countries have for centuries been victims of economic exploitation of natural resources, cheap labor, and location. The case of most states in Africa, Latin America, the Middle East, and Asia support this contention. However, other states that have dealt with these actors and built successful and diversified economies, such as South Korea and Malaysia, point to a different conclusion.

The following articles reflect the deep chasm of opinion regarding the role of IFI's in the global economy. The Global Exchange argues that the IMF is but a "Global Loan Shark" by loaning out money to countries and extracting structural reforms and concessions designed to promote western corporate interests and drain money from the target countries. They contend that the IMF is a secretive, predatory institution that is about profit and extraction and as such acerbates gaps between rich and poor and worsens the economic situation for millions of people. Larsen recognizes the critics of the IMF but argues that it does what it is designed to do, which is stabilize local economies; set them on a sound footing with the world economy; and allow for investment, growth, and wealth creation through private corporate investment and resources.

YES

The Top Ten Reasons to Oppose the IMF

What Is the IMF?

The International Monetary Fund and the World Bank were created in 1944 at a conference in Bretton Woods, New Hampshire, and are now based in Washington, DC. The IMF was originally designed to promote international economic cooperation and provide its member countries with short term loans so they could trade with other countries (achieve balance of payments). Since the debt crisis of the 1980's, the IMF has assumed the role of bailing out countries during financial crises (caused in large part by currency speculation in the global casino economy) with emergency loan packages tied to certain conditions, often referred to as structural adjustment policies (SAPs). The IMF now acts like a global loan shark, exerting enormous leverage over the economies of more than 60 countries. These countries have to follow the IMF's policies to get loans, international assistance, and even debt relief. Thus, the IMF decides how much debtor countries can spend on education, health care, and environmental protection. The IMF is one of the most powerful institutions on Earth—yet few know how it works.

1. **The IMF has created an immoral system of modern day colonialism that SAPs the poor**
   The IMF—along with the WTO and the World Bank—has put the global economy on a path of greater inequality and environmental destruction. The IMF's and World Bank's structural adjustment policies (SAPs) ensure debt repayment by requiring countries to cut spending on education and health; eliminate basic food and transportation subsidies; devalue national currencies to make exports cheaper; privatize national assets; and freeze wages. Such belt-tightening measures increase poverty, reduce countries' ability to develop strong domestic economies and allow multinational corporations to exploit workers and the environment A recent IMF loan package for Argentina, for example, is tied to cuts in doctors' and teachers' salaries and decreases in social security payments. The IMF has made elites from the Global South more accountable to First World elites than their own people, thus undermining the democratic process.

2. **The IMF serves wealthy countries and Wall Street**

   Unlike a democratic system in which each member country would have an equal vote, rich countries dominate decision-making in the IMF because voting power is determined by the amount of money that each country pays into the IMF's quota system. It's a system of one dollar, one vote. The U.S. is the largest shareholder with a quota of 18 percent. Germany, Japan, France, Great Britain, and the US combined control about 38 percent. The disproportionate amount of power held by wealthy countries means that the interests of bankers, investors and corporations from industrialized countries are put above the needs of the world's poor majority.

3. **The IMF is imposing a fundamentally flawed development model**

   Unlike the path historically followed by the industrialized countries, the IMF forces countries from the Global South to prioritize export production over the development of diversified domestic economies. Nearly 80 percent of all mal-nourished children in the developing world live in countries where farmers have been forced to shift from food production for local consumption to the production of export crops destined for wealthy countries. The IMF also requires countries to eliminate assistance to domestic industries while providing benefits for multinational corporations—such as forcibly lowering labor costs. Small businesses and farmers can't compete. Sweatshop workers in free trade zones set up by the IMF and World Bank earn starvation wages, live in deplorable conditions, and are unable to provide for their families. The cycle of poverty is perpetuated, not eliminated, as governments' debt to the IMF grows.

4. **The IMF is a secretive institution with no accountability**

   The IMF is funded with taxpayer money, yet it operates behind a veil of secrecy. Members of affected communities do not participate in designing loan packages. The IMF works with a select group of central bankers and finance ministers to make polices without input from other government agencies such as health, education and environment departments. The institution has resisted calls for public scrutiny and independent evaluation.

5. **IMF policies promote corporate welfare**

   To increase exports, countries are encouraged to give tax breaks and subsidies to export industries. Public assets such as forestland and government utilities (phone, water and electricity companies) are sold off to foreign investors at rock bottom prices. In Guyana, an Asian owned timber company called Barama received a logging concession that was 1.5 times the total amount of land all the indigenous communities were granted. Barama also received a five-year tax holiday. The IMF forced Haiti to open its market to imported, highly subsidized US rice at the same time it prohibited Haiti from subsidizing its own farmers. A US corporation called Early Rice now sells nearly 50 percent of the rice consumed in Haiti.

6. **The IMF hurts workers**

   The IMF and World Bank frequently advise countries to attract foreign investors by weakening their labor laws—eliminating collective bargaining laws and suppressing wages, for example. The IMF's mantra of "labor flexibility" permits corporations to fire at whim and move where wages are cheapest. According to the 1995 UN Trade and Development Report, employers are using this extra "flexibility" in labor laws to shed workers rather than create jobs. In Haiti, the government was told to eliminate a statute in their labor code that mandated increases in the minimum wage when inflation exceeded 10 percent. By the end of 1997, Haiti's minimum wage was only $2.40 a day. Workers in the U.S. are also hurt by IMF policies because they have to compete with cheap, exploited labor. The IMF's mismanagement of the Asian financial crisis plunged South Korea, Indonesia, Thailand and other countries into deep depression that created 200 million "newly poor." The IMF advised countries to "export their way out of the crisis." Consequently, more than 12,000 US steelworkers were laid off when Asian steel was dumped in the US.

7. **The IMF's policies hurt women the most**
SAPs make it much more difficult for women to meet their families' basic needs. When education costs rise due to IMF-imposed fees for the use of public services (so-called "user fees") girls are the first to be withdrawn from schools. User fees at public clinics and hospitals make healthcare unaffordable to those who need it most. The shift to export agriculture also makes it harder for women to feed their families. Women have become more exploited as government workplace regulations are rolled back and sweat-shops abuses increase.

8. **IMF policies hurt the environment**
IMF loans and bailout packages are paving the way for natural resource exploitation on a staggering scale. The IMF does not consider the environmental impacts of lending policies, and environmental ministries and groups are not included in policy making. The focus on export growth to earn hard currency to pay back loans has led to an unsustainable liquidation of natural resources. For example, the Ivory Coast's increased reliance on cocoa exports has led to a loss of two-thirds of the country's forests.

9. **The IMF bails out rich bankers, creating a moral hazard and greater instability in the global economy**
The IMF routinely pushes countries to deregulate financial systems. The removal of regulations that might limit speculation has greatly increased capital investment in developing country financial markets. More than $1.5 trillion crosses borders every day. Most of this capital is invested short-term, putting countries at the whim of financial speculators. The 1995 Mexican peso crisis was partly a result of these IMF policies. When the bubble popped, the IMF and US government stepped in to prop up interest and exchange rates, using taxpayer money to bail out Wall Street bankers. Such bailouts encourage investors to continue making risky, speculative bets, thereby increasing the instability of national economies. During the bailout of Asian countries, the IMF required governments to assume the bad debts of private banks, thus making the public pay the costs and draining yet more resources away from social programs.

10. **IMF bailouts deepen, rather then solve, economic crisis**
During financial crises—such as with Mexico in 1995 and South Korea, Indonesia, Thailand, Brazil, and Russia in 1997—the IMF

stepped in as the lender of last resort. Yet the IMF bailouts in the Asian financial crisis did not stop the financial panic—rather, the crisis deepened and spread to more countries. The policies imposed as conditions of these loans were bad medicine, causing layoffs in the short run and undermining development in the long run. In South Korea, the IMF sparked a recession by raising interest rates, which led to more bankruptcies and unemployment. Under the IMF imposed economic reforms after the peso bailout in 1995, the number of Mexicans living in extreme poverty increased more than 50 percent and the national average minimum wage fell 20 percent.

 **NO**

The IMF's Views and Actions in Dealing with Its Poorest Member Countries

I would like to focus my introductory remarks on how the institution I represent should be viewed from the perspective of the concerns you all have about sustainable development, poverty alleviation, and social justice. These concerns are very much shared by the IMF's member governments, management, and staff.

It is not an easy task to convince critics of the Fund that the IMF's views and actions are driven by ethical considerations and concerns about people's welfare. When thinking about the IMF, our critics tend to associate concepts such as social injustice, austerity, violation of human rights, protection of the interests of financial speculators and transnational enterprises, and of course dominance of U.S. views. That we are often labeled neo-liberal is clearly not meant as a tribute to a compassionate institution.

But before taking a closer look at our policies and actions, remember that the IMF emerged from the ashes of the second world war as one of the cornerstones of a strong economic and financial global economic system, based on cooperation, solidarity, the "rule-of-law" in international economic relations, and instruments to help countries in financial distress. The objective was to help members achieve high living standards, and sustainable economic growth and development, while eschewing the beggar-thy-neighbor policies that had contributed to the Great Depression. The language of our Articles of Agreement, or the IMF's constitution, is somewhat technical, but there can be no doubt that ethical considerations were overarching among the IMF's founding fathers who had all seen the devastating effects of the Great Depression and the two world wars.

Since then the Fund has evolved, and so has the language you will typically find in Fund policy statements. But the objectives of our work have not changed. If anything, our policy statements have become much more precise about the Fund's responsibility to guide member countries on some—but certainly not all—of the essential prerequisites for economic progress, namely macroeconomic and financial stability. This is because economic and financial stability have proven again and again to be absolutely necessary conditions for sustained improvements in living standards. And because economic and financial instability and crises are almost always felt disproportionately by the poor. And finally, with respect to the Fund's important but less than all-encompassing responsibilities, because the membership considers that the Fund is most effective if it concentrates on the core issues that form its mandate.

Safeguarding macroeconomic and financial stability, and restoring it as rapidly as possible after economic crises, are clearly the dominant concerns of the Fund. We in the Fund consider that by fulfilling these responsibilities we are contributing to a better world. We also consider that we are contributing in a major way to protecting those who are the most vulnerable, to reducing poverty, and to enhancing social justice—within countries and across the world. We also help to protect the vulnerable and reduce poverty through advice on the allocation of public spending and the design of social safety nets. While we do not usually frame our objectives in terms of human rights, in my view the IMF's work clearly is geared toward protecting what I consider basic human rights: the right to work, save, invest, engage in enterprise, and secure the future of ones family in a reasonably predictable economic and financial environment characterized by low and stable inflation, and respect for property rights.

It is true of course that there is not a single bridge, road, school or hospital to point to as something the IMF has financed. What we can point to as a success is when a country manages to reduce an unsustainable fiscal deficit and therefore avoids a fiscal crisis. Or when a country rapidly tackles the root causes of a financial crisis and thereby can emerge from a recession relatively quickly and resume solid economic growth. But even in such cases, what we will be remembered for, and blamed for, is typically the fiscal belt-tightening or structural adjustments we are perceived to have "imposed" as a condition for financial assistance. And it does sometimes happen that the government of a country we have assisted is happy joining the chorus blaming the Fund for the "harsh" measures that had to be adopted—overlooking conveniently that it was the government's own policy mistakes or mismanagement that may have contributed importantly to the crisis in the first place.

This is indeed our key public relations challenge: we find it sometimes difficult to get across that the work of the IMF represents a cooperative effort by the world community to assist countries in difficulty. And that we provide this assistance through the best possible advice our professional knowledge and experience suggest is appropriate in a given case, together with short-term financial assistance at much easier terms than the market would provide in a crisis situation. And that our efforts thereby allow the crisis-struck country to recover much faster, and at much lower economic and social costs, than if it had to act on its own. For all of that the IMF is seldom given credit.

Not that we do not make mistakes. We did not sufficiently warn emerging market countries in the 1980s and early 1990s about the requirements for a successful liberalization of short-term capital flows; and we underestimated the fragility of financial systems in some countries—partly because information about the levels of non-performing loans that had been accumulating in the booms preceding the crises, were only revealed—reluctantly—by the authorities after the panics had started. But we also have the capacity to recognize that we have made mistakes, to take into account differences among countries, and to suggest alternative policies-based on a constant effort of research and evaluation. The image of the Fund as a static, dogmatic bureaucracy that always prescribes the same medicine to everybody is totally misplaced.

The IMF is also criticized for what is viewed as our unreserved promotion of globalization. We look at globalization as a trend being propelled by technological progress and the desires of consumers, workers, and holders of financial assets with predominantly, though not only, positive consequences. There is strong evidence, for example, that trade liberalization can be a powerful driving force for economic development, but that the process of liberalization involves adjustment costs for some groups that countries need to deal with. There is nothing new in this: economic progress has always meant economic change with gainers as well as losers. As with other types of economic change, each country needs to face up to the challenges arising from globalization to maximize its benefits and minimize its adverse effects. And countries need to act together, cooperatively, to tackle those challenges that are common but beyond the capacity of any country to deal with on its own. The Fund sees itself as particularly well placed to advise members on some of the conditions necessary for a country to reduce its vulnerability to external shocks as a result of growing economic and financial integration. And we are of course also working with the entire membership to strengthen the global financial architecture, to make it less crisis prone, and to address abuses such as money laundering and financing of terrorism.

The IMF's Role in Low Income Countries

Some critics have gone as far as calling for the IMF to get out of Africa, to stop assisting (the critics would say to stop harming) the poorest countries. The fact is that these countries strongly disagree, as do virtually all of our member countries: they all see a vital role for the Fund in supporting our poorest members. Needless to say, IMF management and staff remain deeply committed to fulfilling this mandate.

But it is important to be clear about the Fund's role, what we reasonably can be expected to accomplish, and what lies beyond our mandate and competencies.

Again, the principal areas in which we provide assistance to low-income countries are fiscal, monetary and exchange rate policies, the stability and soundness of the financial system, and macroeconomic governance and institutions—through policy advice, technical assistance, capacity building, and concessional financial assistance.

In recent years, the World Bank and the IMF have jointly been helping countries develop a framework to make choices about their development strategies—as formulated in the Poverty Reduction Strategy Paper (PRSP). The IMF's principal role is to help ensure that the PRSP is consistent with macroeconomic and financial stability to help justify the Fund's own financial support and also help mobilize financial support from the broader international community. We also offer technical assistance to help establish and strengthen policy-making institutions, in particular to translate medium-term development priorities and action plans into budget allocations and to help establish systems to track execution of expenditures in priority sectors. For the most heavily indebted countries, the IMF and the World Bank are helping to reduce external debt burdens in a durable way through the HIPC initiative; this initiative has already helped to increase the share of public expenditures that can be allocated to education, health spending, and other policies that benefit the poor. But it is not the IMF's business to intervene in defining countries' development strategies. That is beyond our mandate and our competencies. Instead, this task is up to each country in cooperation with its development partners (especially the World Bank) and domestic stakeholders such as the legislature, trade unions, business, religious groups, and NGOs, including representatives of the poor. In this way, by fostering greater transparency around the development strategy and openness in the debate, the PRSP can help to empower advocates for the poor. In the most successful cases, the PRSP can thus become a framework for social cohesion.

There are in fact encouraging signs that many years' efforts to strengthen countries' macroeconomic foundations are beginning to pay off. In the countries that have pursued policies consistent with our advice, economic growth is generally up, inflation is increasingly under control, budget deficits have shrunk, and foreign exchange reserves have increased. All very good, you will probably say, but too little to make a sufficient difference, and far too little progress to allow the Millennium Development Goals—the MDGs—including deep reductions in poverty, to be achieved.

That is absolutely correct. Much more is needed to stimulate private sector development, raise living standards, and improve the quality of life.

Let me focus on one particular dimension, which also appears in the title for this session, which was labeled Wealth Creation and Social Justice.

In fact wealth creation is at the heart of economic development. Without wealth creation there can be no development, no growth in the tax base and therefore in social expenditures, and no social justice. To accelerate wealth creation requires high rates of investment in physical and human capital and increased efficiency in the production process. Strengthening human capital is critical, and the prevalence of diseases such as malaria and HIV/AIDS and inadequate education not only represent formidable challenges to improving the human condition, but also have severe consequences by limiting production capacity. Most PRSPs rightly try to address the severe shortcomings in the supply of public goods and human capital protection and formation. The IMF supports this emphasis by helping countries design their budgets with the objective of increasing the share of public resources allocated to poverty-reducing programs. The World Bank and the IMF are also helping countries strengthen their public expenditure management—to make sure that budgeted resources do in fact reach the recipients. . . .

In the Fund, with our traditional focus on (you might say obsession with) sustainability, we are also paying a great deal of attention to the fact that many low-income countries remain highly vulnerable to natural disasters and external shocks beyond their control, including falling commodity prices. In too many cases, the progress a country has been making can be reversed almost overnight. Much of our advice therefore is geared toward helping countries strengthen their resilience, including by providing temporary financial assistance to assist the country's adjustment effort. These efforts very much benefit the poor since poverty often increases during periods of economic instability.

The Fund's financial assistance to low-income countries is in the form of adjustment financing—typically at concessional terms and with a longer reimbursement period than for middle- and high-income members. Such adjustment financing helps countries stabilize after crises and assists them when they need to adjust their policies to strengthen their fundamentals and hence their growth potential. The Fund also provides financial assistance to countries experiencing temporary export shortfalls or increased import bills for cereals, or facing urgent balance of payments needs in the wake of natural disasters or armed conflicts. In addition, the Executive Board of the IMF is currently considering providing assistance to countries that may face temporary adjustment costs as a result of meeting obligations under the Doha round. However, all of the Fund's assistance is temporary and cannot serve as long-term development finance; this needs to come from bilateral donors and multilateral development banks. Eventually, and hopefully in the foreseeable future, as some low-income countries reach a more mature stage and establish a record of consistent policy formulation and execution, foreign direct investment and other types of private capital flows are expected to begin to supplement, and eventually replace, such official flows.

Finally, I want to emphasize the Fund's advocacy role. We are working closely with the official aid community to encourage debt forgiveness for highly-indebted low-income countries. We are also working to promote higher aid levels, grants instead of loans, untied assistance rather than tied to specific purchases, harmonization of donor practices to reduce the burden for recipient countries, and greater aid predictability. To illustrate how the Fund liaises with donors, I can mention that one of my staff in Paris is a permanent observer to the OECD's Development Assistance Committee. We also remain a strong critic of the industrial countries' agricultural policies and an unwavering proponent for the developing countries' interests in the Doha trade round. There is no better illustration of the IMF's views in this area than the statement issued last week by our Managing Director together with the Heads of the World Bank and the OECD.

POSTSCRIPT

Do Global Financial Institutions and Multinational Corporations Exploit the Developing World?

Determining whether IFIs and MNCs are exploitive often centers on the impact of their actions combined with the degree to which Third World states are complicit or resist such unequal terms of exchange. Evidence regarding such exploitation was more pronounced during the colonial period when states literally dominated and extracted resources and cheap labor from the third world. Today, in the current globalizing economic system, the evidence is more mixed.

From recent history, we can make certain determinations that may help frame this debate for future discussion. First, MNCs are motivated by profit and not altruence. Thus, if unchecked, they may and often do negotiate terms beneficial to themselves and not necessarily beneficial to Third World states and people. Second, IFIs possess a particular economic philosophy that works in some states and regions and apparently does not in others. As a result, some people are adversely effected by structural adjustment reforms and policies enforced by IFIs. Third, Third World governments have an obligation to protect the interests of their people in such dealings, and the extent that exploitation does occur may be a result of their diligence in such exchanges. Fourth, the nature of whether these entities are exploitive will be determined by the long-term impact of trade and investment. Whatever your position, the final arbiters of this question will be the millions of people in the Third World whose poverty will either improve or be exacerbated by the current globalizing capitalist system and its main actors, IFIs and MNCs.

Some key literature on this subject includes a debate between Thomas Friedman and Robert Kaplan, "States of Discord, *Foreign Policy* (March/April 2002), Thomas Friedman, *The Lexus and the Olive Tree* (Farrar, Straus and Giroux, 1999), David C. Korten, *When Corporations Rule the World* (1997), Donald Marsh, "Free Trade and Their Critics: The Need for Education" Washington Council on International Trade, (2000), Caroline Thomas and Melvyn Reader, "Development and Inequality," in *Issues in World Politics*, edited by Brian White, Richard Little, and Michael Smith (2001), Jones Raymond Vreeland and Stephen Kosuck, eds., *Globalization and The Nation-State: The Impact of the IMF World Bank*, Gustav Ranis, (Routledge, 2005), and *The IMF and its Critics: Reform of Global Financial Architecture*, David Vines and Christopher Gilbert, eds. (Cambridge University Press, 2004).

On the Internet . . .

Nuclear Terrorism: How to Prevent It

This site of the Nuclear Control Institute discusses nuclear terrorism and how best to prevent it. Topics include terrorists' ability to build nuclear weapons, the threat of "dirty bombs," and whether or not nuclear reactors are adequately protected against attack. This site features numerous links to key nuclear terrorism documents and Web sites as well as to recent developments and related news items.

http://www.nci.org/nuketerror.htm

CDI Terrorism Project

The Center for Defense Information's (CDI) Terrorism Project is designed to provide insights, in-depth analysis, and facts on the military, security, and foreign policy challenges of terrorism. The project looks at all aspects of fighting terrorism, from near-term issues of response and defense to long-term questions about how the United States should shape its future international security strategy.

http://www.cdi.org/terrorism/

Exploring Global Conflict: An Internet Guide to the Study of Conflict

Exploring Global Conflict: An Internet Guide to the Study of Conflict is an Internet resource designed to provide understanding of global conflict. Information related to specific conflicts in areas such as Northern Ireland, the Middle East, the Great Lakes region in Africa, and the former Yugoslavia is included on this site. Current news and educational resource sites are listed as well.

http://www.uwm.edu/Dept/CIS/conflict/congeneral.html

Central Intelligence Agency

The U.S. government agency with major responsibility for the war on terrorism provides substantial information on its Web site.

http://wwwcia.gov/terrorism/

The Cato Institute

This U.S. public organization conducts research on a wide range of public policy issues. It subscribes to what it terms "basic American principles." One important research issue is civil liberties concerns relating to the war on terrorism.

http://cato.org/current/civil-liberties/

Center for Strategic and International Studies (CSIS)

CSIS now provides substantial information on terrorism in the aftermath of 9/11. The URL noted here leads directly to its thinking on the issue of homeland defense.

http://www.csis.org/homeland

PART 4

The New Global
Security Dilemma

*W*ith the end of the Cold War, the concept of security was freed *from its bipolar constraints of great power calculations. And as a consequence of 9/11, the definition of security and how to achieve it were once again redefined to encompass new kinds of threats from a new group of perpetrators. In short, our concept of security in a post-modern age has broadened considerably.*

These include concerns over cultural and ethnic conflicts, the impact of immigration on a nation's security, the prospects of nuclear terrorism, the dangers of proliferation, and the role of the United Nations in a world of American preeminence.

This section examines some of the key issues shaping the security dilemma of the twenty-first century.

- Does Immigration Policy Affect Terrorism?

- Are We Headed Toward a Nuclear 9/11?

- Are Cultural and Ethnic Wars the Defining Dimensions of Twenty-First Century Conflict?

- Can Nuclear Proliferation Be Stopped?

- Has U.S. Hegemony Rendered the United Nations Irrelevant?

ISSUE 16

Does Immigration Policy Affect Terrorism?

YES: Mark Krikorian, from "Keeping Terror Out," *The National Interest* (Spring 2004)

NO: Daniel T. Griswold, from "Don't Blame Immigrants for Terrorism," *The Cato Institute* (May 24, 2006)

ISSUE SUMMARY

YES: Mark Krikorian argues that immigration and security are directly and inexorably linked. He contends that the nature of terrorism is such that individual and small-group infiltration of our U.S. borders is a prime strategy for terrorists and thus undermines individual calls for relaxed or open immigration.

NO: Daniel T. Griswold argues that by coupling security and immigration, we simplify a complex issue and in fact do little to enhance security while we demonize a huge segment of the population who are by and large law abiding and not a threat.

Immigration has always been a social, economic, and cultural issue. Countries around the world struggle with their attitudes regarding the migration of people both into and out of their countries. Also, attitudes and acceptance of immigration change over time. Certainly this is true in the United States where attitudes regarding immigration have changed dramatically over the decades in response to a number of factors.

Today, with the expansion of terrorism as a form of conflict and with the extension of that into the United States with 9/11, security and immigration have become irrevocably linked. Attitudes regarding immigration policy, whether in favor of relaxed restrictions or greater exclusion, are now intertwined with debates regarding security. Also, who are the immigrants and where do they come from is now central to the debate regarding immigration policy and any person's potential threat to a states' security.

For countries like the United States where immigration has become a central feature of the history and culture of society, this debate is both emotional and complex. Attitudes about immigrants, their status in society, and their rights and responsibilities are intertwined with views regarding certain ethnic groups and their perceived propensity to engage in acts that may be a threat to the nation's security. In the United States this debate has already become heated with extreme views expressed on all sides.

As security has become paramount in the American mindset, concerns over illegal immigration and the large and porous borders that the United States possesses have accelerated. Those who see security as central to our policy see these borders and the illegal immigration as a shield to future terrorists and more 9/11s. They see no choice but to restrict our borders, shut down illegal immigration and begin mass deportations if necessary. Those who see immigration as a cultural socio-economic issue want a more complex approach that meets security needs but does not change the fundamental character of an America committed to sheltering those in search of greater opportunity.

Mark Krikorian places this debate within a security policy mindset and argues that 9/11 changed our notions of security, free movement of people, what the "Home Front" means, and what our obligations are to immigrants. Essentially he contends that a policy designed to thwart 9/11s must take into account that the hijackers were from outside the United States and that this fact illustrates the reality that immigrants can be a threat to U.S. security. Krikorian argues that only through stricter enforcement and a greater acceptance of the potential immigrant threat will we be able to prevent another 9/11.

Daniel Griswold sees immigration and terrorism as separate issues and policy concerns. He sees immigration as a vital part of who America is and what it represents and by demonizing immigrants through a policy of "cracking down" we run the risk of destroying what is a noble legacy of American acceptance of immigrants. He sees security as a matter of making sure we keep out the wrong people without changing our fundamental approach to the vast majority of immigrants who simply want a better life.

YES

Keeping Terror Out

Supporters of open immigration have tried to de-link 9/11 from security concerns. "There's no relationship between immigration and terrorism," said a spokeswoman for the National Council of the advocacy group La Raza. "I don't think [9/11] can be attributed to the failure of our immigration laws," claimed the head of the immigration lawyers' guild a week after the attacks.

President Bush has not gone that far, but in his January 7 speech proposing an illegal alien amnesty and guest worker program, he claimed the federal government is now fulfilling its responsibility to control immigration, thus justifying a vast increase in the flow of newcomers to America. Exploring the role of immigration control in promoting American security can help provide the context to judge the president's claim that his proposal is consistent with our security imperatives, and can help to sketch the outlines of a secure immigration system.

Home Front

The phrase "Home Front" is a metaphor that gained currency during World War I, with the intention of motivating a civilian population involved in total war. The image served to increase economic output and the purchase of war bonds, promote conservation and the recycling of resources and reconcile the citizenry to privation and rationing.

But in the wake of 9/11, "Home Front" is no longer a metaphor. As Deputy Secretary of Defense Paul Wolfowitz said in October 2002,

> *"Fifty years ago, when we said, 'home front,' we were referring to citizens back home doing their part to support the war front. Since last September, however, the home front has become a battlefront every bit as real as any we've known before."*

Nor is this an aberration unique to Al-Qaeda or to Islamists generally. No enemy has any hope of defeating our armies in the field and must therefore resort to asymmetric means.[1] And though there are many facets to asymmetric or "Fourth-Generation" warfare—as we saw in Al-Qaeda's pre-9/11 assaults on our interests in the Middle East and East Africa and as we are seeing today in Iraq. The Holy Grail of such a strategy is mass-casualty attacks on America.

The military has responded to this new threat with the Northern Command, just as Israel instituted its own "Home Front Command" in 1992, after the Gulf War. But our objective on the Home Front is different, for this front is different from other fronts; the goal is defensive, blocking and disrupting the enemy's ability to carry out attacks on our territory. This will then allow offensive forces to find, pin, and kill the enemy overseas.

Because of the asymmetric nature of the threat, the burden of homeland defense is not borne mainly by our armed forces but by agencies formerly seen as civilian entities—mainly the Department of Homeland Security (DHS). And of DHS's expansive portfolio, immigration control is central. The reason is elementary: no matter the weapon or delivery system—hijacked airliners, shipping containers, suitcase nukes, anthrax spores—operatives are required to carry out the attacks. Those operatives have to enter and work in the United States. In a very real sense, the primary weapons of our enemies are not inanimate objects at all, but rather the terrorists themselves—especially in the case of suicide attackers. Thus keeping the terrorists out or apprehending them after they get in is indispensable to victory. As President Bush said recently, "Our country is a battlefield in the first war of the 21st century."

In the words of the July 2002 National Strategy for Homeland Security:

> *Our great power leaves these enemies with few conventional options for doing us harm. One such option is to take advantage of our freedom and openness by secretly inserting terrorists into our country to attack our homeland. Homeland security seeks to deny this avenue of attack to our enemies and thus to provide a secure foundation for America's ongoing global engagement.*

Our enemies have repeatedly exercised this option of inserting terrorists by exploiting weaknesses in our immigration system. A Center for Immigration Studies analysis of the immigration histories of the 48 foreign-born Al-Qaeda operatives who committed crimes in the United States from 1993 to 2001 (including the 9/11 hijackers) found that nearly every element of the immigration system has been penetrated by the enemy.[2] Of the 48, one-third were here on various temporary visas, another third were legal residents or naturalized citizens, one-fourth were illegal aliens, and the remainder had pending asylum applications. Nearly half of the total had, at some point or another, violated existing immigration laws.

Supporters of loose borders deny that inadequate immigration control is a problem, usually pointing to flawed intelligence as the most important shortcoming that needs to be addressed. Mary Ryan, for example, former head of the State Department's Bureau of Consular Affairs (which issues visas), testified in January 2004 before the 9/11 Commission that

> *"Even under the best immigration controls, most of the September 11 terrorists would still be admitted to the United States today . . . because they had no criminal records, or known terrorist connections, and had not been identified by intelligence methods for special scrutiny."*

But this turns out to be untrue, both for the hijackers and for earlier Al-Qaeda operatives in the United States. A normal level of visa scrutiny, for instance, would have excluded almost all the hijackers. Investigative reporter Joel Mowbray acquired copies of 15 of the 19 hijackers' visa applications (the other four were destroyed—yes, destroyed—by the State Department), and every one of the half-dozen current and former consular officers he consulted said every application should have been rejected on its face.[3] Every application was incomplete or contained patently inadequate or absurd answers.

Even if the applications had been properly prepared, many of the hijackers, including Mohammed Atta and several others, were young, single, and had little income—precisely the kind of person likely to overstay his visa and become an illegal alien, and thus the kind of applicant who should be rejected. And, conveniently, those *least* likely to overstay their visas—older people with close family, property and other commitments in their home countries—are also the very people least likely to commit suicide attacks.

9/11 was not the only terrorist plot to benefit from lax enforcement of ordinary immigration controls—every major Al-Qaeda attack or conspiracy in the United States has involved at least one terrorist who violated immigration law. Gazi Ibrahim Abu Mezer, for example, who was part of the plot to bomb the Brooklyn subway, was actually caught three times by the Border Patrol trying to sneak in from Canada. The third time the Canadians would not take him back. What did we do? Because of a lack of detention space, he was simply released into the country and told to show up for his deportation hearing. After all, with so many millions of illegal aliens here already, how much harm could one more do? . . .

Prior to the growth of militant Islam, the only foreign threat to our population and territory in recent history has been the specter of nuclear attack by the Soviet Union. To continue that analogy, since the terrorists are themselves the weapons, immigration control is to asymmetric warfare what missile defense is to strategic warfare. There are other weapons we must use against an enemy employing asymmetric means—more effective international coordination, improved intelligence gathering and distribution, special military operations—but in the end, the lack of effective immigration control leaves us naked in the face of the enemy. This lack of defensive capability may have made sense with regard to the strategic nuclear threat under the doctrine of Mutual Assured Destruction, but it makes no sense with regard to the asymmetric threats we face today and in the future.

Unfortunately, our immigration response to the wake-up call delivered by the 9/11 attacks has been piecemeal and poorly coordinated. Specific initiatives that should have been set in motion years ago have finally begun to be enacted, but there is an *ad hoc* feel to our response, a sense that bureaucrats in the Justice and Homeland Security departments are searching for ways to tighten up immigration controls that will not alienate one or another of a bevy of special interest groups.

Rather than having federal employees cast about for whatever enforcement measures they feel they can get away with politically, we need a strategic assessment of what an effective immigration-control system would look like.

Homeland Security Begins Abroad

To extend the missile defense analogy, there are three layers of immigration control, comparable to the three phases of a ballistic missile's flight: boost, midcourse and terminal. In immigration the layers are overseas, at the borders, and inside the country. But unlike existing missile defense systems, the redundancy built into our immigration control system permits us repeated opportunities to exclude or apprehend enemy operatives.

Entry to America by foreigners is not a right but a privilege, granted exclusively at the discretion of the American people. The first agency that exercises that discretion is the State Department's Bureau of Consular Affairs, whose officers make the all-important decisions about who gets a visa. Consular Affairs is, in effect, America's other Border Patrol.[4] In September 2003, DHS Under Secretary As a Hutchinson described the visa process as "forward-based defense" against terrorists and criminals.

The visa filter is especially important because the closer an alien comes to the United States the more difficult it is to exclude him. There is relatively little problem, practically or politically, in rejecting a foreign visa applicant living abroad. Once a person presents himself at a port of entry, it becomes more difficult to turn him back, although the immigration inspector theoretically has a free hand to do so. Most difficult of all is finding and removing people who have actually been admitted; not only is there no specific chokepoint in which aliens can be controlled, but even the most superficial connections with American citizens or institutions can lead to vocal protests against enforcement of the law. . . .

Despite improvements, the most important flaw in the visa filter still exists: the State Department remains in charge of issuing visas. State has a corporate culture of diplomacy, geared toward currying favor with foreign governments. In the context of visa issuance, this has fostered a "customer-service" approach, which sees the foreign visa applicant as the customer who needs to be satisfied. The attitude in management is summed up by the catchphrase of the former U.S. consul general in Saudi Arabia: "People gotta have their visas!"[5] Such an approach views high visa-refusal rates as a political problem, rather than an indicator of proper vigilance.

Nor will oversight of visa officers by DHS officials be an adequate antidote. As long as the decisions about raises, promotions, and future assignments for visa officers are made by the State Department, the culture of diplomacy will win out over the culture of law enforcement. In the end, the only remedy may be to remove the visa function from the State Department altogether.

Order at the Border

The next layer of immigration security is the border, which has two elements: "ports of entry," which are the points where people traveling by land, sea, or air enter the United States; and the stretches between those entry points. The first are staffed by inspectors working for DHS's Bureau of Customs and Border Protection, the second monitored by the Border Patrol and the Coast Guard, both now also part of DHS.

This is another important chokepoint, as almost all of the 48 Al-Qaeda operatives who committed terrorist acts through 2001 had had contact with immigration inspectors. But here, too, the system failed to do its job. For instance, Mohammed Atta was permitted to reenter the country in January 2001 even though he had overstayed his visa the last time. Also, before 9/11 hijacker Khalid Al-Midhar's second trip to the United States, the CIA learned that he had been involved in the bombing of the U.S.S. *Cole*—but it took months for his name to be placed on the watch list used by airport inspectors, and by then he had already entered the country. And in any case, there still are 12 separate watch lists, maintained by nine different government agencies. . . .

There were also failures *between* the ports of entry. Abdelghani Meskini and Abdel Hakim Tizegha, both part of the Millennium Plot that included Ahmed Ressam, first entered the country as stowaways on ships that docked at U.S. ports. Tizegha later moved to Canada and then returned to the United States by sneaking across the land border. And of course, Abu Mezer, though successfully apprehended by the U.S. Border Patrol, was later released.

And finally, perhaps the biggest defect in this layer of security is the lack of effective tracking of departures. Without exit controls, there is no way to know who has overstayed his visa. This is especially important because most illegal alien terrorists have been overstayers. The opportunities for failure are numerous and the system is so dysfunctional that the INS's own statistics division declared that it was no longer possible to estimate the number of people who have overstayed their visas.

Certainly, there have been real improvements since 9/11. The US-VISIT system has begun to be implemented, with arriving visa-holders being digitally photographed and having their index fingerprints scanned; this will eventually grow into a "check in/check out" system to track them and other foreign visitors. Also, the 45-minute maximum for clearing foreign travelers has been repealed. Lastly, all foreign carriers are now required to forward their passenger manifests to immigration before the plane arrives.

But despite these and other improvements in the mechanics of border management, the same underlying problem exists here as in the visa process: lack of political seriousness about the security importance of immigration control. The Coast Guard, for instance, still considers the interdiction of illegal aliens a "nonsecurity" mission. More importantly, pressure to expedite entry at the expense of security persists; a DHS memo leaked in January outlined how the US-VISIT system would be suspended if lines at airports grew too long. And, to avoid complaints from businesses in Detroit, Buffalo, and elsewhere, most Canadian visitors have been exempted from the requirements of the US-VISIT system. . . .

Safety Through Redundancy

The third layer of immigration security—the terminal phase, in missile defense jargon—is interior enforcement. Here, again, ordinary immigration control can be a powerful security tool. Of the 48 Al-Qaeda operatives, nearly half were either illegal aliens at the time of their crimes or had violated immigration laws at some point prior to their terrorist acts.

Many of these terrorists lived, worked, opened bank accounts, and received driver's licenses with little or no difficulty. Because such a large percentage of terrorists violated immigration laws, enforcing the law would be extremely helpful in disrupting and preventing terrorist attacks.

But interior enforcement is also the most politically difficult part of immigration control. While there is at least nominal agreement on the need for improvements to the mechanics of visas and border monitoring, there is no elite consensus regarding interior enforcement. This is especially dangerous given that interior enforcement is the last fallback for immigration control, the final link in a chain of redundancy that starts with the visa application overseas.

There are two elements to interior enforcement: first, conventional measures such as arrest, detention, and deportation; and second, verification of legal status when conducting important activities. The latter element is important because its goal is to disrupt the lives of illegal aliens so that many will return home on their own (and, in a security context, to disrupt the planning and execution of terrorist attacks).

Inadequacies in the first element of interior enforcement have clearly helped terrorists in the past. Because there is no way of determining which visitors have overstayed their visas, much less a mechanism for apprehending them, this has been a common means of remaining in the United States—of the 12 (out of 48) Al-Qaeda operatives who were illegal aliens when they took part in terrorism, seven were visa overstayers.

Among terrorists who were actually detained for one reason or another, several were released to go about their business inside America because of inadequate detention space. This lack of space means that most aliens in deportation proceedings are not detained, so that when ordered deported, they receive what is commonly known as a "run letter" instructing them to appear for deportation—and 94 percent of aliens from terrorist-sponsoring states disappear instead. . . .

Perhaps the most outrageous phenomenon in this area of conventional immigration enforcement is the adoption of "sanctuary" policies by cities across the country. Such policies prohibit city employees—including police—from reporting immigration violations to federal authorities or even inquiring as to a suspect's immigration status. It is unknown whether any terrorists have yet eluded detention with the help of such policies, but there is no doubt that many ordinary murderers, drug dealers, gang members, and other undesirables have and will continue to do so.

The second element of interior enforcement has been, if anything, even more neglected. The creation of "virtual chokepoints," where an alien's legal status would be verified, is an important tool of immigration control, making it difficult for illegals to engage in the activities necessary for modern life.

The most important chokepoint is employment. Unfortunately, enforcement of the prohibition against hiring illegal aliens, passed in 1986, has all but stopped. This might seem to be of little importance to security, but in fact holding a job can be important to terrorists for a number of reasons. By giving them a means of support, it helps them blend into society. Neighbors might well become suspicious of young men who do not work, but seem able to pay their bills. Moreover, supporting themselves by working would enable terrorists to avoid the scrutiny that might attend the transfer of money from abroad. Of course, terrorists who do not work can still arrive with large sums of cash, but this too creates risks of detection.

That said, the ban on employment by illegal aliens is one of the most widely violated immigration laws by terrorists. Among those who worked illegally at some point were CIA shooter Mir Aimal Kansi; Millennium plot conspirator Abdelghani Meskini; 1993 World Trade Center bombers Eyad Ismoil, Mohammed Salameh and Mahmud and Mohammed Abouhalima.

Other chokepoints include obtaining a driver's license and opening a bank account, two things that most of the 9/11 hijackers had done. It is distressing to note that, while Virginia, Florida and New Jersey tightened their driver's license rules after learning that the hijackers had used licenses from those states, other states have not. Indeed, California's then-Governor Gray Davis signed a bill last year intended specifically to provide licenses to illegal aliens (which was repealed after his recall).

As for bank accounts, the trend is toward making it easier for illegal aliens to open them. The governments of Mexico and several other countries have joined with several major banks to promote the use of consular identification cards (for illegals who can't get other ID) as a valid form of identification, something the U.S. Department of the Treasury explicitly sanctioned in an October 2002 report.

Finally, the provision of immigration services is an important chokepoint, one that provides the federal government additional opportunities to screen the same alien. There is a hierarchy of statuses a foreign-born person might possess, from illegal alien to short-term visitor, long-term visitor, permanent resident (green card holder) and finally, naturalized citizen. It is very beneficial for terrorists to move up in this hierarchy because it affords them additional opportunities to harm us. To take only one example: Mahmud Abouhalima—one of the leaders of the first World Trade Center bombing—was an illegal-alien visa overstayer; but he became a legal resident as part of the 1986 illegal-alien amnesty by falsely claiming to be a farmworker, and he was only then able to travel to Afghanistan for terrorist training and return to the United States. . . .

Former INS Commissioner James Ziglar expressed the general resistance to linking immigration law with homeland security when he said a month after the 9/11 attacks that "We're not talking about immigration, we're talking about evil." It is as if the terrorists were summoned from a magic lamp, rather than moving through our extensive but neglected immigration control system, by applying for visas, being admitted by inspectors, and violating laws with impunity inside America.

Upholding the Law

Such ambivalence about immigration enforcement, at whatever stage in the process, compromises our security. It is important to understand that the security function of immigration control is not merely opportunistic, like prosecuting Al Capone for tax violations for want of evidence on his other numerous crimes. The FBI's use of immigration charges to detain hundreds of Middle Easterners in the immediate aftermath of 9/11 was undoubtedly necessary, but it cannot be a model for the role of immigration law in homeland security. If our immigration system is so lax that it can be penetrated by a Mexican busboy, it can sure be penetrated by an Al-Qaeda terrorist.

Since there is no way to let in "good" illegal aliens but keep out "bad" ones, countering the asymmetric threats to our people and territory requires sustained, across-the-board immigration law enforcement. Anything less exposes us to grave dangers. Whatever the arguments for the president's amnesty and guest worker plan, no such proposal can plausibly be entertained until we have a robust, functioning immigration-control system. And we are nowhere close to that day.

Endnotes

1. See the National Defense University's Institute for National Strategic Studies 1998 Strategic Assessment: "Put simply, asymmetric threats or techniques are a version of not 'fighting fair,' which can include the use of surprise in all its operational and strategic dimensions and the use of weapons in ways unplanned by the United States. Not fighting fair also includes the prospect of an opponent designing a strategy that fundamentally alters the terrain on which a conflict is fought."

2. Steven A. Camarota, "The Open Door: How terrorists entered and remained in the United States, 1993–2001" (Washington, DC: Center for Immigration Studies, 2002).

3. "Visas for Terrorists: They were ill-prepared. They were laughable. They were approved," *National Review,* October 28, 2002.

4. "America's Other Border Patrol: The State Department's Consular Corps and Its Role in U.S. Immigration," by Nikolai Wenzel, Center for Immigration Studies Backgrounder, August 2000, http://www.cis.org/articles/2000/back800.html

5. Joel Mowbray, "Perverse Incentives; The State Department rewards officials responsible for terror visas." *National Review Online,* October 22, 2002.

Daniel T. Griswold  **NO**

Don't Blame Immigrants for Terrorism

In the wake of the September 11 terrorist attacks on the Pentagon and the World Trade Center, the U.S. government must strengthen its efforts to stop terrorists or potential terrorists from entering the country. But those efforts should not result in a wider effort to close our borders to immigrants.

Obviously, any government has a right and a duty to "control its borders" to keep out dangerous goods and dangerous people. The U.S. federal government should implement whatever procedures are necessary to deny entry to anyone with terrorist connections, a criminal record, or any other ties that would indicate a potential to commit terrorist acts.

This will require expanding and upgrading facilities at U.S. entry points so that customs agents and immigration officials can be notified in a timely manner of persons who should not be allowed into the country. Communications must be improved between law enforcement, intelligence agencies and border patrol personnel. Computer systems must be upgraded to allow effective screening without causing intolerable delays at the border. A more effective border patrol will also require closer cooperation from Mexico and Canada to prevent potential terrorists from entering those countries first in an attempt to then slip across our long land borders into the United States.

Long-time skeptics of immigration, including Pat Buchanan and the Federation for American Immigration Reform, have tried in recent days to turn those legitimate concerns about security into a general argument against openness to immigration. But immigration and border control are two distinct issues. Border control is about who we allow to enter the country, whether on a temporary or permanent basis; immigration is about whom we allow to stay and settle permanently.

Immigrants are only a small subset of the total number of foreigners who enter the United States every year. According to the U.S. Immigration and Naturalization Service, 351 million aliens were admitted through INS ports of entry in fiscal year 2000—nearly a million entries a day. That total includes individuals who make multiple entries, for example, tourists and business travelers with temporary visas, and aliens who hold border-crossing cards that allow them to commute back and forth each week from Canada and Mexico.

The majority of aliens who enter the United States return to their homeland after a few days, weeks, or months. Reducing the number of people we allow to reside permanently in the United States would do nothing to protect us from terrorists who do not come here to settle but to plot and commit violent acts. And closing our borders to those who come here temporarily would cause a huge economic disruption by denying entry to millions of people who come to the United States each year for lawful, peaceful (and temporary) purposes.

It would be a national shame if, in the name of security, we were to close the door to immigrants who come here to work and build a better life for themselves and their families. Like the Statue of Liberty, the World Trade Center towers stood as monuments to America's openness to immigration. Workers from more than 80 different nations lost their lives in the terrorist attacks. According to the *Washington Post*, "The hardest hit among foreign countries appears to be Britain, which is estimating about 300 deaths . . . Chile has reported about 250 people missing, Colombia nearly 200, Turkey about 130, the Philippines about 115, Israel about 113, and Canada between 45 and 70. Germany has reported 170 people unaccounted for, but expects casualties to be around 100." Those people were not the cause of terrorism but its victims.

The problem is not that we are letting too many people into the United States but that the government is not keeping out the wrong people. An analogy to trade might be helpful: We can pursue a policy of open trade, with all its economic benefits, yet still exclude goods harmful to public health and safety, such as diseased meat and fruits, explosives, child pornography, and other contraband materials. In the same way, we should keep our borders open to the free flow of people, but at the same time strengthen our ability to keep out those few who would menace the public.

Immigrants come here to realize the American dream; terrorists come to destroy it. We should not allow America's tradition of welcoming immigrants to become yet another casualty of September 11.

POSTSCRIPT

Does Immigration Policy Affect Terrorism?

Immigration policy is a complex issue. It reflects attitudes regarding the socio-cultural fabric and history of a society. It also reaches deep into the core attitudes regarding each person's relationship to the society in which they live. Terrorism, too, is a complex issue. It is a form of warfare that demands dexterous responses, excellent intelligence, and awareness that civilian areas are targets and must be protected in ways heretofore not considered in modern warfare.

As these issues have been linked together in the United States and elsewhere, the debates regarding immigrants and security have become heated, contentious, and filled with polemic on both sides. Those who wish for to build walls and those who wish for open door policies have become entrenched in mindsets that lend little to the complexities of the issues at hand.

If one is to ask whether an open door policy on immigration might allow more terrorists into a country like the United States than a closed door policy, statistical analysis tells us that the former would lead to more terrorists "getting in." However, if one embraces the notion that only a fraction (less than 1/10 of 1%) of people trying to enter the United States are of suspicious motives, then it seems that proper and effective intelligence services could prevent such entry without demonizing thousands of immigrants.

What position you take will often come down to what value you see as more important. Is it the value of a country willing to open its doors to those who are seeking a better life or security and the desire to protect one's own citizens? Clearly, U.S. society has not reached a consensus on which value has greater importance in terms of policy and it may never reach that consensus given our long and storied history as a country welcoming immigration.

Some important literature on this subject includes "Playing Games With Security: Taking Two Steps Back for Every Step Forward on Immigration" Mark Krikorian, *The National Review OnLine* (August 18, 2004), "Keeping Extremists Out: The History of Ideological Exclusion and the Need for Its Revival" James R. Edwards, Jr. *Center for Immigration Studies* (September 2005), *Globalization and the Future of Terrorism: Patterns and Predications,* Brynjar Lia, (Routledge 2005), and *Terrorism as a Challenge for National and International Law: Security Versus Liberty?* Christian Walter, Silva Vwneky, and Volker Rwben (Springer 2004).

ISSUE 17

Are We Headed Toward a Nuclear 9/11?

YES: David Krieger, from "Is a Nuclear 9/11 in Our Future?" *Nuclear Age Peace Foundation* (October 6, 2003)

NO: Graham Allison, from "Preventing Nuclear Terrorism," *Los Angeles World Affairs Council* (September 22, 2005)

ISSUE SUMMARY

YES: David Kreiger, president of the Nuclear Age Peace Foundation, argues that a nuclear 9/11 is very likely in a U.S. city due to the prevalence of nuclear weapons and the failure of nuclear member states to adequately enforce a true non-proliferation regime.

NO: Graham Allison, noted international scholar, argues that a nuclear 9/11 is preventable provided that the United States and other states halt proliferation to states predisposed toward assisting terrorists, particularly North Korea.

Since the terrorist attacks of September 11, 2001, much has been written about the specter of nuclear terrorism and the releasing of a dirty bomb (one loaded with radioactive material) in an urban/civilian setting. The events of September 11 have all but ensured the world's preoccupation with such an event for the foreseeable future. Indeed, the arrest of a U.S. man with dirty bomb materials indicates that such plans may indeed be in the works between Al Qaeda and other terrorist cells. When this horror is combined with the availability of elements of nuclear-related material in places like the states of the former Soviet Union, Pakistan, India, Iraq, Iran, North Korea, and many other states, one can envision a variety of sobering scenarios.

Hollywood feeds these views with such films as *The Sum of All Fears* and *The Peacemaker,* in which nuclear terrorism is portrayed as all too easy to carry out and likely to occur. It is difficult in such environments to separate fact from fiction and to ascertain objectively the probabilities of such events. So many factors go into a successful initiative in this area. One needs to find a committed cadre of terrorists, sufficient financial backing, technological know-how, intense security and secrecy, the means of delivery, and many other variables, including luck. In truth, such acts may have already been advanced and thwarted by governments, security services, or terrorist mistakes and incompetence. We do not know, and we may never know.

Regional and ethnic conflicts of a particularly savage nature in places like Chechnya, Kashmir, Colombia, and Afghanistan help to fuel fears that adequately financed zealots will see in nuclear weapons a swift and cata- strophic answer to their demands and angers. Osama bin Laden's contribu- tion to worldwide terrorism has been the success of money over security and the realization that particularly destructive acts with high levels of coordina- tion can be "successful." This will undoubtedly encourage others with similar ambitions against real or perceived enemies.

Conversely, many argue that fear of the terrorist threat has left us imagining that which is not likely. They point to a myriad of roadblocks to ter- rorist groups' obtaining all of the elements necessary for a nuclear or dirty bomb. They cite technological impediments, monetary issues, lack of sophistication, and inability to deliver. They also cite governments' universal desire to prevent such actions. Even critics of Iraqi leader Saddam Hussein have argued that were he to develop such weapons, he would not deliver them to terrorist groups nor would he use them except in the most dire of circumstances, such as his own regime's survival. They argue that the threat is overblown and, in some cases, merely used to justify increased security and the restriction of civil liberties.

The following selections reflect this dichotomy of views. While both authors see proliferation as key, Krieger feels that a new approach to prolifer- ation must be undertaken to ensure that such an attack won't take place. He takes the U.S. and other nuclear weapons states to task for not setting an example by removing nuclear weapons and as such, this ensures a nuclear terrorist attack eventually.

Allison also believe proliferation is the key to terrorism but argues that more pro-action and stricter proliferation enforcement will ensure that rogue states don't get the bomb and thus transfer it to terrorist groups for use.

YES

David Krieger

Is a Nuclear 9/11 in Our Future?

Sooner or later there will be a nuclear 9/11 in an American city or that of a US ally unless serious program is undertaken to prevent such an occurrence. A terrorist nuclear attack against an American city could take many forms. A worst case scenario would be the detonation of a nuclear device within a city. Depending upon the size and sophistication of the weapon, it could kill hundreds of thousands or even millions of people.

Terrorists could obtain a nuclear device by stealing or purchasing an already created nuclear weapon or by stealing or purchasing weapons-grade nuclear materials and fashioning a crude bomb. While neither of these options would be easy, they cannot be dismissed as beyond the capabilities of a determined terrorist organization.

If terrorists succeeded in obtaining a nuclear weapon, they would also have to bring it into the US, assuming they did not already obtain or create the weapon in this country. While this would not necessarily be easy, many analysts have suggested that it would be within the realm of possibility. An oft-cited example is the possibility of bringing a nuclear device into an American port hidden on a cargo ship.

Another form of terrorist nuclear attack requiring far less sophistication would be the detonation of a radiation weapon or "dirty bomb." This type of device would not be capable of a nuclear explosion but would use conventional explosives to disperse radioactive materials within a populated area. The detonation of such a device could cause massive panic due to the public's appropriate fears of radiation sickness and of developing cancers and leukemias in the future.

A bi-partisan task force of the Secretary of Energy's Advisory Board, headed by former Senate Majority Leader Howard Baker and former White House Counsel Lloyd Cutler, called upon the US in 2001 to spend $30 billion over an eight to ten year period to prevent nuclear weapons and materials in the former Soviet Union from getting into the hands of terrorists or so called "rogue" states. The task force called the nuclear dangers in the former USSR "the most urgent unmet national security threat facing the United States today." At present, the US government is spending only about one-third of the recommended amount, while it pours resources into paying for the invasion, occupation and rebuilding of Iraq as well as programs unlikely to provide effective security to US citizens such as missile defense.

From *Waging Peace*, October 6, 2003, pp. 1. Copyright © 2003 by David Krieger. Reprinted by Permission.

The great difficulty in preventing a nuclear 9/11 is that it will require ending the well-entrenched nuclear double standards that the US and other nuclear weapons states have lived by throughout the Nuclear Age. Preventing nuclear terrorism in the end will not be possible without a serious global program to eliminate nuclear weapons and control nuclear materials that could be converted to weapons. Such a program would require universal agreement in the form of an enforceable treaty providing for the following:

- full accounting and international safeguarding of all nuclear weapons, weapons-grade nuclear materials and nuclear reactors in all countries, including the nuclear weapons states;
- international tracking and control of the movement of all nuclear weapons and weapons-grade materials;
- dismantling and prohibiting all uranium enrichment facilities and all plutonium separation facilities, and the implementation of a plan to expedite the phasing out all nuclear power plants;
- full recognition and endorsement by the nuclear weapons states of their existing obligation pursuant to the Nuclear Non-Proliferation Treaty for an "unequivocal undertaking" to eliminate their nuclear arsenals;
- rapidly dismantling existing nuclear weapons in an orderly and transparent manner and the transfer of nuclear materials to international control sites; and criminalizing the possession, threat or use of nuclear weapons.

While these steps may appear extreme, they are in actuality the minimum necessary to prevent a nuclear 9/11. If that is among our top priorities as a country, as surely it should be, the US government should begin immediately to lead the world in this direction. Now is the time to act, before one or more US cities are devastated by nuclear terrorism.

Preventing Nuclear Terrorism

Even when talking about a subject that on the one hand is so gloomy if realistic, the main message here is actually quite a hopeful message. The most important part of this story is the subtitle of my book: *The Ultimate Preventable Catastrophe*. So, while some part of what I'm going to say at the beginning may seem a little frightening, don't give up before the end of the presentation. For those of you who get the book, make sure to read part two, not just part one, because after part one you might be tempted to do something else rather than your normal day-to-day business.

All of us can remember 9/11, three years ago just this month, when Al Qaeda hijacked airplanes and crashed them into the World Trade Center and the Pentagon. Probably for most of us we can remember where we were, what we were doing, what we thought. My wife was supposed to have been on American Flight 77 coming to Los Angeles that day for a board meeting here in town, but [the meeting] was postponed for a day so she was on the plane for the 12th, not for the 11th. Obviously, the plane never went.

One month to the day after those events—one month to the day—George Tenet, who was the Director of the CIA, walked into the Oval Office for the president's morning intelligence briefing and informed the president that a CIA agent code-named Dragonfire—a wonderful name—had reported that Al Qaeda had acquired a ten-kiloton nuclear weapon. That's a very small nuclear weapon, but one that would make a ten-kiloton blast, and it had this weapon in New York City. There was a stunned silence followed by a series of interrogating questions in which the president was essentially trying to see if this was a real possibility or just another story. Were there ten-kiloton weapons in the former Soviet arsenal? and they say "yes." Are all these weapons accounted for? The answer, "uncertain." Could Al Qaeda have acquired one of these weapons? The answer, "of course." Could Al Qaeda have bought a weapon like this to New York City and have it there without us otherwise knowing about it? The answer "certainly." So on this, Vice President Cheney evacuated, he left Washington. He stayed for some considerable period of time in a secret alternative site for our government called "Site R," which is a site in the hole of a mountain. At that point they set up very rapidly an alternative government consisting of several hundred people who worked in the government

From *Los Angeles World Affairs Council*, September 22, 2005, pp. 1–3, 5–6. Copyright © 2005 by Los Angeles World Affairs Council. Reprinted by permission.

most of the time but who were there in this alternative site in the case that a weapon was in Washington as well. If it were to explode in Washington, all the current government would likely be killed, but the country would still survive and we would need a government to try to see what we could do to put the pieces together thereafter.

The point of this story, which is told in the introduction of the book, is that as the U.S. government confronted this report there was no basis in science, there was no basis in technology, no basis in logic, no basis in politics for dismissing this as a real possibility. Our government took it as a possible fact. Nuclear Emergency Support Teams were dispatched to New York City to search to see if they could find any signals of radioactivity, and other pieces of information in the report by Dragonfire were traced down. After less than a week it was concluded that this was a false alarm. Mayor Giuliani was never informed of this at the time—something that he expressed some considerable dissatisfaction about after he learned about it. But I believe the president made the right decision. If he had told Giuliani, he would undoubtedly have told his Commissioner of Police, and the more people that knew the more likely it would become a fact, and if you turned on your television and heard that the president thought that there might be a nuclear bomb in Los Angeles you wouldn't be here listening to a lecture tonight. I wouldn't be here giving it, so it can have a lot of consequences.

For tonight there are four things to remember. The first one was Dragonfire. The second story I tell in the introduction is four million. What is four million? Four million is the answer to the question, "How many Americans does Bin Laden say Al Qaeda needs to kill?" Four million. Several months after the 9/11 attack, Bin Laden's press spokesman, a fellow named Abu Gheith, put up on the Al Qaeda website Al Qaeda's objective for America in which he said, "our goal is to kill four million Americans, including two million children and to maim an equivalent number." He then goes on to explain in a fascinating, if grotesque, calculus that this number is not picked up out of thin air. This number is actually what is required, he says, to balance the scales of justice for the deaths and destruction that have been caused to Muslims by what they call "Jewish-Christian crusaders," by whom they mean Israel and the United States. He then goes through a whole series of incidents and gives us his body count, and I describe this in the introductory chapter— Chatilla, how many? sanctions against Iraq, how many? And even when I wrote the book, which was just published a month ago, I never quite got the thing about the children. It seemed kind of strange. Four million. I couldn't see how he gets this calculus. I think it's crazy but at least I can see some logic in it. The children point came home to me more vividly recently with this horrible action in Russia with the kids at the school in Beslan where killing children was also part of a conception of what's imagined, in some crazed way, to balance the scales of justice.

So, the second thing to remember is four million, and there's a debate about this among people that try to study Bin Laden and Al Qaeda in the international security community who say, "Oh, no. They're not really serious about this, it's too hard to believe." Actually, when people say they want to kill large numbers of people, most people find it unbelievable, and any of us who study history know that this is not the first time claims about proposals to kill millions of people were discounted. So I myself take this quite seriously. I think it's a serious effort. If you ask yourself how many 9/11 attacks would it take to kill four million Americans, we can do the math. It would be about 1,400. So, you're not going to get this goal by hijacking airplanes and crashing them into buildings. Someone is going to have to go upscale in terms of consequences.

Now, imagine, God forbid, a terrorist nuclear bomb like the bomb that Dragonfire said, and we thought, was in New York City was in Los Angeles. How big is a ten-kiloton bomb out of the former Soviet arsenal? Less than half the size of one of these tables. You could put it in a big wheelie—these huge suitcase-types that you wheel around. And there are a large number of nuclear weapons much smaller than that. People have an idea that nuclear weapons are these huge things that you couldn't possibly haul about. There are some nuclear weapons that are quite small. So, in any case what would a ten-kiloton explosion look like? Think of Hiroshima—that was twelve and a half kilotons, so it's about the same size. In Los Angeles, if you imagine the bomb was at the intersection of Hollywood and Highland, a ten-kiloton explosion would vaporize everything a third of a mile from ground zero. So, that would be the Chinese Theatre, the Walk of Fame, the home of the Academy Awards would look like the Federal Office Building in Oklahoma City. Then you'd have raging fires out to past the Hollywood hill sign. So, in New York City, at Time Square on a work day, you could imagine killing half a million people instantaneously and about that many as well would die over the following several days.

Because nuclear weapons, even for those of you old enough to remember the Cold War, have kind of got out of people's heads, we put up a website called nuclearterrorism.org, which you can go to and you can put in the zip code you're interested in and see the consequences of a ten-kiloton weapon in that neighborhood. Just think of ground zero, a third of a mile gone completely, and out to beyond a mile looking like the federal office building in Oklahoma City. So, for a small nuclear weapon this is a huge consequence.

I go through this much detail in an introductory fashion, not to try to be just doom and gloom, but to say this is a real possibility that we face. I not only believe this is a great threat for us now, today, but President Bush has said that this is our ultimate nightmare. As he says, "the world's most destructive technologies in the hands of the world's most dangerous actors"—that's his bumper sticker for it, and it's a very good bumper sticker, I believe. What's the world's most destructive technology? Nuclear. Who are the world's most dangerous actors? Terrorists, because, as Bin Laden says, there are people who love death more than we love life. They also don't have a return address. A good part of the reason why the Soviet Union never attacked the U.S. was that those of us who are old Cold War warriors, and that would include me, built up a vast arsenal of nuclear weapons so that they would know—any Soviet leader would know—that the moment an attack occurred on the U.S. they had signed a suicide note for their country. Well, that clarifies the mind, but in the case of a nuclear bomb, if it went off tonight in San Francisco or in Boston or in Los

Angeles, who did it? Let's imagine even that Bin Laden says, "Good for us. Five hundred thousand down, three and one-half million to go." So, we would be unhappy, we would be angry, we would be eager to attack somebody but, excuse me, if we knew where Bin Laden was we would be capturing him tonight. So, a person who has no return address is very difficult to deter.

Okay, point two, four million. Dragonfire and four million.

Two more points. The book consists of part one and part two. Part one says "inevitable," part two says "preventable." Let me say a word about each. Part one is for an ordinary citizen who reads the newspapers, not for national security experts. This is kind of speaking to people as citizens and is written for somebody who is running for Congress, but not a national security expert, who wants to play a role and cares about the country. I go through who, what, where, when and how. So, who could want to do this? Al Qaeda? I go through the history of their search for nuclear weapons. Bin Laden says it's their religious duty, but I point out they're not the only game in town. There's a group called Hezbollah. This is a very, very sophisticated terrorist group that operates in conjunction with Iran. They are, as the Deputy Secretary of State Rich Armitage says, the A-Team of terrorists. They're much more sophisticated than Bin Laden and Al Qaeda. They actually blew up the barracks of the U.S. soldiers at Khobar Towers, and they've mounted a number of terrorist attacks upon Israel. This is a sophisticated group, but there are a whole number of others. The Chechens, actually. I pointed out that their most likely target for their first weapon would be Moscow, not the U.S. And when I talk to the Russians about this I tell them, "Wake up. These guys would toast Moscow first." I'm worried about their second nuclear weapon more, that they might actually sell it to Al Qaeda, that it would come to the U.S., but I'm also interested in Moscow. So that's the who.

What might they do with a weapon? There are two versions: A ready-made bomb—like the ten-kiloton bomb that Dragonfire warned about—or a homemade bomb. The ready-made bombs come in many varieties. I discuss in the book the suitcase bombs that the former Soviet arsenal included, some of which you can carry around literally in a suitcase or a backpack and some number of which it is unclear what happened to them.

But the other side of this is homemade nuclear bombs. The book has an appendix of frequently asked questions in which I discuss dirty bombs and attacks on nuclear power plants, but in the book I'm talking about nuclear bombs, which are bombs that create a mushroom cloud, and their vast destructive effect. It's conceivable for a terrorist to make a nuclear bomb if they start with one hundred pounds of highly enriched uranium. The hard thing to do is make highly enriched uranium or plutonium—that is actually beyond the capacity of the terrorist groups. That's a multi-billion dollar investment over many years in a big facility. Iran is now in its 18th year of this project and is just now coming to the finish line. It's possible to do. Pakistan did it over a decade with a very successful effort, but terrorists are not going to do this by themselves. But if terrorists got 100 pounds of highly enriched uranium that had been made by somebody else and stolen or given to them— from that to a nuclear bomb like Hiroshima is a very straight path. As

President Bush said in the run-up to the war with Iraq, if Saddam got a soft-ball-size lump of highly enriched uranium he could make a bomb in a year. That's true, but so could Al Qaeda or any other group. If you start with the highly enriched uranium the rest of the design is unfortunately quite straight-forward and simple. The rest of the material that's required is stuff that's industrially available. So, either a pre-made bomb or a ready-made bomb or a homemade bomb. . . .

Preventable. We have Dragonfire, four million, inevitable and this is the last point—preventable. Unlike other catastrophic terrorism of which there's a number of varieties, and unlike the fact that there will be additional cata-strophic terrorist attacks on America of 9/11 proportions—that is, kills hun-dreds *or* thousands of people—I would say the chances of that are 100 percent, unlike bio-terrorism, where I'm sure there will be additional attacks like the anthrax attacks, the ultimate terrorist weapon, a nuclear bomb, is preventable. How can this be? Because there's fortunately in this issue a strategic narrow [window] to check this issue, this challenge.

There are only two elements in the world from which you can make a fissionable explosion: They are highly enriched uranium and plutonium. Neither of these elements exists in nature. You can't go dig them up. Neither of these can be made in somebody's basement. As I say, it's a multi-billion dollar, multi-year undertaking. So all we have to do, though "all" is big, is prevent terrorists acquiring highly enriched uranium and plutonium and we can prevent nuclear terrorism. Now, what's required to do that? Well, locking down all the stuff that now exists, preventing any more being produced and cleaning it out of the places where you can't lock it down successfully. That is the big picture.

I tried to organize a campaign for doing this under a doctrine of three "nos." Let me say just a word about each of them, because each of them is a lot of stuff, but I'm just going to do it briefly. The nos are: no loose nukes, no new nascent nukes and no new nuclear weapon states. Let's go through them very quickly. No loose nukes means developing with Russia a new gold stan-dard and locking down all weapons and all materials, first in the U.S. and Russia and then every where else on the fastest feasible timetable to this new gold standard.

The U.S. loses how much gold from Fort Knox? Zero. Not an ounce. Russia loses how many treasures from the Kremlin armory? None. So do human beings know how to lock things down that they really care about peo-ple not stealing? Yes. There's no lock that's 100 percent, but relative to the people who want to steal gold, the chances of them getting it out of Ft. Knox is very, very slim—almost nonexistent. I had a debate with a senator whose name I won't mention but who kept saying, "You can't be serious about this! You mean locking down nuclear weapons as good as gold?" I said, "Yeah, I'm absolutely serious about that. Why not? What is gold?" Gold will become a relatively uninteresting substance after a nuclear terrorist attack, I believe. So, no loose nukes.

Second. No new nascent nukes. We haven't appreciated the extent to which if people get highly enriched uranium or plutonium they are about

90 percent of the way to having a nuclear bomb. So, no new national production of highly enriched uranium or plutonium. The specific test case for this today is Iran. Iran is just about to get across the finish line. So, I outline a strategy which I believe could be implemented today, but the window keeps narrowing all the time, for stopping Iran where they are right now and backing them down step-by-step in a verifiable process in which there would be no new production of highly enriched uranium or plutonium in Iran—which means stopping these factories from being completed.

The third one is no new nuclear weapons states. There are eight states that have nuclear weapons in the world today. Five acknowledged. India and Pakistan have tested and say that they're nuclear weapons states, but other people haven't "accepted" them in any official status, and Israel, which is an undeclared nuclear weapons state. So, that's eight. I say draw a bright line there and say simply, "We're not having any more." Yes, it is unfair that these eight should have them and other people shouldn't, to which the answer is we're going to work in the longer run on getting this problem solved—the eight—but in the short run it's not advanced by having more. In any case the current challenge to this is called North Korea. North Korea is *the* most dangerous property on earth. Why? Because with Americans hardly even noticing, since January of 2003—so just the last 18 months— while we've been consumed by Iraq, Kim Jong Il has noticed that we've been giving him a pass and he's been moving rapidly to build additional nuclear weapons or to produce material for additional nuclear weapons. Since January 2003 he withdrew from the non-proliferation treaty, he kicked out the IAE inspectors, turned off the video cams that were watching these fuel rods that had enough stuff for six more nuclear bombs, he put that stuff on trucks and took them off to factories that are reprocessing them to produce more plutonium and at some point he's going to announce "We're finished. We have a nuclear arsenal."

Indeed, when this mysterious explosion occurred in North Korea last week, one of the worries within the intelligence community is that it could be a nuclear bomb. This could be the wake-up call to that fact and the intelligence community tonight is still sitting on their seat thinking, "Is this going to happen and if it didn't happen, when could it happen?" So, if North Korea succeeds with this project they're going to be a nuclear weapon state and they're going to have a nuclear weapons production line for another dozen weapons a year.

What do we know about North Korea? We know that it sells whatever it makes to anybody that will pay for it. So, they are in the business of Missiles-R-Us. They sell missiles to whom? People who pay, Iran, Iraq, Egypt, Libya, and others.

They have two other products: illegal drugs and counterfeit hundred dollar bills. That's it. The rest of the place has no income. Ten percent of the citizens have been starved to death in North Korea in the last half dozen years. So it's a genuine basket case as a country, ruled by a strange fellow, Mr. Kim Jong Il, who, if he has a nuclear arsenal and a nuclear weapons production line, will for sure sell nuclear weapons to other states and terrorist groups and we will not know that the weapon has been sold. So, I think this is the most dangerous site. In the book I outline the strategy for trying to deal with North Korea now. This window is closing, very, very fast, but if North Korea makes its way into this status as a nuclear weapon state with a nuclear weapon production line, then the likelihood of nuclear terrorist attacks on the U.S. goes right up the scale. So, this would be the worst failure in American security policy ever, I believe, if this is allowed to happen and it's just about to happen. But I have a strategy, which some people will regard as slightly crazy, but if I were in charge I would do it tonight. I would have done it two years ago.

So, three nos: no loose nukes, no new nascent nukes, no new nuclear weapons states.

My final point is just a question. There's a strategy group that met in the summer in Colorado, Republicans and Democrats from the national security community. There were two or three former secretaries of defense, a couple of former secretaries of state, three or four former directors of CIA, and several former national security advisors. A broad base of sensible people in my view, plus some people in the academic community. The subject this past summer was nuclear terrorism and my book was some part of the argument about it and at the end of the conversation Bill Perry, who's one of the genuine wise men, in my view, in the national security world, who was Secretary of Defense under Clinton and a very calm man, he hardly ever raises his blood pressure, said, "We are racing towards unprecedented catastrophe. I see no sense of urgency in the public about this threat. What in the world can we do to awaken the public and energize the administration?" I would say that's the question before the house. I don't have a very good answer to that. I saw Bill in San Francisco Monday night, and he's still worrying about that question.

Thank you.

POSTSCRIPT

Are We Headed Toward
a Nuclear 9/11?

There are many arguments to support the contention that nuclear and dirty bombs are hard to obtain, difficult to move and assemble, and even harder to deliver. There is also ample evidence to suggest that most, if not all, of the U.S. government's work is in one way or another designed to thwart such actions because of the enormous consequences were such acts to be carried out. These facts should make Americans rest easier and allay fears if only for the reasons of probability.

However, Allison's contention that failure to assume the worst may prevent the thwarting of such terrorist designs is persuasive. Since September 11 it is clear that the world has entered a new phase of terrorist action and a new level of funding, sophistication, and motivation. The attitude that because something is difficult it is unlikely to take place may be too dangerous to possess. The collapse of the USSR has unleashed a variety of forces, some positive and some more sinister and secretive. The enormous prices that radioactive material and nuclear devices can command on the black market make the likelihood of temptation strong and possibly irresistible.

If states are to err, perhaps they should err on the side of caution and preventive action rather than on reliance on the statistical probability that nuclear terrorism is unlikely. We may never see a nuclear terrorist act in this century, but it is statistically likely that the reason for this will not be for lack of effort on the part of motivated terrorist groups.

Some important research and commentary on nuclear terrorism can be found in Elaine Landau, *Osama bin Laden: A War Against the West* (Twenty-First Century Books, 2002); Jan Lodal, *The Price of Dominance: The New Weapons of Mass Destruction and Their Challenge to American Leadership* (Council on Foreign Relations Press, 2001); Jessica Stern, *The Ultimate Terrorists* (Harvard University Press, 1999); Graham Allison, *Nuclear Terrorism: The Ultimate Presentable Catastrophe* (Times Books, 2004); and Zbigniew Brzezinski, *The Choice: Global Domination or Global Leadership* (Basic Books, 2005).

ISSUE 18

Are Cultural and Ethnic Wars the Defining Dimensions of Twenty-First Century Conflict?

YES: Samuel P. Huntington, from "The Clash of Civilizations?" *Foreign Affairs* (Summer 1993)

NO: Wendell Bell, from "Humanity's Common Values: Seeking a Positive Future," *The Futurist* (September/October 2004)

ISSUE SUMMARY

YES: Political scientist Samuel P. Huntington argues that the emerging conflicts of the twenty-first century will be cultural and not ideological. He identifies the key fault lines of conflict and discusses how these conflicts will reshape global policy.

NO: Wendell Bell, professor emeritus of sociology at Yale University, argues that by emphasizing our common humanity and shared values, cultural divisions will not be the defining wave but rather shared mission and vision will characterize human experience in this century.

Ethnic conflicts seem to be flaring up around the world with greater and greater frequency. The last few years have witnessed ethnic fighting in places such as Northern Ireland, the Middle East, Southeast Asia, and southern Africa. Ethnic clashes have also broken out in places like Bosnia, Kosovo, Rwanda, East Timor, and Chechnya. Certainly, such ethnic conflicts have flared up throughout the centuries in various places. Yet is it possible that ethnic conflict and clashes between cultures are on the rise and will dominate understanding of conflict in the twenty-first century?

For most of the twentieth century, ideological battles between nations took center stage. From the growth of communism and fascism in the 1920s and 1930s came ideological battles centered on notions of governance, economics, race, and the role of people in relation to the state. In that major battle communism and capitalist democracy won out. In the subsequent years the ideological bipolar conflict dominated, as the United States, the USSR, and their surrogates fought on battlefields from Angola to El Salvador and from Vietnam to Afghanistan. At least in part, the battle centered over which system and method of governance would achieve preeminence. In the end the United States won, and capitalist democracy triumphed over communism.

Today scholars and policymakers grapple with the new dynamics of the global system and search for unifying elements that will help to explain why and how groups engage in conflict. Since ideology has lost its zest, and since no apparent philosophies stand ready to directly challenge capitalism, other rallying cries are being uttered and adopted. This period has witnessed the rise or reinvigoration of such philosophies as Islamic fundamentalism, environmentalism, national self-determination (omnipresent in global politics), and ethnic identity as movements and rallying cries for groups to challenge perceived or real oppressors.

Combined with this development is a technological revolution that has brought people in various parts of the world closer than ever before. Ethnic groups traditionally divided by closed borders or reduced contact are now increasingly thrust together by political, economic, and social factors. While this development is benign enough on the surface, many who feel that their cultures are threatened, their identities challenged, and their rights usurped have reacted with disdain and distrust for "the other." As a result, ethnic conflicts have flared in the Balkans, the Middle East, the former Soviet Union, the Great Lakes region of Africa, South America, and the Indian subcontinent.

With increased tensions, greater amounts of weaponry, and less restraint offered by a bipolar world, these conflicts have raged with devastating human consequences. Issues of ethnic cleansing, genocide, land mines, nuclear proliferation, and narcoterrorism have all sprung from or been fueled by these conflicts.

In the following selection, Samuel P. Huntington argues that cultural rivalries are the wave of the coming age. He contends that fault lines between civilizations (where dominant cultures meet) will be "the flash points for crisis and bloodshed" in the coming decades. He contends that cultures will predominate in the battle for hearts and minds and that groups will engage in conflict to defend against challenges to their cultures as they perceive them.

In the second selection, Wendell Bell contends that the notion of cultural divisions being the centerpiece of human conflict ignores the many common values and ideals that humans share. He also intimates that holding to the notion of cultural clashes perpetuates such conflicts. He argues that by emphasizing the common values we share and by taking actions to promote them such conflicts will be mitigated and as such the twenty-first century can be an era of reduced, not increased, tensions.

YES

Samuel P. Huntington

The Clash of Civilizations?

The Next Pattern of Conflict

World politics is entering a new phase, and intellectuals have not hesitated to proliferate visions of what it will be—the end of history; the return of traditional rivalries between nation states, and the decline of the nation state from the conflicting pulls of tribalism and globalism, among others. Each of these visions catches aspects of the emerging reality. Yet they all miss a crucial, indeed a central, aspect of what global politics is likely to be in the coming years.

It is my hypothesis that the fundamental source of conflict in this new world will not be primarily ideological or primarily economic. The great divisions among humankind and the dominating source of conflict will be cultural. Nation states will remain the most powerful actors in world affairs, but the principal conflicts of global politics will occur between nations and groups of different civilizations. The clash of civilizations will dominate global politics. The fault lines between civilizations will be the battle lines of the future.

Conflict between civilizations will be the latest phase in the evolution of conflict in the modern world. For a century and a half after the emergence of the modern international system with the Peace of Westphalia [1648], the conflicts of the Western world were largely among princes—emperors, absolute monarchs and constitutional monarchs attempting to expand their bureaucracies, their armies, their mercantilist economic strength and, most important, the territory they ruled. In the process they created nation states, and beginning with the French Revolution the principal lines of conflict were between nations rather than princes. In 1973, as R. R. Palmer put it, "The wars of kings were over; the wars of peoples had begun." This nineteenth-century pattern lasted until the end of World War I. Then, as a result of the Russian Revolution and the reaction against it, the conflict of nations yielded to the conflict of ideologies, first among communism, fascism-Nazism and liberal democracy, and then between communism and liberal democracy. During the Cold War, this latter conflict became embodied in the struggle between the two superpowers, neither of which was a nation state in the classical European sense and each of which defined its identity in terms of its ideology.

From Samuel P. Huntington, "The Clash of Civilizations?" *Foreign Affairs,* vol. 72, no. 3 (Summer 1993). Copyright © 1993 by The Council on Foreign Relations, Inc. Reprinted by permission of *Foreign Affairs.*

These conflicts between princes, nation states and ideologies were primarily conflicts within Western civilization, "Western civil wars," as William Lind has labeled them. This was as true of the Cold War as it was of the world wars and the earlier wars of the seventeenth, eighteenth and nineteenth centuries. With the end of the Cold War, international politics moves out of its Western phase, and its centerpiece becomes the interaction between the West and non-Western civilizations and among non-Western civilizations. In the politics of civilizations, the peoples and governments of non-Western civilizations no longer remain the objects of history as targets of Western colonialism but join the West as movers and shapers of history.

The Nature of Civilizations

During the Cold War the world was divided into the First, Second and Third Worlds. Those divisions are no longer relevant. It is far more meaningful now to group countries not in terms of their political or economic systems or in terms of their level of economic development but rather in terms of their culture and civilization.

What do we mean when we talk of a civilization? A civilization is a cultural entity. Villages, regions, ethnic groups, nationalities, religious groups, all have distinct cultures at different levels of cultural heterogeneity. The culture of a village in southern Italy may be different from that of a village in northern Italy, but both will share in a common Italian culture that distinguishes them from German villages. European communities, in turn, will share cultural features that distinguish them from Arab or Chinese communities. Arabs, Chinese and Westerners, however, are not part of any broader cultural entity. They constitute civilizations. A civilization is thus the highest cultural grouping of people and the broadest level of cultural identity people have short of that which distinguishes humans from other species. It is defined both by common objective elements, such as language, history, religion, customs, institutions, and by the subjective self-identification of people. People have levels of identity: a resident of Rome may define himself with varying degrees of intensity as a Roman, an Italian, a Catholic, a Christian, a European, a Westerner. The civilization to which he belongs is the broadest level of identification with which he intensely identifies. People can and do redefine their identities and, as a result, the composition and boundaries of civilizations change.

Civilizations may involve a large number of people, as with China, ("a civilization pretending to be a state," as Lucian Pye put it), or a very small number of people, such as the Anglophone Caribbean. A civilization may include several nation states, as is the case with Western, Latin American and Arab civilizations, or only one, as is the case with Japanese civilization. Civilizations obviously blend and overlap, and may include subcivilizations. Western civilization has two major variants, European and North American, and Islam has its Arab, Turkic and Malay subdivisions. Civilizations are nonetheless meaningful entities, and while the lines between them are seldom sharp, they are real. Civilizations are dynamic; they rise and fall; they divide and merge. And, as any student of history knows, civilizations disappear and are buried in the sands of time.

Westerners tend to think of nation states as the principal actors in global affairs. They have been that, however, for only a few centuries. The broader reaches of human history have been the history of civilizations. In *A Study of History,* Arnold Toynbee identified 21 major civilizations; only six of them exist in the contemporary world.

Why Civilizations Will Clash

Civilization identity will be increasingly important in the future, and the world will be shaped in large measure by the interactions among seven or eight major civilizations. These include Western, Confucian, Japanese, Islamic, Hindu, Slavic-Orthodox, Latin American and possibly African civilization. The most important conflicts of the future will occur along the cultural fault lines separating these civilizations from one another.

Why will this be the case?

First, differences among civilizations are not only real; they are basic. Civilizations are differentiated from each other by history, language, culture, tradition and, most important, religion. The people of different civilizations have different views on the relations between God and man, the individual and the group, the citizen and the state, parents and children, husband and wife, as well as differing views of the relative importance of rights and responsibilities, liberty and authority, equality and hierarchy. These differences are the product of centuries. They will not soon disappear. They are far more fundamental than differences among political ideologies and political regimes. Differences do not necessarily mean conflict, and conflict does not necessarily mean violence. Over the centuries, however, differences among civilizations have generated the most prolonged and the most violent conflicts.

Second, the world is becoming a smaller place. The interactions between peoples of different civilizations are increasing; these increasing interactions intensify civilization consciousness and awareness of differences between civilizations and commonalities within civilizations. North African immigration to France generates hostility among Frenchmen and at the same time increased receptivity to immigration by "good" European Catholic Poles. Americans react far more negatively to Japanese investment than to larger investments from Canada and European countries. Similarly, as Donald Horowitz has pointed out, "An Ibo may be . . . an Owerri Ibo or an Onitsha Ibo in what was the Eastern region of Nigeria. In Lagos, he is simply an Ibo. In London, he is a Nigerian. In New York, he is an African." The interactions among peoples of different civilizations enhance the civilization-consciousness of people that, in turn, invigorates differences and animosities stretching or thought to stretch back deep into history.

Third, the processes of economic modernization and social change throughout the world are separating people from longstanding local identities. They also weaken the nation state as a source of identity. In much of the world religion has moved in to fill this gap, often in the form of movements that are labeled "fundamentalist." Such movements are found in Western Christianity, Judaism, Buddhism and Hinduism, as well as in Islam. In most countries and most religions the people active in fundamentalist movements are young, college-educated, middle-class technicians, professionals and business persons. The "unsecularization of the world," George Weigel has remarked, is one of the dominant social facts of life in the late twentieth century." The revival of religion, "la revanche de Dieu," as Gilles Kepel labeled it, provides a basis for identity and commitment that transcends national boundaries and unites civilizations.

Fourth, the growth of civilization-consciousness is enhanced by the dual role of the West. On the one hand, the West is at a peak of power. At the same time, however, and perhaps as a result, a return to the roots phenomenon is occurring among non-Western civilizations. Increasingly one hears references to trends toward a turning inward and "Asianization" in Japan, the end of the Nehru legacy and the "Hinduization" of India, the failure of Western ideas of socialism and nationalism and hence "re-Islamization" of the Middle East, and now a debate over Westernization versus Russianization in Boris Yeltsin's country. A West at the peak of its power confronts non-Wests that increasingly have the desire, the will and the resources to shape the world in non-Western ways.

In the past, the elites of non-Western societies were usually the people who were most involved with the West, had been educated at Oxford, the Sorbonne or Sandhurst, and had absorbed Western attitudes and values. At the same time, the populace in non-Western countries often remained deeply imbued with the indigenous culture. Now, however, these relationships are being reversed. A de-Westernization and indigenization of elites is occurring in many non-Western countries at the same time that Western, usually American, cultures, styles and habits become more popular among the mass of the people.

Fifth, cultural characteristics and differences are less mutable and hence less easily compromised and resolved than political and economic ones. In the former Soviet Union, communists can become democrats, the rich can become poor and the poor rich, but Russians cannot become Estonians and Azeris cannot become Armenians. In class and ideological conflicts, the key question was "Which side are you on?" and people could and did choose sides and change sides. In conflicts between civilizations, the question is "What are you?" That is a given that cannot be changed. And as we know, from Bosnia to the Caucasus to the Sudan, the wrong answer to that question can mean a bullet in the head. Even more than ethnicity, religion discriminates sharply and exclusively among people. A person can be half-French and half-Arab and simultaneously even a citizen of two countries. It is more difficult to be half-Catholic and half-Muslim.

Finally, economic regionalism is increasing. The proportions of total trade that were intraregional rose between 1980 and 1989 from 51 percent to 59 percent in Europe, 33 percent to 37 percent in East Asia, and 32 percent to 36 percent in North America. The importance of regional economic blocs is likely to continue to increase in the future. On the one hand, successful economic regionalism will reinforce civilization-consciousness. On the other hand, economic regionalism may succeed only when it is rooted in a common civilization. The European Community rests on the shared foundation of

European culture and Western Christianity. The success of the North American Free Trade Area depends on the convergence now underway of Mexican, Canadian and American cultures. Japan, in contrast, faces difficulties in creating a comparable economic entity in East Asia because Japan is a society and civilization unique to itself. However strong the trade and investment links Japan may develop with other East Asian countries, its cultural differences with those countries inhibit and perhaps preclude its promoting regional economic integration like that in Europe and North America.

Common culture, in contrast, is clearly facilitating the rapid expansion of the economic relations between the People's Republic of China and Hong Kong, Taiwan, Singapore and the overseas Chinese communities in other Asian countries. With the Cold War over, cultural commanalities increasingly overcome ideological differences, and mainland China and Taiwan move closer together. If cultural commonality is a prerequisite for economic integration, the principal East Asian economic bloc of the future is likely to be centered on China. This bloc is, in fact, already coming into existence. As Murray Weidenbaum has observed,

> Despite the current Japanese dominance of the region, the Chinese-based economy of Asia is rapidly emerging as a new epicenter for industry, commerce and finance. This strategic area contains substantial amounts of technology and manufacturing capability (Taiwan), outstanding entrepreneurial, marketing and services acumen (Hong Kong), a fine communications network (Singapore), a tremendous pool of financial capital (all three), and very large endowments of land, resources and labor (mainland China). . . . From Guangzhou to Singapore, from Kuala Lumpur to Manila, this influential network—often based on extensions of the tranditional clans—has been described as the beckbone of the East Asian economy.[1]

Culture and religion also form the basis of the Economic Cooperation Organization, which brings together ten non-Arab Muslim countries: Iran, Pakistan, Turkey, Azerbaijan, Kazakhstan, Kyrgyzstan, Turkmenistan, Tadjikistan, Uzbekistan and Afghanistan. One impetus to the revival and expansion of this organization, founded originally in the 1960s by Turkey, Pakistan and Iran, is the realization by the leaders of several of these countries that they had no chance of admission to the European Community. Similarly, Caricom, the Central American Common Market and Mercosur rest on common cultural foundations. Efforts to build a broader Caribbean-Central American economic entity bridging the Anglo-Latin divide, however, have to date failed.

As people define their identity in ethnic and religious terms, they are likely to see an "us" versus "them" relation existing between themselves and people of different ethnicity or religion. The end of ideologically defined states in Eastern Europe and the former Soviet Union permits traditional ethnic identities and animosities to come to the fore. Differences in culture and religion create differences over policy issues, ranging from human rights to immigration to trade and commerce to the environment. Geographical propinquity gives rise to conflicting territorial claims from Bosnia to Mindanao. Most important, the efforts of the West to promote its values of democracy and liberalism as universal values, to maintain its military predominance and to advance its economic interests engender countering responses from other civilizations. Decreasingly able to mobilize support and form coalitions on the basis of ideology, governments and groups will increasingly attempt to mobilize support by appealing to common religion and civilization identity.

The clash of civilizations thus occurs at two levels. At the micro-level, adjacent groups along the fault lines between civilizations struggle, often violently, over the control of territory and each other. At the macro-level, states from different civilizations compete for relative military and economic power, struggle over the control of international institutions and third parties, and competitively promote their particular political and religious values.

The Fault Lines Between Civilizations

The fault lines between civilizations are replacing the political and ideological boundaries of the Cold War as the flash points for crisis and bloodshed. The Cold War began when the Iron Curtain divided Europe politically and ideologically. The Cold War ended with the end of the Iron Curtain. As the ideological division of Europe has disappeared, the cultural division of Europe between Western Christianity, on the one hand, and Orthodox Christianity and Islam, on the other, has reemerged. The most significant dividing line in Europe, as William Wallace has suggested, may well be the eastern boundary of Western Christianity in the year 1500. This line runs along what are now the boundaries between Finland and Russia and between the Baltic states and Russia, cuts through Belarus and Ukraine separating the more Catholic western Ukraine from Orthodox eastern Ukraine, swings westward separating Transylvania from the rest of Romania, and then goes through Yugoslavia almost exactly along the line now separating Croatia and Slovenia from the rest of Yugoslavia. In the Balkans this line, of course, coincides with the historic boundary between the Hapsburg and Ottoman empires. The peoples to the north and west of this line are Protestant or Catholic; they shared the common experiences of European history—feudalism, the Renaissance, the Reformation, the Enlightenment, the French Revolution, the Industrial Revolution; they are generally economically better off than the peoples to the east; and they may now look forward to increasing involvement in a common European economy and to the consolidation of democratic political systems. The peoples to the east and south of this line are Orthodox or Muslim; they historically belonged to the Ottoman or Tsarist empires and were only lightly touched by the shaping events in the rest of Europe; they are generally less advanced economically; they seem much less likely to develop stable democratic political systems. The Velvet Curtain of culture has replaced the Iron Curtain of ideology as the most significant dividing line in Europe. As the events in Yugoslavia show, it is not only a line of difference; it is also at times a line of bloody conflict.

Conflict along the fault line between Western and Islamic civilizations has been going on for 1,300 years. After the founding of Islam, the Arab and Moorish surge west and north only ended at Tours in 732. From the eleventh to the thirteenth century the Crusaders attempted with temporary success to bring Christianity and Christian rule to the Holy Land. From the fourteenth to the seventeenth century, the Ottoman Turks reversed the balance, extended their sway over the Middle East and the Balkans, captured Constantinople, and twice laid siege to Vienna. In the nineteenth and early twentieth centuries as Ottoman power declined Britain, France, and Italy established Western control over most of North Africa and the Middle East.

After World War II, the West, in turn, began to retreat; the colonial empires disappeared; first Arab nationalism and then Islamic fundamentalism manifested themselves; the West became heavily dependent on the Persian Gulf countries for its energy; the oil-rich Muslim countries became money-rich and, when they wished to, weapons-rich. Several wars occurred between Arabs and Israel (created by the West). France fought a bloody and ruthless war in Algeria for most of the 1950s; British and French forces invaded Egypt in 1956; American forces went into Lebanon in 1958; subsequently American forces returned to Lebanon, attacked Libya, and engaged in various military encounters with Iran; Arab and Islamic terrorists, supported by at least three Middle Eastern governments, employed the weapon of the weak and bombed Western planes and installations and seized Western hostages. This warfare between Arabs and the West culminated in 1990, when the United States sent a massive army to the Persian Gulf to defend some Arab countries against aggression by another. In its aftermath NATO planning is increasingly directed to potential threats and instability along its "southern tier."

This centuries-old military interaction between the West and Islam is unlikely to decline. It could become more virulent. The Gulf War left some Arabs feeling proud that Saddam Hussein had attacked Israel and stood up to the West. It also left many feeling humiliated and resentful of the West's military presence in the Persian Gulf, the West's overwhelming military dominance, and their apparent inability to shape their own destiny. Many Arab countries, in addition to the oil exporters, are reaching levels of economic and social development where autocratic forms of government become inappropriate and efforts to introduce democracy become stronger. Some openings in Arab political systems have already occurred. The principal beneficiaries of these openings have been Islamist movements. In the Arab world, in short, Western democracy strengthens anti-Western political forces. This may be a passing phenomenon, but it surely complicates relations between Islamic countries and the West.

Those relations are also complicated by demography. The spectacular population growth in Arab countries, particularly in North Africa, has led to increased migration to Western Europe. The movement within Western Europe toward minimizing internal boundaries has sharpened political sensitivities with respect to this development. In Italy, France and Germany, racism is increasingly open, and political reactions and violence against Arab and Turkish migrants have become more intense and more widespread since 1990.

On both sides the interaction between Islam and the West is seen as a clash of civilizations. The West's "next confrontation," observes M. J. Akbar, an Indian Muslim author, "is definitely going to come from the Muslim world. It is in the sweep of the Islamic nations from the Maghreb to Pakistan that the struggle for a new world order will begin." Bernard Lewis comes to a similar conclusion:

> We are facing a mood and a movement far transcending the level of issues and policies and the governments that pursue them. This is no less than a clash of civilizations—the perhaps irrational but surely historic reaction of an ancient rival against our Judeo-Christian heritage, our secular present, and the worldwide expansion of both.[2]

Historically, the other great antagonistic interaction of Arab Islamic civilization has been with the pagan, animist, and now increasingly Christian black peoples to the south. In the past, this antagonism was epitomized in the image of Arab slave dealers and black slaves. It has been reflected in the ongoing civil war in the Sudan between Arabs and blacks, the fighting in Chad between Libyan-supported insurgents and the government, the tensions between Orthodox Christians and Muslims in the Horn of Africa, and the political conflicts, recurring riots and communal violence between Muslims and Christians in Nigeria. The modernization of Africa and the spread of Christianity are likely to enhance the probability of violence along this fault line. Symptomatic of the intensification of this conflict was the Pope John Paul II's speech in Khartoum in February 1993 attacking the actions of the Sudan's Islamist government against the Christian minority there.

On the northern border of Islam, conflict has increasingly erupted between Orthodox and Muslim peoples, including the carnage of Bosnia and Sarajevo, the simmering violence between Serb and Albanian, the tenuous relations between Bulgarians and their Turkish minority, the violence between Ossetians and Ingush, the unremitting slaughter of each other by Armenians and Azeris, the tense relations between Russians and Muslims in Central Asia, and the deployment of Russian troops to protect Russian interests in the Caucasus and Central Asia. Religion reinforces the revival of ethnic identities and restimulates Russian fears about the security of their southern borders. This concern is well captured by Archie Roosevelt:

> Much of Russian history concerns the struggle between the Slavs and the Turkic peoples on their borders, which dates back to the foundation of the Russian state more than a thousand years ago. In the Slavs' millennium-long confrontation with their eastern neighbors lies the key to an understanding not only of Russian history, but Russian character. To understand Russian realities today one has to have a concept of the great Turkic ethnic group that has preoccupied Russians through the centuries.[3]

The conflict of civilizations is deeply rooted elsewhere in Asia. The historic clash between Muslim and Hindu in the subcontinent manifests itself now not only in the rivalry between Pakistan and India but also in intensifying religious strife within India between increasingly militant Hindu groups and India's substantial Muslim minority. The destruction of the Ayodhya mosque in December 1992 brought to the fore the issue of whether India will remain a secular democratic state or become a Hindu one. In East Asia, China has outstanding territorial disputes with most of its neighbors. It has pursued a ruthless policy toward the Buddhist people of Tibet, and it is pursuing an increasingly ruthless policy toward its Turkic-Muslim minority. With the Cold War over, the underlying differences between China and the United States have reasserted themselves in areas such as human rights, trade and weapons proliferation. These differences are unlikely to moderate. A "new cold war," Deng Xiaoping reportedly asserted in 1991, is under way between China and America. . . .

Civilization Rallying: The Kin-Country Syndrome

Groups or states belonging to one civilization that become involved in war with people from a different civilization naturally try to rally support from other members of their own civilization. As the post–Cold War world evolves, civilization commonality, what H. D. S. Greenway has termed the "kin-country" syndrome, is replacing political ideology and traditional balance of power considerations as the principal basis for cooperation and coalitions. It can be seen gradually emerging in the post–Cold War conflicts in the Persian Gulf, the Caucasus and Bosnia. None of these was a full-scale war between civilizations, but each involved some elements of civilizational rallying, which seemed to become more important as the conflict continued and which may provide a foretaste of the future.

First, in the Gulf War one Arab state invaded another and then fought a coalition of Arab, Western and other states. While only a few Muslim governments overtly supported Saddam Hussein, many Arab elites privately cheered him on, and he was highly popular among large sections of the Arab publics. Islamic fundamentalist movements universally supported Iraq rather than the Western-backed governments of Kuwait and Saudi Arabia. Forswearing Arab nationalism, Saddam Hussein explicitly invoked an Islamic appeal. He and his supporters attempted to define the war as a war between civilizations. "It is not the world against Iraq," as Safar Al-Hawali, dean of Islamic Studies at the Umm Al-Qura University in Mecca, put it in a widely circulated tape. "It is the West against Islam." Ignoring the rivalry between Iran and Iraq, the chief Iranian religious leader, Ayatollah Ali Khamenei, called for a holy war against the West: "The struggle against American aggression, greed, plans and policies will be counted as a jihad, and anybody who is killed on that path is a martyr." "This is a war," King Hussein of Jordan argued, "against all Arabs and all Muslims and not against Iraq alone."

The rallying of substantial sections of Arab elites and publics behind Saddam Hussein caused those Arab governments in the anti-Iraq coalition to moderate their activities and temper their public statements. Arab governments opposed or distanced themselves from subsequent Western efforts to apply pressure on Iraq, including enforcement of a no-fly zone in the summer of 1992 and the bombing of Iraq in January 1993. The Western-Soviet-Turkish-Arab anti-Iraq coalition of 1990 had by 1993 become a coalition of almost only the West and Kuwait against Iraq.

Muslims contrasted Western actions against Iraq with the West's failure to protect Bosnians against Serbs and to impose sanctions on Israel for violating U.N. resolutions. The West, they alleged, was using a double standard. A world of clashing civilizations, however, is inevitably a world of double standards: people apply one standard to their kin-countries and a different standard to others.

Second, the kin-country syndrome also appeared in conflicts in the former Soviet Union. Armenian military successes in 1992 and 1993 stimulated Turkey to become increasingly supportive of its religious, ethnic and linguistic brethren in Azerbaijan. "We have a Turkish nation feeling the same sentiments as the Azerbaijanis," said one Turkish official in 1992. "We are under pressure. Our newspapers are full of the photos of atrocities and are asking us if we are still serious about pursuing our neutral policy. Maybe we should show Armenia that there's a big Turkey in the region." President Turgut Özal agreed, remarking that Turkey should at least "scare the Armenians a little bit." Turkey, Özal threatened again in 1993, would "show its fangs." Turkish Air Force jets flew reconnaissance flights along the Armenian border; Turkey suspended food shipments and air flights to Armenia; and Turkey and Iran announced they would not accept dismemberment of Azerbaijan. In the last years of its existence, the Soviet government supported Azerbaijan because its government was dominated by former communists. With the end of the Soviet Union, however, political considerations gave way to religious ones. Russian troops fought on the side of the Armenians, and Azerbaijan accused the "Russian government of turning 180 degrees" toward support for Christian Armenia.

Third, with respect to the fighting in the former Yugoslavia, Western publics manifested sympathy and support for the Bosnian Muslims and the horrors they suffered at the hands of the Serbs. Relatively little concern was expressed, however, over Croatian attacks on Muslims and participation of the dismemberment of Bosnia-Herzegovina. In the early stages of the Yugoslav breakup, Germany, in an unusual display of diplomatic initiative and muscle, induced the other 11 members of the European Community to follow its lead in recognizing Slovenia and Crotia. As a result of the pope's determination to provide strong backing to the two Catholic countries, the Vatican extended recognition even before the Community did. The United States followed the European lead. Thus the leading actors in Western civilization rallied behind their coreligionists. Subsequently Croatia was reported to be receiving substantial quantities of arms from Central European and other Western countries. Boris Yeltsin's government, on the other hand, attempted to pursue a middle course that would be sympathetic to the Orthodox Serbs but not alienate Russia from the West. Russian conservative and nationalist groups, however, including many legislators, attacked the government for not being more forthcoming in its support for the Serbs. By early 1993 several hundred Russians apparently were serving with the Serbian forces, and reports circulated of Russian arms being supplied to Serbia.

Islamic governments and groups, on the other hand, castigated the West for not coming to the defense of the Bosnians. Iranian leaders urged Muslims from all countries to provide help to Bosnia; in violation of the U.N. arms embargo, Iran supplied weapons and men for the Bosnians; Iranian-supported Lebanese groups sent guerrillas to train and organize the Bosnian forces. In 1993 up to 4,000 Muslims from over two dozen Islamic countries were reported to be fighting in Bosnia. The governments of Saudi Arabia and other countries felt under increasing pressure from fundamentalist groups in their own societies to provide more vigorous support for the Bosnians. By the end of 1992, Saudi Arabia had reportedly supplied substantial funding for weapons and supplies for the Bosnians, which significantly increased their military capabilities vis-à-vis the Serbs.

In the 1930s the Spanish Civil War provoked intervention from countries that politically were fascist, communist and democratic. In the 1990s the Yugoslav conflict is provoking intervention from countries that are Muslim, Orthodox and Western Christian. The parallel has not gone unnoticed. "The war in Bosnia-Herzegovina has become the emotional equivalent of the fight against fascism in the Spanish Civil War," one Saudi editor observed. "Those who died there are regarded as martyrs who tried to save their fellow Muslims."

Conflicts and violence will also occur between states and groups within the same civilization. Such conflicts, however, are likely to be less intense and less likely to expand than conflicts between civilizations. Common membership in a civilization reduces the probability of violence in situations where it might otherwise occur. In 1991 and 1992 many people were alarmed by the possibility of violent conflict between Russia and Ukraine over territory, particularly Crimea, the Black Sea fleet, nuclear weapons and economic issues. If civilization is what counts, however, the likelihood of violence between Ukrainians and Russians should be low. They are two Slavic, primarily Orthodox peoples who have had close relationships with each other for centuries. As of early 1993, despite all the reasons for conflict, the leaders of the two countries were effectively negotiating and defusing the issues between the two countries. While there has been serious fighting between Muslims and Christians elsewhere in the former Soviet Union and much tension and some fighting between Western and Orthodox Christians in the Baltic states, there has been virtually no violence between Russians and Ukrainians.

Civilization rallying to date has been limited, but it has been growing, and it clearly has the potential to spread much further. As the conflicts in the Persian Gulf, the Caucasus and Bosnia continued, the positions of nations and the cleavages between them increasingly were along civilizational lines. Populist politicians, religious leaders and the media have found it a potent means of arousing mass support and of pressuring hesitant governments. In the coming years, the local conflicts most likely to escalate into major wars will be those, as in Bosnia and the Caucasus, along the fault lines between civilizations. The next world war, if there is one, will be a war between civilizations.

The West Versus the Rest

The West is now at an extraordinary peak of power in relation to other civilizations. Its superpower opponent has disappeared from the map. Military conflict among Western states is unthinkable, and Western military power is unrivaled. Apart from Japan, the West faces no economic challenge. It dominates international political and security institutions and with Japan international economic institutions. Global political and security issues are effectively settled by a directorate of the United States, Britain and France, world economic issues by a directorate of the United States, Germany and Japan, all of which maintain extraordinarily close relations with each other to the exclusion of lesser and largely non-Western countries. Decisions made at the U.N. Security Council or in the International Monetary Fund [IMF] that reflect the interests of the West are presented to the world as reflecting the desires of the world community. The very phrase "the world community" has become the euphemistic collective noun (replacing "the Free World") to give global legitimacy to actions reflecting the interests of the United States and other Western powers.[4] Through the IMF and other international economic institutions, the West promotes its economic interests and imposes on other nations the economic policies it thinks appropriate. In any poll of non-Western peoples, the IMF undoubtedly would win the support of finance ministers and a few others, but get an overwhelmingly unfavorable rating from just about everyone else who would agree with Georgy Arbatov's characterization of IMF officials as "neo-Bolsheviks who love expropriating other people's money, imposing undemocratic and alien rules of economic and political conduct and stifling economic freedom."

Western domination of the U.N. Security Council and its decisions, tempered only by occasional abstention by China, produced U.N. legitimation of the West's use of force to drive Iraq out of Kuwait and its elimination of Iraq's sophisticated weapons and capacity to produce such weapons. It also produced the quite unprecedented action by the United States, Britain and France in getting the Security Council to demand that Libya hand over the Pan Am 103 bombing suspects and then to impose sanctions when Libya refused. After defeating the largest Arab army, the West did not hesitate to throw its weight around in the Arab world. The West in effect is using international institutions, military power and economic resources to run the world in ways that will maintain Western predominance, protect Western interests and promote Western political and economic values.

That at least is the way in which non-Westerners see the new world, and there is a significant element of truth in their view. Differences in power and struggles for military, economic and institutional power are thus one source of conflict between the West and other civilizations. Differences in culture, that is basic values and beliefs, are a second source of conflict. V. S. Naipaul has argued that Western civilization is the "universal civilization" that "fits all men." At a superficial level much of Western culture has indeed permeated the rest of the world. At a more basic level, however, Western concepts differ fundamentally from those prevalent in other civilizations. Western ideas of individualism, liberalism, constitutionalism, human rights, equality, liberty, the rule of law, democracy, free markets, the separation of church and state, often have little resonance in Islamic, Confucian, Japanese, Hindu, Buddhist or Orthodox cultures. Western efforts to propagate such ideas produce instead a reaction against "human rights imperialism" and a reaffirmation of indigenous values, as can be seen in the support for religious fundamentalism by the younger generation in non-Western cultures. The very notion that there could be a "universal civilization" is a Western idea, directly at odds with the particularism of most Asian societies and their emphasis on what distinguishes one people from another. Indeed, the author of a review of 100 comparative studies of values in different societies concluded that "the values that are most important in the West are least important worldwide."[5] In the political realm, of course, these differences are most manifest in the efforts of the United States and other Western powers to induce other peoples to adopt Western ideas concerning democracy and human rights. Modern democratic government originated in the West. When it has developed in non-Western societies it has usually been the product of Western colonialism or imposition.

The central axis of world politics in the future is likely to be, in Kishore Mahbubani's phrase, the conflict between "the West and the Rest" and the responses of non-Western civilizations to Western power and values.[6] Those responses generally take one or a combination of three forms. At one extreme, non-Western states can, like Burma and North Korea, attempt to pursue a course of isolation, to insulate their societies from penetration or "corruption" by the West, and, in effect, to opt out of participation in the Western-dominated global community. The costs of this course, however, are high, and few states have pursued it exclusively. A second alternative, the equivalent of "band-wagoning" in international relations theory, is to attempt to join the West and accept its values and institutions. The third alternative is to attempt to "balance" the West by developing economic and military power and cooperating with other non-Western societies against the West, while preserving indigenous values and institutions; in short, to modernize but not to Westernize. . . .

The Confucian-Islamic Connection

The obstacles to non-Western countries joining the West vary considerably. They are least for Latin American and East European countries. They are greater for the Orthodox countries of the former Soviet Union. They are still greater for Muslim, Confucian, Hindu and Buddhist societies. Japan has established a unique position for itself as an associate member of the West: it is in the West in some respects but clearly not of the West in important dimensions. Those countries that for reason of culture and power do not wish to, or cannot, join the West compete with the West by developing their own economic, military and political power. They do this by promoting their internal development and by cooperating with other non-Western countries. The most prominent form of this cooperation is the Confucian-Islamic connection that has emerged to challenge Western interests, values and power.

Almost without exception, Western countries are reducing their military power; under Yeltsin's leadership so also is Russia. China, North Korea and several Middle Eastern states, however, are significantly expanding their military capabilities. They are doing this by the import of arms from Western and non-Western sources and by the development of indigenous arms industries. One result is the emergence of what Charles Krauthammer has called "Weapon

States," and the Weapon States are not Western states. Another result is the redefinition of arms control, which is a Western concept and a Western goal. During the Cold War the primary purpose of arms control was to establish a stable military balance between the United States and its allies and the Soviet Union and its allies. In the post–Cold War world the primary objective of arms control is to prevent the development by non-Western societies of military capabilities that could threaten Western interests. The West attempts to do this through international agreements, economic pressure and controls on the transfer of arms and weapons technologies.

The conflict between the West and the Confucian-Islamic states focuses largely, although not exclusively, on nuclear, chemical and biological weapons, ballistic missiles and other sophisticated means for delivering them, and the guidance, intelligence and other electronic capabilities for achieving that goal. The West promotes non-proliferation as a universal norm and nonproliferation treaties and inspections as means of realizing that norm. It also threatens a variety of sanctions against those who promote the spread of sophisticated weapons and proposes some benefits for those who do not. The attention of the West focuses, naturally, on nations that are actually or potentially hostile to the West.

The non-Western nations, on the other hand, assert their right to acquire and to deploy whatever weapons they think necessary for their security. They also have absorbed, to the full, the truth of the response of the Indian defense minister when asked what lesson he learned from the Gulf War: "Don't fight the United States unless you have nuclear weapons." Nuclear weapons, chemical weapons and missiles are viewed, probably erroneously, as the potential equalizer of superior Western conventional power. China, of course, already has nuclear weapons; Pakistan and India have the capability to deploy them. North Korea, Iran, Iraq, Libya and Algeria appear to be attempting to acquire them. A top Iranian official has declared that all Muslim states should acquire nuclear weapons, and in 1988 the president of Iran reportedly issued a directive calling for development of "offensive and defensive chemical, biological and radiological weapons."

Centrally important to the development of counter-West military capabilities is the sustained expansion of China's military power and its means to create military power. Buoyed by spectacular economic development, China is rapidly increasing its military spending and vigorously moving forward with the modernization of its armed forces. It is purchasing weapons from the former Soviet states; it is developing long-range missiles; in 1992 it tested a one-megaton nuclear device. It is developing power-projection capabilities, acquiring aerial refueling technology, and trying to purchase an aircraft carrier. Its military buildup and assertion of sovereignty over the South China Sea are provoking a multilateral regional arms race in East Asia. China is also a major exporter of arms and weapons technology. It has exported materials to Libya and Iraq that could be used to manufacture nuclear weapons and nerve gas. It has helped Algeria build a reactor suitable for nuclear weapons research and production. China has sold to Iran nuclear technology that American officials believe could only be used to create weapons and apparently has shipped components of 300-mile-range missiles to Pakistan. North Korea has had a nuclear weapons program under way for some while and has sold advanced missiles and missile technology to Syria and Iran. The flow of weapons and weapons technology is generally from East Asia to the Middle East. There is, however, some movement in the reverse direction; China has received Stinger missiles from Pakistan.

A Confucian-Islamic military connection has thus come into being, designed to promote acquisition by its members of the weapons and weapons technologies needed to counter the military power of the West. It may or may not last. At present, however, it is, as Dave McCurdy has said, "a renegades' mutual support pact, run by the proliferators and their backers." A new form of arms competition is thus occurring between Islamic-Confucian states and the West. In an old-fashioned arms race, each side developed its own arms to balance or to achieve superiority against the other side. In this new form of arms competition, one side is developing its arms and the other side is attempting not to balance but to limit and prevent that arms build-up while at the same time reducing its own military capabilities.

Implications for the West

This article does not argue that civilization identities will replace all other identities, that nation states will disappear, that each civilization will become a single coherent political entity, that groups within a civilization will not conflict with and even fight each other. This paper does set forth the hypotheses that differences between civilizations are real and important; civilization-consciousness is increasing; conflict between civilizations will supplant ideological and other forms of conflict as the dominant global form of conflict; international relations, historically a game played out within Western civilization, will increasingly be de-Westernized and become a game in which non-Western civilizations are actors and not simply objects; successful political, security and economic international institutions are more likely to develop within civilizations than across civilizations; conflicts between groups in different civilizations will be more frequent, more sustained and more violent than conflicts between groups in the same civilization; violent conflicts between groups in different civilizations are the most likely and most dangerous source of escalation that could lead to global wars; the paramount axis of world politics will be the relations between "the West and the Rest;" the elites in some torn non-Western countries will try to make their countries part of the West, but in most cases face major obstacles to accomplishing this; a central focus of conflict for the immediate future will be between the West and several Islamic-Confucian states.

This is not to advocate the desirability of conflicts between civilizations. It is to set forth descriptive hypotheses as to what the future may be like. If these are plausible hypotheses, however, it is necessary to consider their implications for Western policy. These implications should be divided between short-term advantage and long-term accommodation. In the short term it is clearly in the interest of the West to promote greater cooperation and unity within its own civilization, particularly between its European and North American components; to incorporate into the West societies in Eastern

Europe and Latin America whose cultures are close to those of the West; to promote and maintain cooperative relations with Russia and Japan; to prevent escalation of local intercivilization conflicts into major inter-civilization wars; to limit the expansion of the military strength of Confucian and Islamic states; to moderate the reduction of Western military capabilities and maintain military superiority in East and Southwest Asia; to exploit differences and conflicts among Confucian and Islamic states; to support in other civilizations groups sympathetic to Western values and interests; to strengthen international institutions that reflect and legitimate Western interests and values and to promote the involvement of non-Western states in those institutions.

In the longer term other measures would be called for. Western civilization is both Western and modern. Non-Western civilizations have attempted to become modern without becoming Western. To date only Japan has fully succeeded in this quest. Non-Western civilizations will continue to attempt to acquire the wealth, technology, skills, machines and weapons that are part of being modern. They will also attempt to reconcile this modernity with their traditional culture and values. Their economic and military strength relative to the West will increase. Hence the West will increasingly have to accommodate these non-Western modern civilizations whose power approaches that of the West but whose values and interests differ significantly from those of the West. This will require the West to maintain the economic and military power necessary to protect its interests in relation to these civilizations. It will also, however, require the West to develop a more profound understanding of the basic religious and philosophical assumptions underlying other civilizations and the ways in which people in those civilizations see their interests. It will require an effort to identify elements of commonality between Western and other civilizations. For the relevant future, there will be no universal civilization, but instead a world of different civilizations, each of which will have to learn to coexist with the others.

Notes

1. Murray Weidenbaum, *Greater China: The Next Economic Superpower?*, St. Louis: Washington University Center for the Study of American Business, Contemporary Issues, Series 57, February 1993, pp. 2–3.

2. Bernard Lewis, "The Roots of Muslim Rage," *The Atlantic Monthly*, vol. 266, September 1990, p. 60; *Time*, June 15, 1992, pp. 24–28.

3. Archie Roosevelt, *For Lust of Knowing*, Boston; Little, Brown, 1988, pp. 332–333.

4. Almost invariably Western leaders claim they are acting on behalf of "the world community." One minor lapse occurred during the run-up to the Gulf War. In an interview on "Good Morning America," Dec. 21, 1990, British Prime Minister John Major referred to the actions "the West" was taking against Saddam Hussein. He quickly corrected himself and subsequently referred to "the world community." He was, however, right when he erred.

5. Harry C. Triandis, *The New York Times*, Dec. 25, 1990, p. 41, and "Cross-Cultural Studies of Individualism and Collectivism," Nebraska Symposium on Motivation, vol. 37, 1989, pp. 41–133.

6. Kishore Mahbubani, "The West and the Rest," *The National Interest*, Summer 1992, pp. 3–13.

Wendell Bell

Humanity's Common Values: Seeking a Positive Future

Some commentators have insisted that the terrorist attacks of September 11, 2001, and their aftermath demonstrate Samuel P. Huntington's thesis of "the clash of civilizations," articulated in a famous article published in 1993. Huntington, a professor at Harvard University and director of security planning for the National Security Council during the Carter administration, argued that "conflict between groups from differing civilizations" has become "the central and most dangerous dimension of the emerging global politics."

Huntington foresaw a future in which nation-states no longer play a decisive role in world affairs. Instead, he envisioned large alliances of states, drawn together by common culture, cooperating with each other. He warned that such collectivities are likely to be in conflict with other alliances formed of countries united around a different culture.

Cultural differences do indeed separate people between various civilizations, but they also separate groups within a single culture or state. Many countries contain militant peoples of different races, religions, languages, and cultures, and such differences do sometimes provoke incidents that lead to violent conflict—as in Bosnia, Cyprus, Northern Ireland, Rwanda, and elsewhere. Moreover, within many societies today (both Western and non-Western) and within many religions (including Islam, Judaism, and Christianity) the culture war is primarily internal, between fundamentalist orthodox believers on the one hand and universalizing moderates on the other. However, for most people most of the time, peaceful accommodation and cooperation are the norms.

Conflicts between groups often arise and continue not because of the differences between them, but because of their similarities. People everywhere, for example, share the capacities to demonize others, to be loyal to their own group (sometimes even willing to die for it), to believe that they themselves and those they identify with are virtuous while all others are wicked, and to remember past wrongs committed against their group and seek revenge. Sadly, human beings everywhere share the capacity to hate and kill each other, including their own family members and neighbors.

Discontents of Globalization

Huntington is skeptical about the implications of the McDonaldization of the world. He insists that the "essence of Western civilization is the Magna Carta not the Magna Mac." And he says further, "The fact that non-Westerners may bite into the latter has no implications for accepting the former."

His conclusion may be wrong, for if biting into a Big Mac and drinking Coca-Cola, French wine, or Jamaican coffee while watching a Hollywood film on a Japanese TV and stretched out on a Turkish rug means economic development, then demands for public liberties and some form of democratic rule may soon follow where Big Mac leads. We know from dozens of studies that economic development contributes to the conditions necessary for political democracy to flourish.

Globalization, of course, is not producing an all-Western universal culture. Although it contains many Western aspects, what is emerging is a global culture, with elements from many cultures of the world, Western and non-Western.

Local cultural groups sometimes do view the emerging global culture as a threat, because they fear their traditional ways will disappear or be corrupted. And they may be right. The social world, after all, is constantly in flux. But, like the clean toilets that McDonald's brought to Hong Kong restaurants, people may benefit from certain changes, even when their fears prevent them from seeing this at once.

And local traditions can still be—and are—preserved by groups participating in a global culture. Tolerance and even the celebration of many local variations, as long as they do not harm others, are hall-marks of a sustainable world community. Chinese food, Spanish art, Asian philosophies, African drumming, Egyptian history, or any major religion's version of the Golden Rule can enrich the lives of everyone. What originated locally can become universally adopted (like Arabic numbers). Most important, perhaps, the emerging global culture is a fabric woven from tens of thousands—possibly hundreds of thousands—of individual networks of communication, influence, and exchange that link people and organizations across civilizational boundaries. Aided by electronic communications systems, these networks are growing stronger and more numerous each day.

Searching for Common, Positive Values

Global religious resurgence is a reaction to the loss of personal identity and group stability produced by "the processes of social, economic, and cultural modernization that swept across the world in the second half of the twentieth century," according to Huntington. With traditional systems of authority disrupted, people become separated from their roots in a bewildering maze of new rules and expectations. In his view, such people need "new sources of identity, new forms of stable community, and new sets of moral precepts to provide them with a sense of meaning and purpose." Organized religious groups, both mainstream and fundamentalist, are growing today precisely to meet these needs, he believes.

Although uprooted people may need new frameworks of identity and purpose, they will certainly not find them in fundamentalist religious groups, for such groups are not "new sources of identity." Instead, they recycle the past. Religious revival movements are reactionary, not progressive. Instead of facing the future, developing new approaches to deal with perceived threats of economic, technological, and social change, the movements attempt to retreat into the past.

Religions will likely remain among the major human belief systems for generations to come, despite—or even because of—the fact that they defy conventional logic and reason with their ultimate reliance upon otherworldly beliefs. However, it is possible that some ecumenical accommodations will be made that will allow humanity to build a generally accepted ethical system based on the mainly similar and overlapping moralities contained in the major religions. A person does not have to believe in supernatural beings to embrace and practice the principles of a global ethic, as exemplified in the interfaith declaration, "Towards a Global Ethic," issued by the Parliament of the World's Religions in 1993.

Interfaith global cooperation is one way that people of different civilizations can find common cause. Another is global environmental cooperation seeking to maintain and enhance the life-sustaining capacities of the earth. Also, people everywhere have a stake in working for the freedom and welfare of future generations, not least because the future of their own children and grandchildren is at stake.

Many more examples of cooperation among civilizations in the pursuit of common goals can be found in every area from medicine and science to moral philosophy, music, and art. A truly global commitment to the exploration, colonization, and industrialization of space offers still another way to harness the existing skills and talents of many nations, with the aim of realizing and extending worthy human capacities to their fullest. So, too, does the search for extraterrestrial intelligence. One day, many believe, contact will be made. What, then, becomes of Huntington's "clash of civilizations"? Visitors to Earth will likely find the variations among human cultures and languages insignificant compared with the many common traits all humans share.

Universal human values do exist, and many researchers, using different methodologies and data sets, have independently identified similar values. Typical of many studies into universal values is the global code of ethics compiled by Rushworth M. Kidder in *Shared Values for a Troubled World* (Wiley, 1994). Kidder's list includes love, truthfulness, fairness, freedom, unity (including cooperation, group allegiance, and oneness with others), tolerance, respect for life, and responsibility (which includes taking care of yourself, of other individuals, and showing concern for community interests). Additional values mentioned are courage, knowing right from wrong, wisdom, hospitality, obedience, and stability. . . .

The Search for Global Peace and Order

Individuals and societies are so complex that it may seem foolhardy even to attempt the ambitious task of increasing human freedom and well-being. Yet what alternatives do we have? In the face of violent aggressions, injustice, threats to the environment, corporate corruption, poverty, and other ills of our present world, we can find no satisfactory answers in despair, resignation, and inaction.

Rather, by viewing human society as an experiment, and monitoring the results of our efforts, we humans can gradually refine our plans and actions to bring closer an ethical future world in which every individual can realistically expect a long, peaceful, and satisfactory life.

Given the similarity in human values, I suggest three principles that might contribute to such a future: inclusion, skepticism, and social control.

1. The Principle of Inclusion

Although many moral values are common to all cultures, people too often limit their ethical treatment of others to members of their own groups. Some, for example, only show respect or concern for other people who are of their own race, religion, nationality, or social class.

Such exclusion can have disastrous effects. It can justify cheating or lying to people who are not members of one's own ingroup. At worst, it can lead to demonizing them and making them targets of aggression and violence, treating them as less than human. Those victimized by this shortsighted and counterproductive mistreatment tend to pay it back or pass it on to others, creating a nasty world in which we all must live.

Today, our individual lives and those of our descendants are so closely tied to the rest of humanity that our identities ought to include a sense of kinship with the whole human race and our circle of caring ought to embrace the welfare of people everywhere. In practical terms, this means that we should devote more effort and resources to raising the quality of life for the worst-off members of the human community; reducing disease, poverty, and illiteracy; and creating equal opportunity for all men and women. Furthermore, our circle of caring ought to include protecting natural resources, because all human life depends on preserving the planet as a livable environment.

2. The Principle of Skepticism

One of the reasons why deadly conflicts continue to occur is what has been called "the delusion of certainty." Too many people refuse to consider any view but their own. And, being sure that they are right, such people can justify doing horrendous things to others.

As I claimed in "Who Is Really Evil?" (*The Futurist,* March–April 2004), we all need a healthy dose of skepticism, especially about our own beliefs. Admitting that we might be wrong can lead to asking questions, searching for better answers, and considering alternative possibilities.

Critical realism is a theory of knowledge I recommend for everyone, because it teaches us to be skeptical. It rests on the assumption that knowledge is never fixed and final, but changes as we learn and grow. Using evidence and reason, we can evaluate our current beliefs and develop new ones in response to new information and changing conditions. Such an approach is essential to future studies, and indeed to any planning. If your cognitive maps of reality are wrong, then using them to navigate through life will not take you where you want to go.

Critical realism also invites civility among those who disagree, encouraging peaceful resolution of controversies by investigating and discussing facts. It teaches temperance and tolerance, because it recognizes that the discovery of hitherto unsuspected facts may overturn any of our "certainties," even long-cherished and strongly held beliefs.

3. **The Principle of Social Control**

Obviously, there is a worldwide need for both informal and formal social controls if we hope to achieve global peace and order. For most people most of the time, informal social controls may be sufficient. By the end of childhood, for example, the norms of behavior taught and reinforced by family, peers, school, and religious and other institutions are generally internalized by individuals.

Yet every society must also recognize that informal norms and even formal codes of law are not enough to guarantee ethical behavior and to protect public safety in every instance. Although the threats we most often think of are from criminals, fanatics, and the mentally ill, even "normal" individuals may occasionally lose control and behave irrationally, or choose to ignore or break the law with potentially tragic results. Thus, ideally, police and other public law enforcement, caretaking, and rehabilitation services protect us not only from "others," but also from ourselves.

Likewise, a global society needs global laws, institutions to administer them, and police/peacekeepers to enforce them. Existing international systems of social control should be strengthened and expanded to prevent killing and destruction, while peaceful negotiation and compromise to resolve disputes are encouraged. A global peacekeeping force with a monopoly on the legitimate use of force, sanctioned by democratic institutions and due process of law, and operated competently and fairly, could help prevent the illegal use of force, maintain global order, and promote a climate of civil discourse. The actions of these global peacekeepers should, of course, be bound not only by law, but also by a code of ethics. Peacekeepers should use force as a last resort and only to the degree needed, while making every effort to restrain aggressors without harming innocent people or damaging the infrastructures of society.

Expanding international law, increasing the number and variety of multinational institutions dedicated to controlling armed conflict, and strengthening efforts by the United Nations and other organizations to encourage the spread of democracy, global cooperation, and peace, will help create a win-win world.

Conclusion: Values for a Positive Global Future

The "clash of civilizations" thesis exaggerates both the degree of cultural diversity in the world and how seriously cultural differences contribute to producing violent conflicts.

In fact, many purposes, patterns, and practices are shared by all—or nearly all—peoples of the world. There is an emerging global ethic, a set of shared values that includes:

- Individual responsibility.
- Treating others as we wish them to treat us.
- Respect for life.
- Economic and social justice.
- Nature-friendly ways of life.
- Honesty.
- Moderation.
- Freedom (expressed in ways that do not harm others).
- Tolerance for diversity.

The fact that deadly human conflicts continue in many places throughout the world is due less to the differences that separate societies than to some of these common human traits and values. All humans, for example, tend to feel loyalty to their group, and may easily overreact in the group's defense, leaving excluded "outsiders" feeling marginalized and victimized. Sadly, too, all humans are capable of rage and violent acts against others.

In past eras, the killing and destruction of enemies may have helped individuals and groups to survive. But in today's interconnected world that is no longer clearly the case. Today, violence and aggression too often are blunt and imprecise instruments that fail to achieve their intended purposes, and frequently blow back on the doers of violence.

The long-term trends of history are toward an ever-widening definition of individual identity (with some people already adopting self-identities on the widest scale as "human beings"), and toward the enlargement of individual circles of caring to embrace once distant or despised "outsiders." These trends are likely to continue, because they embody values—learned from millennia of human experience—that have come to be nearly universal: from the love of life itself to the joys of belonging to a community, from the satisfaction of self-fulfillment to the excitement of pursuing knowledge, and from individual happiness to social harmony.

How long will it take for the world to become a community where every human everywhere has a good chance to live a long and satisfying life? I do not know. But people of good will can do much today to help the process along. For example, we can begin by accepting responsibility for our own life choices: the goals and actions that do much to shape our future. And we can be more generous and understanding of what we perceive as mistakes and failures in the choices and behavior of others. We can include all people in our circle of concern, behave ethically toward everyone we deal with, recognize that every human being deserves to be treated with respect, and work to raise minimum standards of living for the least well-off people in the world.

We can also dare to question our personal views and those of the groups to which we belong, to test them and consider alternatives. Remember that knowledge is not constant, but subject to change in the light of new information and conditions. Be prepared to admit that anyone—even we ourselves—can be misinformed or reach a wrong conclusion from the limited evidence available. Because we can never have all the facts before us, let us admit to ourselves, whenever we take action, that mistakes and failure are possible. And let us be aware that certainty can become the enemy of decency.

In addition, we can control ourselves by exercising self-restraint to minimize mean or violent acts against others. Let us respond to offered friendship with honest gratitude and cooperation; but, when treated badly by another person, let us try, while defending ourselves from harm, to respond not with anger or violence but with verbal disapproval and the withdrawal of our cooperation with that person. So as not to begin a cycle of retaliation, let us not overreact. And let us always be willing to listen and to talk, to negotiate and to compromise.

Finally, we can support international law enforcement, global institutions of civil and criminal justice, international courts and global peacekeeping agencies, to build and strengthen nonviolent means for resolving disputes. Above all, we can work to ensure that global institutions are honest and fair and that they hold all countries—rich and poor, strong and weak—to the same high standards.

If the human community can learn to apply to all people the universal values that I have identified, then future terrorist acts like the events of September 11 may be minimized, because all people are more likely to be treated fairly and with dignity and because all voices will have peaceful ways to be heard, so some of the roots of discontent will be eliminated. When future terrorist acts do occur—and surely some will—they can be treated as the unethical and criminal acts that they are.

There is no clash of civilizations. Most people of the world, whatever society, culture, civilization, or religion they revere or feel a part of, simply want to live—and let others live—in peace and harmony. To achieve this, all of us must realize that the human community is inescapably bound together. More and more, as Martin Luther King Jr. reminded us, whatever affects one, sooner or later affects all.

POSTSCRIPT

Are Cultural and Ethnic Wars the Defining Dimensions of Twenty-First Century Conflict?

Examining the causes and trends in conflict is a difficult enterprise. Perhaps no area in world politics has been studied and analyzed more than conflict. Arguing that certain forms of conflict are on the upswing or downswing is difficult because we do not have the benefit of historical context. Looking back on the twentieth century, for example, we can clearly see the increasing scope and impact of war; deaths and weapon destructiveness increased at virtually every stage. Determining the likelihood that ethnic conflicts will dominate the security agenda in the coming years, however, is difficult.

The sheer ferocity of ethnic conflicts in the past decade in places like Rwanda, Bosnia, Kosovo, Chechnya, Angola, Afghanistan, and Israel and Palestine seems to support Huntington's thesis that ethnicity and culture will define the battle lines between groups. Yet is culture at work or simply political power? The passion of the combatants may mask the more Machiavellian goals of leaders and individuals.

One element is certain. As the centralized power and influence of large imperial states like the USSR and others diminish, fault lines of conflict that may exist between people in various regions are more readily exploitable, like those in Yugoslavia by former Serb president Slobodan Milosovich. This means that such conflicts are indeed more likely, as leaders take advantage of paranoia, animosity, and fear to create an "other" against which a people may rally, at least for a short time. Whether this is part of a major global trend in ethnic conflict and cultural divides or merely political opportunism is still open to question.

Literature on this topic includes Farhang Rajaee, *Globalization on Trial: The Human Condition and the Information Civilization* (Kumarian Press, 2000); Daniel Patrick Moynihan, *Pandemonium: Ethnicity in International Politics* (Oxford University Press, 1994); Stephen M. Walt, "Building Up a New Bogeyman," *Foreign Policy* (Spring 1997), Edward A. Tiryakian and Said Arjomand, eds., *Rethinking Civilizational Analyses* (Sage, 2004), and David Carment, *Conflict Prevention: Path to Peace or Grand Illusion* (UN University Press, 2003).

ISSUE 19

Can Nuclear Proliferation Be Stopped?

YES: Ira Straus, from "Reversing Proliferation," *National Interest* (Fall 2004)

NO: Mirza Aslam Beg, from "Outside View: Nuclear Proliferators Can't Be Stopped," *SpaceDaily.com* (March 2005)

ISSUE SUMMARY

YES: Ira Straus, U.S. coordinator of the Independent International Committee on Eastern Europe and Russia, argues that non-proliferation has historically been a key component of U.S. policy but that it has been subsumed with terrorism of late, and as such, we have lost precious ground in the fight to maintain non-proliferation. As a result, the threat of states such as North Korea and Iran is dangerous and runs the risk of breaking the prolifera-tion regime's back.

NO: Mirza Aslam Beg, former chief of staff of the Pakistani army, argues that proliferation combined with deterrence leads to stabil-ity and not instability. He contends that proliferation will happen naturally and therefore, the world community should deal with its realities and not try and prevent it, thus creating a set of regional balances that will lead to peace.

In 2006, the world is watching the emergence of two nuclear weapons states, Iran and North Korea. The nuclear club, which currently has eight members (the USA, Russia, Great Britain, France, China, India, Pakistan, Israel) will surely grow by the end of this decade. For fifty years the nuclear weapons states along with others have tried to fashion strategies for containing and eliminating the spread of nuclear weapons. The nuclear Non-Proliferation Treaty of 1968 did much to incentivize states from pursuing this option and it also provided a basis for superpower agreement and cooperation.

However, the fall of the USSR created a power vacuum in that regime while also focusing proliferation strategies as anti-American strategies. With superpower agreement and cooperation, states across the political spectrum were moderated in their pursuit of the nuclear option. Since 1989, many states have focused energies on the acquisition of nuclear weapons as a counter balancing strategy against perceived U.S. hegemony. Iran and North Korea are prime examples of this approach.

Many states during this period pursued this option to varying degrees including Libya, Argentina, Brazil, Iran, and Iraq among others. The capability to produce such weapons technology and then produce the delivery capability requires great expertise, specialized scientific knowledge, and large amounts of cash, billions of dollars. Obviously for many countries this is not an option. However, for some states, this has become a crusade for regional power, national prestige, regime security, and perceived self defense.

With these and other states actively pursuing this option and with greater amounts of expertise and cash to fuel such actions, it seems problematic that proliferation can be halted.

In our first article, Ira Straus calls for a reinvigoration of the non-proliferation regime in the wake of lost opportunities in the previous fifteen years. He contends that now is the time for the western world and other states to enforce non-proliferation and make the acquisition of such weapons counterproductive to regimes that would pursue this option.

Conversely, former General Beg of Pakistan argues that proliferation will occur; it is fine that it does happen and that it is a legitimate response to asymmetries of power in regions such as South Asia and the Middle East. He also sees it as potentially stabilizing to those areas where regional conflicts loom.

YES

Ira Straus

Reversing Proliferation

Of the three interwoven threats to America—terrorists, rogue states and the proliferation of WMD—the third has provoked the least public debate since 9/11. This is curious, since the invasion of Iraq was intended as an exercise in counter-proliferation and the administration has announced a major program to deal with other cases of the spread of WMD. But public debate has focused on the prudence of pre-emptive war and unilateralism, and on whether Iraq had stockpiled WMD in the first place, not on the ways the momentum can be and is being used to overcome further WMD threats in Libya and Pakistan and to strengthen the anti-proliferation regime more generally. The Bush Administration's ongoing program has received little serious attention outside of expert circles, despite eye-catching measures such as the Proliferation Security Initiative, which empowers the United States to board ships suspected of carrying WMD contraband.[1]

Meanwhile Iran and North Korea demonstrate that the old non-proliferation regime is still in crisis. Yet at the same time, wider diplomatic conditions are better than at any point since the 1940s for a realistic policy of not only non- but even counter-proliferation—that is, the use of force to stop proliferation. Measures can be taken in the near future to reverse the impending crisis, and a short capsule history will explain why.

The history of non-proliferation diplomacy falls into three periods. The first was the period of U.S. nuclear monopoly that lasted from 1945 to 1949, during which robust plans for an anti-proliferation regime were proposed. In the late 1940s, the Baruch Plan proposed global management of uranium and UN enforcement actions that would have been exempt from a veto in the Security Council.

. . . The second period of non-proliferation, lasting from 1949 to 1989, was marked by the fact that there was no single power center capable of enforcing the rules. In the early 1960s, John Strachey, an anticommunist British Labour politician and strategist, called on America and the Soviet Union to constitute such a power center together and to enforce tough inspections—through preventive war if necessary. As with the Baruch Plan, however, Cold War rivalry prevented the Strachey Plan from getting off the ground.

. . . During the same period, having created an alliance system within the West designed, among other things, to limit proliferation, the United States

From *The National Interest*, no. 77, Fall 2004, pp. 63–70. Copyright © 2004 by National Interest. Reprinted by permission.

proposed a specific roll-back in the 1960s: a Multilateral Nuclear Force for NATO. This would have integrated French and British nuclear forces with America's and provided a potential core for an ultimate reversion to unity of nuclear forces worldwide. Charles de Gaulle vetoed the idea, however, preferring an independent nuclear deterrent for France. Some Gaullist strategists argued that global proliferation would lead to stability—a view shared by Maoist China and Castroite Cuba and later taken up by the likes of Kenneth Waltz and, of course, by subsequent generations of rogue states.

Thus, while proliferation was not reversed, neither was it allowed to spread unchecked. The NPT-IAEA regime provided regular mechanisms to slow down proliferation while making further proliferation among its signatories illegal—except for the loophole that the treaty could be renounced (as North Korea has recently done). It divided signatories into those acceding as nuclear states and those acceding as non-nuclear states. It perpetuated the discrimination in their status already established. And it enlisted the self-interest of the established nuclear powers in protecting the importance of their nuclear status by limiting the size of their club.

Great Power Responsibilities

It must be emphasized here that the NPT regime reflected and built upon the traditional conception of the special responsibilities of great powers, which goes back at least to the Treaty of Westphalia. The great powers mutually attributed special rights and duties to one another. The small powers accepted the reality, as well as the unwritten prerogative of the great powers to amend occasionally the rules in order to preserve the system's fundamental principles. Indeed, international law was largely the creature of great power practice.

The responsibilities of the great powers naturally grew more urgent with the advent of nuclear weapons. Five powers were formally attributed special rights in the Security Council and given almost unlimited authority to do as they chose—when unanimous—in the name of preserving international peace and security. Other states, accepting the brutal lessons of the two world wars, legally endorsed this extraordinary abrogation of their sovereignty and ratified the UN Charter. Yet when the Cold War broke out, and as long as the Security Council remained divided, the great-power prerogatives were recognized informally as belonging unilaterally to the superpowers. It was the only possible basis for successful nonproliferation.

Yet this was a realpolitik that dared not speak its name. Both superpowers were formally committed to anti-imperialist ideologies derived from their respective revolutions. They took this seriously enough to cooperate symbiotically in ending Europe's overseas empires. They competed for the loyalty of these excolonies and knew no greater accusation to hurl at each other than "imperialist." It would have been hard for them to jointly discuss such an imperialist subject as forceful counter-proliferation.

As long as the Cold War continued, the superpowers were enemies competing for clients, not allies forming a cohesive anti-proliferation core. The reality of competition repeatedly trumped the need for cooperation. Any nuclear-seeking regime would lean to one or the other side, so a preventive attack on it would change the strategic balance between the superpowers. It would have required something close to a miracle to get the superpowers to agree on mutually compensatory terms for proceeding with such a proposed attack. For example, the Soviets reportedly suggested a joint preemptive attack on China in 1969 and 1970; the United States refused, exploiting the situation instead to form its own links with China against the USSR.

The Third Phase

Today the cold war is over—and the third phase of non-proliferation is unfolding before our eyes.

Its most important features are the abatement of Soviet-American mutual hostility; the replacement of bipolarity with unipolarity (allowing tighter multilateral negotiations and more effective enforcement); and the substitution of small states for great powers as the potential targets of counter-proliferation. At the same time, the urgency of curbing proliferation has also grown. Nuclear knowledge and material are spreading, and rogue regimes and Islamist terrorists are actively seeking to acquire them.

But standards, once lowered, are not easy to raise again. A major jolt was needed to send them back upward. The Iraq War supplied the jolt. Libya, a long-intractable case, finally moved to renounce its nuclear (and great power) ambitions. Diplomacy alone would not have achieved this.

Nor would war alone have sufficed. In fact, the Gulf War already supplied an appropriate jolt in 1991, but that opening was not sufficiently exploited. With the Cold War drawing to a close and the trend toward cohesion on the Security Council growing, there could have been a joint Russian-American-led effort to clean up the rubbish they had strewn around during the Cold War. The moral energy was available for a new post-Cold War partnership. Global opinion would have welcomed almost anything that could have been presented as getting rid of "Cold War relics." But it was not to be.

The Gulf War revealed the holes in the old IAEA inspections regime. It led to creation of a new and more intrusive regime of inspection (UNSCOM, for the special case of Iraq). It introduced further discrimination into non-proliferation policy. It applied not to everyone but to a specific known offender. It was asymmetrical. It was put in the terms of truce for a defeated small country, thus perpetuating the war's abridgement of Iraq's sovereignty, and it was applied on that country by the Security Council making use of its unique legal authority over all other states. This enabled UNSCOM to be unusually successful in uncovering and destroying Iraq's WMD programs—although only with the help of further Anglo-American bombings and high-level defectors.

As time passed, the disunity of the Security Council and new forms of great-power rivalry came back into play. The Security Council's unit-veto structure made it ill-suited for carrying out a complex and costly program consistently, for states came to view their respective national interests differently.

In a sense, UNSCOM was too discriminatory. Remedial anti-proliferation was needed in more places than Iraq. Discrimination needed to be raised to the level of a norm applied to all nuclear-seeking rogue states. Several countries grew tired of enforcing the sanctions regime against Iraq in the course of the 1990s. As the UNSCOM regime applied only to a single case, it had differential impact on the great powers of the Security Council. The Iraqi government was able to play upon this, as well as upon resentment of America as the de facto leader and arbiter of the process that the Security Council and UNSCOM were in effect fronting.

Had an UNSCOM-type regime been extended to all the "states of concern", it would have aimed to curb proliferation globally and thus brought the national interests of great powers closer together. It would have recognized, for example, Russia's list of states of concern—Pakistan, Afghanistan, Saudi Arabia-a list whose validity the United States came belatedly to recognize after 9/11— alongside Iraq, Iran, Syria, Sudan, Libya and North Korea on the American list. Such bargaining could have made for mutual support in dealing with both sets of rogue states, instead of inspiring Russia to launch resentful polemics against America's unilateral decisions on which countries counted as "rogues."

But this was not done in the 1990s. The decade passed. The problems grew worse.

Today, as in 1991, we can afford neither diplomacy without war nor war without diplomacy. Where there has been significant success, it is because the coercive action on Iraq has been supplemented with semi-coercive diplomacy, and because that coercive diplomacy has supplemented the work of multilateral treaties and institutions. And there are signs that the United States—despite its alleged propensity to "unilateralism"— has taken these lessons to heart.

President Bush announced a six-pronged initiative to strengthen the NPT regime in February 2004. Taken together, its six points add up to a substantial strengthening of the non-proliferation system. It combines recent ad hoc efforts such as the Proliferation Security Initiative. It tacks on multilateral measures such as the "additional protocol" with the IAEA, which can be agreed to one country at a time, and it joins them to a series of unilateral measures and voluntary collective measures (through the forty-nation Nuclear Suppliers Group). Thanks to this realistic mesh of different levels of work, progress can be made before all countries have signed on—a reality that actually provides an incentive for countries to sign on. It is also realistically discriminatory—indeed, it creates a third tier in the nuclear hierarchy: countries that do not enrich uranium. It does not go nearly as far as the salto mortale Baruch Plan, but it could be described as its partial revival, adapted in light of the realities of the intervening half-century.

The six points received immediate endorsements from men like Mohamed ElBaradei, Director of the IAEA, and from the administration's inveterate critic at the Carnegie Endowment, Joseph Cirincione. The endorsements indicate a broad base of support for follow-through. At the same time, Cirincione called for greater balance between non-proliferation spending (very small), counter-proliferation spending (huge, counting Iraq), and consequence management spending (huge also, in ballistic missile defense and Homeland Security). This is a useful conceptualization. It suggests that a more fullservice approach would help in consolidating elite support for the administration's counter-proliferation innovations. Carnegie has also put forth its own collective draft plan, entitled. Universal Compliance: A Strategy for Nuclear Security, for following up on the anti-proliferation momentum. It calls for taking the successful UNSCOM-UNMOVIC experience of intrusive inspections and extending it to other countries. It acknowledges that the threat of military coercion has been necessary for these successes. It draws heavily on the supranational rights of the Security Council to impose non-proliferation obligations on all countries, urging the use of this to prohibit withdrawal from the NPT and make its norms binding on all. It offers a comprehensive "strategy synthesiz[ing] some innovative approaches of the George W. Bush administration, the benefits of the traditional treaty-based regime, and many new elements."

Both sides of the debate thus agree that more should be done to generalize the recent advances based on war, threat and intrusive inspections. The degree of consensus is encouraging, as was the momentum of action after the Iraq War. Without it, the jolt of the 2003 war could have been mostly wasted, as was the jolt of the 1991 war.

If there is to be an effective and durable non- and counter-proliferation effort, diplomatic gains must be embedded permanently in agreements and institutions. Existing organizations have to be reinforced and supplemented where necessary. Skeptics will point out that such bodies can become obstacles to action—as the United States feared was true in the run up to the Iraq War. Such fears are not always wrong. When an international body is crippled by great-power rivalry, it is not scandalous for other international bodies, or core countries, to act in its stead. More use can be made of devices such as coalitions of the willing inside NATO, and the (little known) absence of a legal requirement of consensus in NATO decision-making.

Cohesion is not consensus; it needs to be understood in the future more in terms of getting the job done for the common good and less in terms of waiting for agreement by every member. Cohesion is deepened when decisions can be taken on a timely basis and implemented by those willing to do so, as long as the minority pledges not to undermine the majority. Such coalition actions do not bypass the alliance but enhance it. Bypass operations are also sometimes necessary, but should not be overused. The core cohesion among the advanced technological allies (the NATO/G-8 plus group) has to be maintained, indeed upgraded; otherwise we could never keep up consistent economic and political pressures on would-be proliferators—and so keep down the cases of military enforcement to a workable remainder.

We also need to modernize our general thinking on proliferation in the light of these new realities. The NPT's leveling demand for universal disarmament was always unrealistic; it is now obsolete. The superpowers have drastically reduced their arsenals, and the danger from them is not in their numbers but in their security. As the former Russian parliamentarian and non-proliferation think-tanker Alexei Arbatov pointed out in June, only a supranational world government would make it possible to pursue general nuclear disarmament without destabilizing effects along the way. Today the NPT's disarmament demand could be replaced with an obligation for the nuclear powers to manage their arsenals better, take them off hair-trigger alert, coordinate them with joint planning, and even—though this is a very long-term goal indeed—aim at their ultimate integration. This is the only way back to the Baruch Plan, which aimed at unified control of nuclear resources and dangers.

For the moment, this ultimate vision seems almost as utopian as disarmament itself. But some aspects of it, as we have seen, are realistic and necessary now. It provides the right orientation for practical steps. General acceptance of Western global leadership is a condition for the success of non-proliferation. In any social order, norms can be efficiently enforced from above only if they are broadly supported and most of the time mutually enforced horizontally. Fortunately, the leadership role of the NATO/OECD/G-8 countries is more widely accepted than it may seem at first sight. They have long been the leaders in global modernization and in providing the public goods of global security and efficient trade. Awareness of this reality has grown as the Western countries have ended their former centuries of internecine warfare. This awareness is in turn illustrated by the very fact that they have come to organize themselves as a core of global cohesion. Legitimacy has accrued accordingly. The old Cold War controversies have dissipated. The UNT has come to accept NATO's regional and global roles. Other nations seek nowadays to be included in their joint structures, rather than trying to overturn them.

The period from the end of Soviet communism to 9/11 was a lost decade for non-proliferation. With the Gulf War, the process of reversing proliferation seemed to be getting underway, but little was done afterwards. Two administrations let opportunities slip through their fingers. Clients of the two sides of the Cold War proceeded to spin out of control and develop independent strategic identities. Islamism got a second wind. Terrorists gained global reach and WMD ambitions. The dangers metastasized.

September 11 finally changed that—with preventive war, in cooperation with Russia (among other states), against the Taliban, coercive diplomacy over Iraq, and then war. Despite embarrassments over Iraq, the process has continued with semi-coercive diplomacy against proliferation elsewhere—diplomacy that has borne unprecedented if uncompleted fruit in Libya and Pakistan. But Iran and North Korea remain as serious unsolved problems and potential crises. If both countries are persuaded by some combination of diplomacy and the use of force to abandon their nuclear programs, then the rollback of nuclear proliferation will have begun. If not, the other countries in those neighborhoods will inevitably acquire nuclear weapons for self-defense. Europe may instinctively prefer diplomacy to force in countering proliferation; the United States may reason that force is needed to help diplomacy work. Whatever their differences, both Europe and the United States jointly have the best chance since 1965 to institute a serious regime of counter-proliferation. It may also be their last.

Note

1. A notable exception to this is Graham Allison & Andrei Kokoshin, "The New Containment," *The National Interest* (Fall 2002).

Mirza Aslam Beg

 **NO**

Outside View: Nuclear Proliferators Can't Be Stopped

Early last year when Abdul Qadeer Khan was targeted for alleged nuclear proliferation, I was also implicated and remained under the world media's focus.

During an NBC TV network interview, I was asked the question whether I would like my future generations to live in this part of the world, which is threatened by nuclear holocaust.

I said: Yes, certainly, I would like my future generations to live in South Asia where I see no threat of nuclear war, because perfect nuclear deterrence holds between India and Pakistan. But certainly I would not like my future generations to live in the neighborhood of "nuclear capable Israel."

He questioned: In that case would you like to pass on the nuclear capability to Iran, which considers itself threatened by Israel? I said no. Countries acquire the capability on their own, as we have done it.

Iran will do the same, because they are threatened by Israel. The media hype and the consequences of the reported nuclear proliferation, led to the tormenting treatment meted out to Khan.

For a long time, Americans and Europeans have been engaged in nuclear proliferation, as part of a concept, called "outsourcing nuclear capability," to friendly countries as a measure of defense against nuclear strikes. The concept is interesting as well as regrettable.

The Natural Resource Defense Council of the United States reveals in its report that: "A specific number of nuclear warheads, under U.S. and NATO war plans, will be transferred to America's non-nuclear allies to be delivered to targets by their warplanes.

"Preparations for delivering 180 nuclear bombs are taking place in peacetime, and equipping non-nuclear countries with the means to conduct nuclear warfare, (which) is inconsistent with today's international efforts to dissuade other countries from obtaining nuclear weapons.

"The arsenal is being kept at eight Air Force bases in Belgium, Germany, Italy, the Netherlands, Turkey and Britain. The strike plans' potential targets are Russia and countries in the Middle East—most likely Iran and Syria."

It would be appropriate to call this concept as "enlightened nuclear proliferation" being implemented by those who are responsible for nuclear non-proliferation regime.

After the breakup of the Soviet Union, nuclear scientists and nuclear material of all kinds proliferated: "Half of the nuclear materials, pieces and parts of it, are unaccounted for by the Russians—and a lot of them, are at places in rural areas, which is more threatening to the world right now."

India, "according to international media—February 2004, reported 25 cases of 'missing' or 'stolen' *radioactive* material from its labs to the International Atomic *Energy* Agency.

Fifty-two percent of the cases were attributed to "theft" and 48 percent to "missing mystery." India claimed to have recovered lost material in 12 of the total 25 cases." How innocently simple, is the way of "innocent *nuclear* proliferation."

For quite a while, North Korea has been complaining about nuclear warheads placed in South Korea by the United States, which prompted North Korea to develop its own *nuclear* weapon capability.

In Pakistan's neighborhood, Iran is under tremendous pressure, for allegedly attempting to develop the nuclear weapons, which Iran has denied. The war of nerves between the United States and Iran thus has been going on for quite sometime.

On Feb. 16 very disturbing news was splashed on a Pakistani private TV channel, picked up from Tehran Radio, that 12 of the suspected Iranian nuclear sites had been hit by missiles.

The news was really alarming but gradually it transpired that, some rock blasting occurred in the southern region of Iran, which was taken as missile attacks.

Whether the news was fake or prompted, it did help Iran test the nerves of United States and Israel, because both promptly denied that any such strike was carried out. Thus deterrence between Iran and Israel now has appears to have crossed the threshold of ambiguity, which, indeed is significant.

While Iran has tested the nerves of its adversaries, North Korea has corrected the imbalance in South East Asia by declaring its capability. Since both are termed "rogue states," it would be proper to call it "rogue nuclear proliferation."

Nuclear deterrence between Iran and Israel has crossed the psychological barrier. Nuclear deterrence between South and North Korea has already been established.

Therefore, the nuclear fault line of the 21st century, now extends from Israel to Iran, Pakistan to India and South Korea to North Korea, while the strategic balance is held by the United States and Europe on one side and Russia and China on the other.

The nuclear non-proliferation regime, therefore, is dying its natural death at the hands of those who are the exponents of the nuclear non-proliferation regime.

How the new balance of terror will be maintained from Mediterranean to Pacific is a task for those who, having themselves violated the nuclear proliferation regime, are now responsible for maintaining global nuclear peace.

The world now has to wait and see how objectives of the utopian nuclear non-proliferation regime are achieved.

At the beginning of the new era, the emerging multipolar world order is facing the formidable challenge of a dangerous global nuclear security paradigm.

Fortunately, the emerging multipolar global order is expected to be less confrontational than the bipolar world order and less brutal and tyrannical than the unipolar world order of today.

With at least six competing geo-economic centers of power, the new world order would be more democratic in nature as it would be governed by forces of globalization and integrative economic demands.

Such democratization of the world order will bring sanity into the entire gambit of nuclear proliferation. The "enlightened proliferators" together, with the "innocent," and the "rogue proliferators," would democratize the global nuclear non-proliferation order. This may be the only hope for all living beings inhabiting this wretched earth.

POSTSCRIPT

Can Nuclear Proliferation Be Stopped?

Nuclear weapons represent real power in the world of international relations. The possession of such weapons gives the possessor real and perceived status as a regional and global player. In addition, these weapons give a regime a potential trump card against overthrow or invasion by external powers, be they regional adversaries or the United States. This makes the acquisition of such weapons tempting, albeit extremely expensive and time consuming.

The non-proliferation regime of the past forty years has worked in that many states with such a capability have not pursued the nuclear option. The partnership between the United States and the Soviet Union maintained that regime because they could provide a series of incentives and disincentives with client states to avoid pursuing the nuclear option. Now that the bipolar cold war is over the dynamics of international relations and the non-proliferation regime have changed.

States that see themselves as adversaries of the United States or at least are leery of its motives are exploring ways of solidifying their power against U.S. action. The Iraq invasion, while aimed at overthrowing an avowed U.S. enemy, may have actually created greater impetus for other adversary states (Iran and North Korea) to pursue nuclear weapons as a balance against perceived American hegemony. Consequently, the non-proliferation regime of the past forty years has been compromised. Whether increased proliferation will result is still a guess but the trends are not positive. And as many critics have pointed out, more weapons mean a greater chance of nuclear action by states or clients of states.

Some important literature on this subject includes *The Nuclear Tipping Point*, edited by Kurt M. Campbell, Robert Einhorn and Mitchell Reiss (Brookings Institution Press, 2004), *War No More: Eliminating Conflict in the Nuclear Age*, Robert Hindle and Joseph Rotblat, (Pluto Press, October 2003), *The Gravest Danger*, Sidney Drell and James Goodby, (Hoover Institution Press, 2003) and *The New Nuclear Danger*, Helen Caldicott, (New Press, 2002)

ISSUE 20

Has U.S. Hegemony Rendered the United Nations Irrelevant?

YES: Tom DeWeese, from The Time Is Now . . . The United Nations—Irrelevant and Dangerous," *American Policy Center* (March 28, 2003)

NO: Shashi Tharoor, from "Is the United Nations Still Relevant?" *Asia Society* (June 14, 2004)

ISSUE SUMMARY

YES: Tom DeWeese argues that the United Nations is principally responsible for much of the chaos in the world and that its ability to positively impact change is negligible. Therefore, it is irrelevant to U.S. policy, and thus global interests, and should be dismantled.

NO: Shashi Tharoor argues that the United Nations does a great deal of good, and in this age of terrorism and U.S. global hegemony, it is necessary to peacefully negotiate and be a force for reason around the world.

The United Nations is an international organization created in the aftermath of World War II to promote peace, security, and global cooperation. Since its beginnings it has forged into myriad areas of global concern from health care, communication, children's rights, genocide, human conflict, and regional and civil conflicts. During its long history one element remains a constant: there are real disagreements about its utility and effectiveness in the world community.

Proponents of the United Nations have argued that it provides a forum for disagreements to be aired along with aiding the poor and sick through many effective programs. They contend that the United Nations allows all nations to be heard and helps promote greater awareness of issues such as landmines, river blindness, and the Darfur conflict in the Sudan. Proponents state that without the United Nations, power politics and wealth will guide problem solving and thus marginalize many states and millions of people.

The United Nations critics argue that it does little to stem conflicts, halt nefarious governments from exploiting their people, and serves to inhibit action in the world community against a variety of injustices. These people see the United Nations as another form of big, bureaucratized government intent on perpetuating itself rather than acting for the goals that it is purported to be committed to.

With the post-cold war world and the elimination of the communist-capitalist dichotomy, many in the west and especially the United States see the UN as merely an impediment to America's application of power for democratic interests. This was most pronounced on the Iraq issue where pre-emption and unilateralism took hold in the Bush foreign policy despite UN protests and this amplified the debate.

The selections that follow tackle the issue of the UN's relevance. DeWeese argues clearly that the UN has served its purpose and needs to be removed. In his view, the UN inhibits global action to promote freedom and free markets and as such is an impediment to the march of history. As an American, he sees the UN as anti-American and thus not useful to a new world order where American hegemony is relatively unquestioned.

Tharoor, Under-Secretary-General of the UN argues that the UN's long history of constructive action and moderation makes it more important than ever with both new forms of conflict and American hegemony looming. He argues that success is not measured in whether the UN agrees with any one country on any one issue but rather on the body of work that they engage in over time across many issues.

YES

Tom DeWeese

The Time Is Now . . . The United Nations—Irrelevant and Dangerous

The world is in chaos and, quite frankly, it's the United Nations' fault. It gives validity to zealots and petty bigots. It helps to keep tyrannical dictators in power. It provides money and aid to international terrorists. And it sets itself up as the international economic and environmental standard to which all nations are to mirror. The truth is, the United Nations is the root of international trouble, not the answer.

Saddam Hussein is in power, able to threaten world peace today because the United States allowed the United Nations to dictate the terms for the finish to the Gulf War after an American-organized coalition all but annihilated Iraq.

In the intervening ten years, Iraq has time and again broken the terms of that treaty. The UN's response has been to pass 17 resolutions to demand that Iraq behave itself. Those resolutions are toothless and irrelevant because other nations that have economic dealings with Iraq refuse to take action for fear of losing money.

When President Bush went before the UN to make his case against Iraq, he simply was asking the Security Council if it agreed with the U.S. assessment that Iraq "has been and remains in material breach," of the terms of the Gulf War disarmament provisions.

True to form, the UN was unable to act on so clear of a position. Instead, France, Russia and China, all nations who have Iraqi economic ties to lose in a war, and all members of the Security Council, used their power to delay and deny U.S. action.

As a result, the UN resolution #17 does not allow for U.S. action if a breach is found. Instead it demands that the UN basically issue resolution #18.

Delay. Negotiate. Recommend. Study. Reconsider. Do nothing. This is the same game the UN has played in nearly every international crisis. It is the reason North Korea remains a threat and its violent dictator remains in power after 50 years. It's the reason why Zimbabwe's murderous dictator, Robert Mugabe, is able to steal his election and then steal the land of white property owners and still have a voice at the UN's Sustainable Development Conference a few months ago in South Africa. It's the reason why the Communist Chinese are able to ignore any UN rules not to their liking while growing as an international military and economic threat. It's the reason why a terrorist nation like Syria can be given a seat on the UN's Human Rights Council while the United States is removed.

The United Nations' main purpose is to provide voice and power to irrelevant or vicious nations to counter the United States. The UN will never approve U.S. military action against Iraq or any tin horn dictators who threaten the peace, because they are the very dictators who now control UN decisions.

It is, however, the United States that is to blame for this situation, because we allow this circus on the East River to exist. The only credibility the UN possesses comes from recognition by the United States. The only financial security the UN enjoys comes from funds provided by U.S. taxpayers. The only military punch controlled by the UN comes from American military might.

The United States is dutifully providing an elegant clubhouse in which pouting and jealous bureaucrats and self-inflated international diplomats can pretend to matter. As long as the United States allows them to exist and as long as this nation goes along with their demands, they *do* matter. The United Nations is nothing more than a house of cards, but it's a very dangerous house of cards.

The UN is dangerous because its most vocal membership stands in opposition of the American values of controlled representative government, justice, free enterprise, privacy of individuals and private property rights. Most of the UN's membership comes from nations controlled either by communist regimes, kingdoms or mad dictators where American values are either unknown or viewed as a threat.

Those same UN members are busy working to implement plans for UN global governance. Already, the UN's International Criminal Court is in place. The UN has held an international meeting to discuss the possibilities and methods of implementing global taxes. More plans are under consideration to establish a UN global army or police force. Most member states participating in these planning sessions are from brutal dictatorships like China and Cuba and a number of brutal fundamental Islamic states like Syria and Iran. Can any clear thinking American honestly believe that the ideas coming out of this group would have a possibility of favoring ideals readily accepted as rights in the United States?

Many Americans simply do not believe that the United States would voluntarily give up its sovereignty to the United Nations. They say our people would never stand for it. Those who think this way somehow seem to believe the issue will come up for a national vote to determine if the United States will agree to dismantle its federal government in acceptance of UN global governance. If it were that simple, even UN opponents wouldn't worry about the UN threat. Such a thing will never happen.

It happens in increments of well meaning, innocent-sounding policies and treaties. The North American Free Trade Agreement (NAFTA) was sold as simply a way for American producers to broaden their markets to the international level. Instead, many have found that details of the treaty dictate rules and regulations, particularly of the environmental kind, that tilt the playing field to other nations. As a result, American markets are flooded with foreign goods as American businesses and jobs head out of the country. As a result of NAFTA, the American sheep industry has all but disappeared. Other industries may soon follow as the United States continues to cling to this discredited policy.

The European Union was originally sold as another NAFTA through which nations could join together to compete with the United States on the international market. Now, once-proud nations have given up their national sovereignty, ancient currencies like the Italian Lira and the French Franc have disappeared in place of the Euro. Would the citizens of France, Italy or Greece ever have agreed to such a move had the whole plan been put on a ballot?

Now there is discussion of an African Union, a South American Union and a North American Union in which the United States would meld its borders with Canada and Mexico. The move will be easy since NAFTA has already set the precedent.

How long will it be after the establishment of all of these geographical unions before the world moves towards one international union? It will be easy to complete after all of these nations have been through the process of letting go of their national identity. And the United Nations is already putting the pieces together with its World Court, global tax schemes and military planning.

Imagine a world run by the justice of China, with the economics of Cuba and the military might of the United States. Such is the world of the future under the United Nations. The United States holds all of the cards, but it has only one vote in this cesspool of Socialism.

The United Nations exists simply because there is an established mindset that says it's supposed to. It is the established mindset of too many Americans who simply view the United Nations through the televised images of the UN's Security Council, as the President of the United States, hat in hand, asks that body to allow the United States to defend itself against murderers. It is the mindset that they somehow have a legitimacy to grant or deny that request. It is the mind set that Resolution #17 means anything. It is the established mindset of NAFTA, the European Union and the International Criminal Court that are already part of our lives. It is the mindset that accepts the movement toward a North American Union and UN global governance as simply the next natural step. These are the reasons why the United Nations is so dangerous.

The United States can end it all now if it wishes. The mindset must be changed. President Bush has proven that we don't need the United Nations to grant us permission to protect our national interests. The United States can and will fight its own war on terrorism. It can and will organize its own coalition of allies, use its own money, it's own weapons and its own troops to defeat an enemy who threatens us.

The United Nations is irrelevant as a body to deliver world peace. But it is United States' participation in propping up the circus that makes the UN dangerous. Seventeen resolutions should be enough to prove we don't need to spend another dime playing this game.

As the new 108th Congress opens in January, Congressman Ron Paul will once again introduce H.R. 1146, the American Sovereignty Restoration Act. His bill calls for the United States to withdraw from the United Nations. It also calls for the United Nations to remove its headquarters from our shores. H.R. 1146 would relieve the United States from participating in UNESCO and UN environmental policies that endanger our economy and property rights. It would end U.S. participation in UN peace keeping missions, meaning we would no longer be helping to prop up criminal governments and enemies who seek our demise.

As the UN's irrelevance becomes clearer to the nation; as it drags its feet, delays and passes yet anther meaningless resolution, the time has never been better to change the national mindset to say, "Get us out of the UN." The time is now.

 **NO**

Is the United Nations Still Relevant?

It really is a pleasure to be back in this vibrant and dynamic city. For me, it is a gap of 23 years since I was last here and I think a few things have changed but it is certainly a pleasure to speak to this distinguished gathering and not just to be here under the auspices of the Asia Society.

(I have been thrown (the) challenge of trying to justify the relevance of the UN.) Well, a little over a year ago, in fact in March of 2003 as the debates were raging in the UN Security Council with Iraq, a BBC interviewer rather glibly asked me, "So how does the UN feel about being seen as the "I" word, "irrelevant"? And he was about to go on when I interrupted him. "As far as we are concerned", I retorted, "the 'I' word is indispensable."

It was not just a debate in point. Those of us who toil everyday at the headquarters of the United Nations and even more our colleagues on the front lines in the field have become a little exasperated at seeing our institutional obituaries in the press.

Last year's debates over Iraq led some to evoke a parallel to The League of Nations, a body created with great hopes at the end of the First World War but a body which was reduced to debating the standardization of European railway gauges, the day the Germans marched into Poland and began the Second World War.

But such comparisons are, to say the least, grossly overstated. As Mark Twain put it when he woke up in the morning and read the newspaper to discover he was supposed to be dead, the report's of the UN's demise are somewhat exaggerated and yet I have to admit we live with a paradox. In the world's sole remaining superpower, the United States, influential sections of political opinion are all too ready to write us off.

Ironically, independent public opinion polls over many years have consistently found that ordinary Americans have great faith in the UN and in multi-lateral solutions to world problems, but no one reads, shall we say, the Asian Wall Street Journal. Could doubts that US leaders and legislators have not always shared their constituent's faith.

On a visit to Washington DC last year, I asked a distinguished Washingtonian what lay behind all the hostility and criticism that we kept hearing towards the UN in that little capital. I said, "Did the critics not understand what we were all about? What was the problem here", I said, "was it ignorance or was it apathy?" And he replied, "I do not know and I do not care", which exactly sums up the problem.

But of course I am assuming you are all here either because you do know or because you want to, and certainly your presence shows that you care. Let me say that I am not here just to pick on the United States. God knows that would be a contest or unequals. One has to accept that most states act both unilaterally and multi-laterally. The former in defence of the their national security or in their immediate backyard, the latter in pursuit of global causes. For the larger a country's backyard, the greater the temptation to act unilaterally across it, and this I think in recent years has been rather acute in the case of the United States.

But I would argue today that even the United States with its enormous economic and military reach needs to act in concert with other states to achieve its real objectives. I want to suggest to you today that the United Nations remains the best means to achieve multilateral co-operation in so many important areas.

But before rising to answer the challenge, the question that Ronnie put to me, let me briefly, perhaps at least for the sake of the students in the room, venture briefly back into history because I have always believed the best crystal ball is a rear view mirror.

The United Nations was founded during a period when the world had known almost nothing but war and strife, book-ended by its two savage World Wars that began within 25 years of each other. In the first half of the 20th century, people in most parts of the world scarcely had the luxury of deciding whether they were interested in world politics. World politics took a thoroughly intrusive interest in them. Horror succeeded horror until in 1945 the world was brought face to face with the terrible tragedies brought by war, fascism, nuclear bombing and attempted genocide. Hiroshima and the holocaust. Had things gone on like that, the future of the human race would have been bleak indeed.

But happily we did not go on like that. We all know that the second half of the 20th century was far from perfect. Lots of things have gone wrong including, in this corner of the world—in Asia, we can all recount the horrors of the rule of the Khmer Rouge in Cambodia, the partition of India; we can each pick own examples. We also know that in other parts of the world there are still a lot to be done. Billions of people living, for example, in extreme and degrading poverty.

But sitting here in Hong Kong, I think you would accept the proposition that the overall record of the second half of the 20th century is one of amazing advances. The world economy not only recovered from the devastation of 1945, it expanded as never before. There was astonishing technological progress. Many in the industrialized world now enjoy a level of prosperity and have access to a range of experiences that their grandparents could scarcely have dreamed of.

Even in the developing world there has been spectacular economic growth, child mortality has been reduced, literacy has spread, the peoples of the so-called third world threw off the yoke of colonialism and those of the Soviet Bloc want political freedom. We know that democracy and human right are not yet universal but they are much more the norm than the exception.

My question to you is: did all this happen by accident? No, it happened quite simply because in and after 1945, the group of far-sighted leaders, statesmen and stateswomen, were determined to make the second half of the 20th century different from the first. They drew rules to govern international behavior and they founded institutions for which different nations could cooperate for the common good to foster international relations, to elaborate consensual global laws and to establish predictable universally applicable rules for the benefit of all. This was precisely to avoid what had happened in the first half the 20th century when these norms did not exist and were not entrenched. The keystone of the arch, so to speak, charged with keeping the peace between all nations and bringing them together in the quest for freedom and prosperity was the United Nations themselves.

It was very much explicitly in the vision of the UN's founders, particularly the great American president Franklin Delano Roosevelt and an explicit alternative to the disastrous experiences of the first half of the century. And in the view people like Roosevelt, the UN stood for a world in which people of different nations and cultures looked on each other not as objects of fear and suspicion but as potential partners, able to exchange goods and ideas to their mutual benefit.

When the successor, President Harry Truman, signed the United Nations charter in San Francisco just 59 years ago this month, he said very clearly: "If you seek to use this instrument selfishly for the advantage of any one nation or any one small group of nations, we shall be betraying the ideals for which the United Nations has been founded".

That was then of course and today, 59 years later, it may be difficult to recognize in the voices of the critics of the United Nations, the voice of an American president like Harry Truman. In fact let me quote something else he said in that same speech: "We all have to recognize", Truman declared, "no matter how great our strength but we must deny ourselves the license to do always as we please. No one nation can or should expect any special privilege which harms any other nation. Unless we are willing to pay that price, no organization for world peace can accomplish its purpose and what a reasonable price that is."

Now, that is an American president in 1945, and I suspect there are many in Washington today who would not agree that this is indeed a reasonable price for the world's only superpower to pay. In the interest of something as war versus world peace, especially in an era of terrorism. Indeed especially after 9/11, the very viability of the organization has suffered most in the United States.

The American critic Charles Krauthammer has even described the UN as a bunch of Lilliputians tying down an American Gulliver with a thousand strings. Others like the American political scientist Robert Kagan has suggested that the UN offers a naively Kantiant aura, an inadequate response to a Hobbesian world that calls for a leviathan, not a weak-kneed, peace organization like the UN.

I must reply to them that the UN is a response to a Hobbesian world, a world in which we have seen the horrors of the First and Second World Wars. The UN charter was the product of those who won that victory at the end of the Second World War, converting where war-time alliance into a peace-time organization, and they saw the horrors and vowed never again.

But their solution was not to create a single power to be a leviathan. It was to create a system of laws that would ensure that the world of the second half of the 20th century and indeed now, the world we have inherited in the 21st would be a better place than the one that had preceded the creation of the United Nations.

I have to admit right away that in making the comment Ronnie made, we are all influenced, as I am sure you were, by what happened in the course of last year when the Security Council debated Iraq, did not agree on the case for war and the US and Britain and a few other allies went ahead with the war anyway. That was when the talk of the UN's irrelevance mounted to a crescendo.

Indeed a Pew Poll—an independent polling organization—taken in 20 countries last summer showed that the UN had suffered perhaps the greatest collateral damage over Iraq. According to this poll, the image of the UN, the credibility of the UN was down in all 20 countries. Down in the US because we did not support the administration in the war and down in the 19 other countries because the UN could not prevent the war. So we were hit from both sides in this debate. I have to say that even more recent polls taken this year confirm that we have indeed taken a battering. Our supporters are fewer than perhaps ever since the founding of the organization.

Next year is the 60th anniversary of the United Nations. Ronnie has been talking about 70 years old today. There are of course in many organizations including the UN, a retirement age of 60 and the question comes up: does the UN too have to start contemplating retirement, have we fallen from grace? No, far from it.

First of all and perhaps most important, even the US has come back to the UN and on Iraq. You all followed last week's unanimous Security Council vote in the news. Well, it is the culmination of a long process. Soon after the end of the war, in fact in May last year the Security Council unanimously adopted a resolution which called on the UN Secretary General to establish a presence in Iraq, to act independently but also in coordination with the occupying powers.

Then in August it passed another resolution giving the UN significant tasks in post-war reconstruction. The horrors of the bombing in Baghdad later prevented us from doing much more immediately after those resolutions on the ground. But the very submission of these resolutions by the US to the Security Council was an acknowledgment by Washington that in Secretary General Kofi Annan's words, there is no substitute for the unique legitimacy provided by the United Nations.

The in fact has been thrown into stark relief this year by the confidence and trust the US then placed on the shoulders of the Secretary General's special adviser, Lakhdar Brahimi who went out and mid-wifed the birth of the new Iraqi government. He identified those who would lead Iraq when the American occupation ends on the 30th June.

The unanimous adoption of last week's resolution endorsing Mr Brahimi's work and his recommendations and like the other resolutions, once supported even by those council members who opposed the US intervention, demonstrates their understanding of the importance of collective action.

In fact, we used to have a nasty little joke about the agreements of the Security Council on all these matters where it was said there was an argument between an American diplomat and a French diplomat and a particular problem, and the American diplomat said, "You now how we can solve this? We can do this and this and this and we can solve it", and the French diplomat replied, "Yes, yes, yes, that will work in practice but will it work in theory?" You can say that now the theory has been agreed. There really is consensus on the basic principles on how to proceed. The divisiveness of last year has been put behind us. The great challenge is: will it work in practice? So on that the jury is still out but I think we have grounds for cautious optimism that we did not have before.

I think that it is also worth mentioning that the tired old League of Nations analogy does not apply because by the late 1930s when the League was falling apart, two of the three most powerful countries in the world, the US and Germany, the third being Great Britain of course, did not even belong to the League of Nations, which therefore had no influence on their actions. The League died because it had become truly irrelevant to the global geopolitics of the era.

By contrast, every country on earth belongs to the United Nations including the world's only superpower. Every newly independent state seeks entry into the UN almost as its first order of government business. Its seats in the UN is its most fundamental confirmation of its membership in the committee of nations.

The United Nations is now even as so essential to the future of the world that even Switzerland, long a holdout because of its fierce neutrality decided by referendum in 2002 to end its isolation and join the UN. No club that attracts every possible eligible member can easily be described as irrelevant.

Third, the authorization or not, as it happens, of war in Iraq is not the only gauge of the Security Council's relevance to that situation. Just 5 years ago, for example, the NATO alliance bombed the former Yugoslavia over its government's conduct in Kosovo, without the approval or even reference to the Security Council. My interviewer's "I" word was heard widely in those days also. Kosovo, it was also said, had demonstrated the UN's irrelevance.

But as soon as the bombing was over Kosovo returned immediately to the Security Council when arrangements had to be found to administer the territory after the war. Who but the United Nations could confer international legitimacy on the arrangements that were required and encourage other nations to support and resource the enterprise.

Iraq in other words is not the first time that the United Nations has been written off by some during the war, only to be found essential to the ensuing peace. The UN offers a legitimacy that no country or ad hoc coalition can muster for itself.

Many member states with a long history of committing troops to support international peacekeeping efforts, countries like Canada or India, even Pakistan, declined US requests to provide soldiers for Iraq because they were not prepared to act without a protective shield of a UN mandate.

In any case, shortly after the war, Washington discovered and has continued to discover in Iraq that the US is better able to win wars alone than to construct peace alone.

Military strength always has its limitations and in the area of nation-building. The great French statesman Talleyrand once said, the one thing you cannot do with a bear net is to sit on it. So I am convinced that the rebuilding of Iraq would increasingly be an international project.

And whatever happens in Iraq, let us also not forget that the relevance of the United Nations does not stand or fall on its conduct on any one issue alone. When this crisis has passed, the world will still be facing, to use Secretary General Kofi Annan's phrase, innumerable problems without passports. Problems cross frontiers uninvited. Problems of the proliferation of weapons of mass destruction certainly but also of the degradation of our common environment, of contagious disease and chronic starvation, of human rights and human wrongs. The mass illiteracy and massive displacement.

Robert Kagan's famous and rather fatuous proposition that Americans are from Mars and Europeans are from Venus has gained wide currency these days. But even if we were for one moment to set aside Asians, it begs the question, where are Africans from, Pluto? And yet, as I was just saying to President Turner a little earlier over lunch, the tragic confluence of AIDS, famine, drought and poverty in parts of Africa threatens far more human lives than the crisis in Iraq ever did. And without the UN the world would not tackle these problems.

All of these are problems that no one country, however powerful, can solve on its own and which are the shared responsibility of humankind. They cry out for solutions but like the problems themselves, also cross frontiers.

The United Nations exists to find these solutions through the common endeavor of all states. That is why I am proud to assert that it is the one indispensable organization we have in our globalizing world. And no, it is not perfect, it has acted unwisely at times, it has failed to act in others. When we only think of what happened in Bosnia or genocide in Rwanda for instances of each kind of failure. It has sometimes been too divided to succeed, as was the case with Iraq.

But of course the UN is at its best a mirror of the world. It reflects our divisions and disagreements as well as our hopes and convictions. It is folly to discourage an entire institution on the basis of a few occasions where it does not succeed. Because in fact, the UN is both a stage and an actor. It is a stage on which the member States play their parts, claiming their differences and their convergences, but it is also an actor in the shape of a Secretary General, his operations in the field, humanitarian agencies, staff like myself who are going out there and executing the policies that have been made on that stage.

So when you do not like the policies, it is sometimes convenient to blame the actor for getting what was discussed on the stage or it was not agreed sometimes on the stage.

Kofi Annan sometimes jokes that the acronym by which he is known inside the organization as Secretary General, "SG", should actually stand for scapegoat. That is something that we sometimes see as a role that the UN is required to play. But the fact is that the UN's records of success and failure is indeed better than that of many national institutions.

Those who criticize the UN appear to be working on the basis of a yard-stick, that it must succeed all the time. Well, sometimes we only muddle through but as Dag Hammarskjold, the UN's great second Secretary General put it, the United Nations was not created to take humanity to heaven but to save it from hell. That it has innumerable times.

How quickly we forget that during the Cold War the UN played the indispensable role of preventing regional crises from igniting a superpower conflagration and even while they were disagreeing on Iraq, the member States of the Security Council, the same ambassadors during the same few weeks were agreeing on a host of other vital issues, from Congo to Cypress to Afghanistan, East Timor, all of these were challenges or problems that matter. Why do we only focus on the one thing that the western media decides should get the headlines: Iraq.

Again, I was saying to Mrs Turner that part of my job in trying to promote the UN around the world and explain its policies is to draw attention to stories that are not in the headlines. I launched last month a list of 10 stories the world ought to hear more about but which were being ignored because of the obsession with Iraq. Well, to no great surprise, that list of 10 stories was also ignored by and large by the mainstream media, though I must say we did get a few minutes on CNN and on some of the more high-winded public service broadcasters in the US.

But the simple message I want to end with is that since I mentioned the US at the very beginning, when the people of the US and elsewhere ask why the world's sole superpower should even need an organization like the UN, the answer is simple: global challenges demand global solutions.

Somebody once said about water pollution, we all live downstream. There is no way that we can escape the challenges around the world.

So I am going to really close by saying that the UN fundamentally helps establish the form norms that we would all like to see the world live by, that we have tried to entrench since the Second World War. People and nations around the globe have tried through these last six decades to strengthen the foundations of stability and to unite around common values and in the process the UN has brought humanitarian relief to millions in need, it has helped people to rebuild their countries from the ruins of armed conflict, it has challenged poverty, it has fought apartheid, it has protected the rights of children, it has promoted de-colonization and its placed environmental and gender issues at the top the world's agenda. So in other words it becomes the one place where we can actually work together to dream the same dreams together.

You now there is a dreadful old story about Adam and Eve in the garden of Eden, and it is said that Adam, when he found that Eve was becoming a bit indifferent to him—Adam said to Eve, "Eve, is there someone else?" You think about that for a minute, because you could ask the same question about the United Nations. Is there another institution that brings together all the countries of the world to work together for common objectives in all our collective interests; there clearly is not.

We may not be perfect but in the name of our common humanity, let us work together to make this the one United Nations we have, the best possible United Nations there can be.

Thank you very much.

POSTSCRIPT

Has U.S. Hegemony Rendered the United Nations Irrelevant?

The world has changed since 1989. The fall of the Soviet Union not only created a profound change in international relations but it changed how numerous states interact with reach other and with the lone remaining superpower, the United States. Also, the collapse of the bipolar world changed how the United States views itself, its allies, regional issues and the uses and applications of power. The United States is both less restrained and also more confused as to where and how force and influence should be applied for its interests and for international concerns. The Iraq war accentuates this tension and dynamic.

The United Nations represents a global entity that, while clearly contentious, has a global voice on numerous issues. This at times places it in alliance with and against U.S. interests. Critics are quick to point out that it "gets in the way" of the applications of U.S. power for good and thus is irrelevant or at worse an obstacle. Supporters argue that it is both a counter weight to U.S. hegemony and a venue for the exploration of global issues designed to serve global and not just western-U.S. interests.

The question of relevance is clearly a matter of where you stand on western/U.S. policy. If you are skeptical of such policies and fearful of U.S. hegemony then the UN becomes very relevant. If you support American activism in foreign affairs, for example Iraq, then you tend to see the UN as irrelevant. Whichever side you may be on, one thing is certain, the United Nations long ago institutionalized itself in the global community and a dismantling of that entity is not likely under any scenario.

Some important literature on this subject includes, *Global Deception: The United Nations and Stealth Assault on American Freedom*, Joseph Klein (World Affairs Publishers, 2006), *The Parliament of Man: The Past, Present and Future of the United Nations*, Paul Kennedy, (Random House, 2006), *The United Nations Exposed: How the United Nations Sabotages American Security and Fails the World*, Eric Shawn, 2006 and Terrell E. Arnold, *"On the Relevance of the United Nations,"* (Rense.com, 2003).

Contributors to This Volume

EDITORS

JAMES E. HARF is a professor of government and world affairs at the University of Tampa where he also serves as director of International Academic Programs. He spent most of his career at The Ohio State University, where he holds the title of professor emeritus. He is coeditor of *The Unfolding Legacy of 9/11* (University Press of America, 2004) and coauthor of *World Politics and You: A Student Companion to International Politics on the World Stage*, 5th ed. (Brown & Benchmark, 1995) and *The Politics of Global Resources* (Duke University Press, 1986). He recently published his first novel, *Memories of Ivy* (Ivy House Publishing Group, 2005) about life as a university professor. He also coedited a four-book series on the global issues of population, food, energy, and environment, as well as three other book series on national security education, international studies, and international business. His current research interests include the world population problem and tools for addressing international conflict. As a staff member of the Presidential Commission on Foreign Language and International Studies, he was responsible for undergraduate education. He also served 15 years as executive director of the Consortium for International Studies Education.

MARK OWEN LOMBARDI is president and chief executive officer at The College of Santa Fe. He is coeditor and author of *The Unfolding Legacy of 9/11* (University Press of America, 2004) and coeditor of *Perspectives of Third World Sovereignty: The Post-Modern Paradox* (Macmillan, 1996). Dr. Lombardi has authored numerous articles on such topics as African political economy,

STAFF

Larry Loeppke	Managing Editor
Jill Peter	Senior Developmental Editor
Susan Brusch	Senior Developmental Editor
Beth Kundert	Production Manager
Jane Mohr	Project Manager
Tara McDermott	Design Coordinator
Nancy Meissner	Editorial Assistant
Julie Keck	Senior Marketing Manager
Mary Klein	Marketing Communications Specialist
Alice Link	Marketing Coordinator
Tracie Kammerude	Senior Marketing Assistant
Lori Church	Pemissions Coordinator

U.S. foreign policy, the politics of the cold war, and curriculum reform at colleges and universities. Dr. Lombardi is the former executive director of the U.S.-Africa Education Foundation. Dr. Lombardi is a member of numerous national and local civic organizations and boards in Santa Fe, and he has given over 125 speeches to local and national groups on topics ranging from higher education to international affairs and U.S. foreign policy. He is also a frequent contributor to local news programs, speaking on national and international issues.

AUTHORS

GRAHAM ALLISON is director of the Belfer Center for Science and International Affairs at Harvard University. He has authored several books and numerous articles on international affairs and security studies.

RONALD BAILEY is science correspondent for *Reason* magazine and author of *ECOSCAM: The False Prophets of Ecological Apocalypse*, as well as an adjunct scholar with the Competitive Enterprise Institute.

MIRZA ASLAM BEG is the former Chief of Staff of the Pakistani army.

WENDELL BELL is professor emeritus of sociology at Yale University's Center for Comparative Research. He is the author of numerous articles.

JÉRÔME BINDÉ is deputy assistant director for Social and Human Sciences at UNESCO and director of its Division of Foresight, Philosophy, and Human Sciences.

SYLVIE BRUNEL, a geographer and a specialist on development issues, is a former president of Action Contre la Faim (Action Against Hunger).

XIMING CAI is assistant professor of civil and environmental engineering at the University of Illinois at Urbana-Champaign. He earned his M.S. in hydrology and water resources at Tsinghua University in Beijing, China, and his Ph.D. in environmental and water resources at the University of Texas at Austin.

RED CAVANEY is president and chief executive officer of the American Petroleum Institute.

JANIE CHUANG is an international legal expert and practitioner-in-residence at the American University Washington College of Law.

SARAH A. CLINE is a former senior research assistant in the Environment and Production Technology Division of the International Food Policy Research Institute. She has also served as research assistant at Resources for the Future. She is currently a Ph.D. candidate in agricultural and resource economics at Colorado State University.

TOM DeWEESE is a conservative columnist and activist who is editor and chief of the *DeWeese Report.*

CHRISTOPHER ESSEX is a professor of applied mathematics at the University of Western Ontario. He previously served as a NSERC visiting fellow at the Canadian Climate Centre and an Alexander von Humboldt Research Fellow.

H.T. GORANSON is the lead scientist of Sirius–Beta Corp. and was senior scientist with the U.S. Defense Advanced Research Projects Agency.

DANIEL T. GRISWOLD is assistant director of Trade Policy Studies for the Cato Institute.

DINA FRANCESCA HAYNES is an associate professor of law at the New England School of Law. She has published in the areas of international law, immigration law, human rights law, and human trafficking.

SAMUEL P. HUNTINGTON is the holder of the Albert J. Weatherhead III Chair at Harvard University and the author of numerous books and articles—most recently *The Challenge to America's National Identity.*

INTERGOVERNMENTAL PANEL ON CLIMATE CHANGE (IPCC) was established in 1988 by the World Meteorological Organization (WMO) and the United Nations Environment Programme (UNEP). The role of the IPCC is to assess the scientific, technical, and socioeconomic information relevant to understanding the scientific basis of risk of human-induced climate change, its potential impacts, and options for adaptation and mitigation.

DAVID KRIEGER is president of the Nuclear Age Peace Foundation, and he is the co-author of *Choose Hope, Your Role in Waging Peace in the Nuclear Age.*

MARK KRIKORIAN is the executive director of the Center for Immigration Studies and is also a regular contributor to *The National Review.* Prior to his current position he was editor of *The Winchester Star.*

JAMES HOWARD KUNSTLER is an urban planner and author of *The Long Emergency.*

FLEMMING LARSEN was director of the International Monetary Fund's European offices from 2000–2005.

PHILLIPE LEGRAIN is chief economist of *Britain in Europe, the Campaign for Britain to Join the Euro.* He is the author of *Open World: The Truth about Globalization* (Abacus, 2002).

BJØRN LOMBORG is an associate professor of statistics in the Department of Political Science at the University of Aarhus in Denmark and a frequent participant in topical coverage in the European media. His areas of interest include the use of surveys in public administration and the use of statistics in the environmental arena. In February 2002 Lomborg was named director of Denmark's National Environmental Assessment Institute. He earned his Ph.D. from the University of Copenhagen in 1994.

MIA MacDONALD is a policy analyst and Worldwatch Institute senior fellow.

FEDERICO MAYOR is former director-general of UNESCO. He has served as Spain's minister of education and science, as president of the University of Granada, and as a member of the European Parliament.

ROBERT McDONALD is a postdoctoral fellow at Harvard University.

ROSS McKITRICK is an associate professor of economics at the University of Guelph in Canada and a senior fellow of the Fraser Institute in Vancouver, B.C.

ROBERT S. McNAMARA was president of the World Bank Group of Institutions until his retirement in 1981. A former lieutenant colonel in the U.S. Air Force, he has also taught business administration at Harvard University, where he earned his M.B.A., and he served as secretary of defense from 1961 until 1968. He has received many awards, including the Albert Einstein Peace Prize, and he is the author of *In Retrospect: The Tragedy and Lessons of Vietnam* (Random House, 1995).

MICHAEL MEYER is senior editor for *Newsweek International.*

GEORGE MONBIOT is a weekly columnist for *The Guardian.*

STEVEN W. MOSHER is president of the Population Research Institute.

JIM MOTAVALLI is editor of *E/The Environmental Magazine* and author of 3 books, including *Feeling the Heat: Dispatches from the Front Lines of Climatic Change* (2004). He is also a frequent contributor to major newspapers on issues of the environment.

DANIELLE NIERENBERG is a research associate at the Worldwatch Institute.

JOHAN NORBERG is a fellow at the Swedish group Timbro, and he is the author of *In Defense of Capitalism.*

DAVID PIMENTEL is a professor of insect ecology and agricultural sciences at Cornell University, and author of numerous related works.

JANET RALOFF is a writer for *Science News.*

MARK W. ROSEGRANT is director of the International Food Policy Research Institute's Environment and Production Technology Division. He has over 24 years of experience in research and policy analysis in agriculture and economic development, with an emphasis on critical water issues as they impact world food security and environmental sustainability. He earned his Ph.D. in public policy from the University of Michigan.

IRA STRAUS is the U.S. Coordinator of the Independent International Committee on Eastern Europe and Russia for NATO. He is a former Fulbright Professor at Moscow State University and senior associate at the Program on Transitions to Democracy.

SHASHI THAROOR is Undersecretary General for Communications and Public Information and is in charge of the Department of Public Information for the United Nations.

J. SCOTT TYNES is a staff writer for the *Star.* This paper is an English weekly publication in Jordan covering politics, economics, and culture.